Erich Folkhard

Welding Metallurgy of Stainless Steels

In collaboration with

Günther Rabensteiner
Ernst Perteneder
Heinz Schabereiter
Josef Tösch

Springer-Verlag Wien New York

a. o. Univ.-Prof. Dipl.-Ing. Dr. mont. Erich Folkhard, Dir. ret.

Dipl.-Ing. Dr. mont. Günther Rabensteiner,
Head of Research, Development and Technology Department

Dipl.-Ing. Dr. mont. Ernst Perteneder,
Head of Development Department for Filler Metals

Heinz Schabereiter,
Head of Welding Technology Department

Ing. (HTL) Josef Tösch,
Development Engineer for Filler Metals

Böhler Ges. m. b. H.
Welding Division
A-8605 Kapfenberg, Austria

Revised Translation of
Metallurgie der Schweißung nichtrostender Stähle
Wien New York: Springer-Verlag 1984

ISBN 3-211-81803-0 Springer-Verlag Wien New York
ISBN 0-387-81803-0 Springer-Verlag New York Wien

Printed in Austria

With 115 partly coloured Figures

Library of Congress Cataloging-in-Publication Data: Folkhard, Erich. [Metallurgie der Schweißung nichtrostender Stähle. English] Welding metallurgy of stainless steels / Erich Folkhard in collaboration with Günther Rabensteiner ... [et al.]. p. cm. Translation of: Metallurgie der Schweißung nichtrostender Stähle. Bibliography: p. Includes index. ISBN 0-387-82043-4 (U.S.) 1. Steel, Stainless-Welding. 2. Steel, Stainless-Metallurgy. I. Rabensteiner, Günther. II. Title. TS227.2.F64 1988. 672.5'2-dc 19.88-16021.

ISBN 3-211-82043-4 Springer-Verlag Wien New York
ISBN 0-387-82043-4 Springer-Verlag New York Wien

Preface to the German Edition

When considering the operational performance of stainless steel weldments the most important points to consider are corrosion resistance, weld metal mechanical properties and the integrity of the welded joint. Mechanical and corrosion resistance properties are greatly influenced by the metallurgical processes that occur during welding or during heat treatment of welded components. This book is aimed, therefore, at providing information on the metallurgical problems that may be encountered during stainless steel welding. In this way we aim to help overcome a certain degree of insecurity that is often encountered in welding shops engaged in the welding of stainless steels and is often the cause of welding problems which may in some instances lead to the premature failure of the welded component.

The metallurgical processes that occur during the welding of stainless steel are of a highly intricate nature. The present book focuses in particular on the significance of constitution diagrams, on the processes occurring during the solidification of weld metal and on the recrystallization and precipitation phenomena which take place in the area of the welds. There are specific chapters covering the hot cracking resistance during welding and the practical welding of a number of different stainless steel grades. In addition, recommendations are given as to the most suitable procedures to be followed in order to obtain maximum corrosion resistance and mechanical properties from the weldments.

During the compilation and writing this book, an attempt has been made to combine a sense of practical usefulness together with a respect for theoretical soundness, arriving thus at a discussion of the highly complex mechanisms that occur during the welding of stainless steel which is designed to be both thorough and easy to understand for those involved in welding engineering at all levels. In order to accomplish this, complicated metallurgical problems have sometimes been represented in a somewhat simplified manner. However a certain basic knowledge of the fundamentals of metallurgy, e.g. the use of constitution diagrams and the evaluation of time-temperature-transformation and precipitation diagrams is required in order to understand the metallurgical processes that are discussed. The survey of basic metallurgical literature that is given at the beginning of the comprehensive list of supplementary literature may help to complete or renew any existing gaps.

There is a wealth of comprehensive literature covering the fields of stainless steel and stainless steel welding. The shop operater often has restricted access to this literature due to the fact that it is spread over a great number of technical or scientific journals dealing with metallurgy, material technology and welding technology problems as well as the question of corrosion behaviour and chemical plant engineering. In the list of supplementary literature, which includes 442 entries, the author has attempted to compile the most important publications from the vast field of pertinent literature which were known to the author and to arrange them for the reader in a way that is easy to survey.

Obviously, the first edition of this book leaves room for further improvement. For this reason, the author appeals to all readers and colleagues to kindly let him share their knowledge in the field of stainless steel welding metallurgy and supply him with suggestions as to expansion, changes or wise restrictions.

I wish to sincerely thank G. Rabensteiner, E. Perteneder, H. Schabereiter and J. Tösch for their valuable assistance given during the compilation and editing of the individual chapters of this book. They have helped the author with unfailing enthusiasm during the completion and editing of the manuscript and have contributed important improvements to the final draft. My thanks also go to H. Kohl, E. Krainer and A. Kulmburg for reading the manuscript or parts of it, to Vereinigte Edelstahlwerke Aktiengesellschaft (VEW) for supporting the publication of this book and for the courtesy of supplying technical and metallurgical data, and to Springer, Verlag, Vienna, for their kind cooperation and the excellent way in which they have printed and illustrated this book.

Kapfenberg, June 1984 Erich Folkhard

Preface to the English Edition

The widespread interest shown in the first edition of this book encouraged us to publish an English translation. In principle, the text corresponds to the German edition but individual chapters have been reviewed and updated. Whenever possible, the designation of individual steel and weld metal grades are derived from the internationally well known US Standards. In addition, all abbreviations used in this book are listed in alphabetic order and explained in the appendix at the end of the book. The list of supplementary literature has been updated and expanded to a total of 487 entries. I would like to thank H. Daurer, Vienna, for translating the book into English and G. T. Finnerty, Birmingham, England, for reviewing the translation. My thanks also go to the co-authors for their indispensable assistance in editing the English edition of this book, to the Böhler Welding Division in Kapfenberg and to Springer-Verlag, Vienna, for their active support of the English edition.

Kapfenberg, November 1987 Erich Folkhard

Contents

1 Significance of Constitution Diagrams for the Understanding of Welding Phenomena[1]

In order to understand the properties of iron based stainless steel weld metals and the associated phenomena that take place in the heat affected zone a knowledge of the constitution diagrams that relate to the alloying elements used in these alloys is required. Many welding phenomena can be explained and in many cases even predicted with the aid of constitution diagrams.

There is one restriction, however. Speaking in metallurgical terms, constitution diagrams are equilibrium diagrams, i.e. they represent the situation at very slow heating and cooling rates. In welding operations, however, cooling rates are normally rather high and a certain degree of deviation from constitution diagrams must be expected. Such deviations may appear, for example, in the form of super-cooling of transformation phenomena or even complete suppression of the latter. Approximate equilibrium is achieved only during subsequent heat treatment. Nevertheless, many metallurgical reactions take place during welding and phenomena such as type of crystallization, segregations, phase transformations, precipitations and embrittlement can be directly related to the constitution diagrams.

The knowledge of constitution diagrams of stainless steels and stainless steel weld metals is also of importance for an additional reason. A major proportion of the stainless steels are made up from the austenitic iron-chromium-nickel alloys, with their best known representative being the so called 18/8 steel with about 18% chromium and 8% nickel (AISI 302), developed according to Strauß and Maurer [1] in 1912 and marketed in the twenties of this century. Austenitic materials maintain their structure even at ambient temperature, i.e. γ-α (gamma-alpha) transformation does not take place. For welding operations, this means that the metallurgical changes occurring during solidification or crystallization often have a much stronger impact on the properties of the stainless steel weld metal than in the case of unalloyed or low alloyed steels and weld metals. In case of the latter, changes during solidification are often superseded by the subsequent γ-α transformation and the influence on weld metal properties is not as strong. A good example in this respect, regarding austenitic materials, is the direct influence that the type of solidification or crystallization structure has on the cracking resistance of the weld metal. But

---

[1] To facilitate practical application, the alloying contents of all constitution diagrams are stated in weight (mass) percent. For information we also call constitution diagrams phase diagrams or equilibrium diagrams.

segregation processes and precipitations during cooling and during subsequent heat treatment may also have a direct influence on cracking resistance, toughness and corrosion properties of welded joints. With a proper understanding of constitution diagrams it is possible, however, to make satisfactory predictions as to possible metallurgical reactions that may take place in the weld metal and the heat affected zone. This knowledge is successfully complemented by an understanding of the kinetics of transformation and precipitation processes as well as possible segregations, particularly in the weld metal.

We would like to refer here to the great wealth of supplementary literature which represents an important aid in the understanding of metallurgical phenomena in binary and ternary alloys and commercial steel grades. A great number of constitution diagrams (binary and ternary phase diagrams) are contained in the books of Hansen and Anderko [2], Elliot [3], Shunk [4], Kubaschewski [5], Rhines [6], Schumann [7], Rapatz [8], Houdremount [9] and in the Metals Handbook, volume 8 [18].

1.1 Iron-Chromium, Iron-Nickel and Chromium-Nickel Constitution Diagrams

Fig. 1 shows the iron-chromium constitution diagram according to Kubaschewski [5]. The element chromium constricts the γ crystal range (austenite) with increasing alloying content until it disappears completely from approx. 12% chromium onwards.

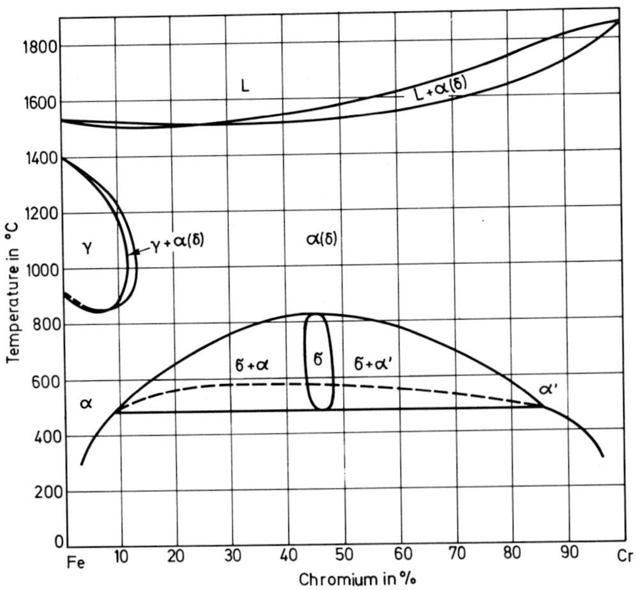

Fig. 1. Iron-chromium constitution diagram to Kubaschewski [5]

This means that alloys with more than about 12% chromium do not show any γ-α transformation, thus ruling out any accompanying grain refinement and the possibility of steel hardening. Chromium strongly promotes ferrite formation to the point where only ferrite crystals (δ [α] phase, see also footnote 3, page 9) are precipitated from the melt over the full alloying range. With higher chromium contents, the brittle sigma phase starts to precipitate from δ ferrite at about 820° C. It consists of approx. 45% Cr and represents a metallic compound of iron and chromium[2]. Due to its high chromium content, embrittlement will occur. It may be accompanied by a chromium depletion of the matrix at the grain boundaries of the precipitated sigma phase. This can have a detrimental effect on corrosion resistance. Below 600° C, we are confronted with a segregation of the α [δ] phase into low chromium α crystals and high chromium α' crystals. This is the cause of the so called 475° C embrittlement of stainless steels.

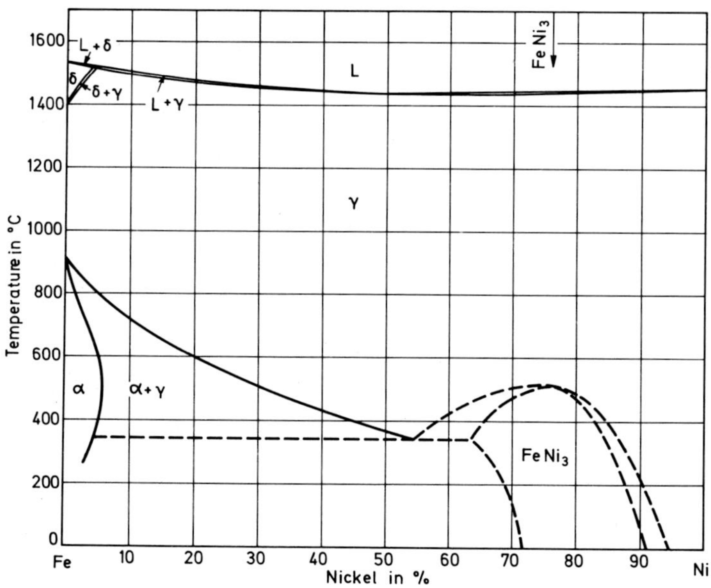

Fig. 2. Iron-nickel constitution diagram to Kubaschewski [5]

Fig. 2 shows the iron-nickel constitution diagram according to Kubaschewski [5]. Contrary to chromium, nickel has a strong tendency towards austenite formation (γ phase), i.e. from about 5% Ni onwards, the melt does not solidify to δ but γ

---

[2] Sigma phases are metal compounds of tetragonal structure and a regular arrangement of atoms of which the compound consists. Of about 50 different sigma phases, the iron-chromium sigma phase represents the best known one. It shows a tetragonal crystal lattice with an alternating arrangement of 1 atom iron and 1 atom chromium. Whenever the term "sigma phase" is being used in the context of this book, it always stands exclusively for this metal compound with the formula FeCr (see also supplementary literature [256]).

crystals. Ferrite formation from the melt is limited to the small δ ferrite corner (upper left in Fig. 2), with this ferrite being changed back to austenite when cooling passes the temperature range of 1400–1450° C. This transformation is peritectic. The area of solidification to γ crystals is rather flat, a feature which sometimes may favour the formation of segregations.

Contrary to chromium, nickel does not constrict but substantially expands the area of austenite. With increasing nickel content, the area of γ-α transformation is shifted to lower temperatures from about 900° C to 350° C. The austenitic structure is stabilized to such a degree that it is still maintained at lower temperatures and even down to ambient temperature with faster cooling rates. The austenitic chromium-nickel steels are based on this effect. Such steel grades are no longer hardenable, since γ-α transformation is completely suppressed. Due to their austenitic structure, they are paramagnetic (nonmagnetic) and can easily be distinguished from ferritic steels with the aid of a magnet. There are no brittle phases in iron-nickel systems.

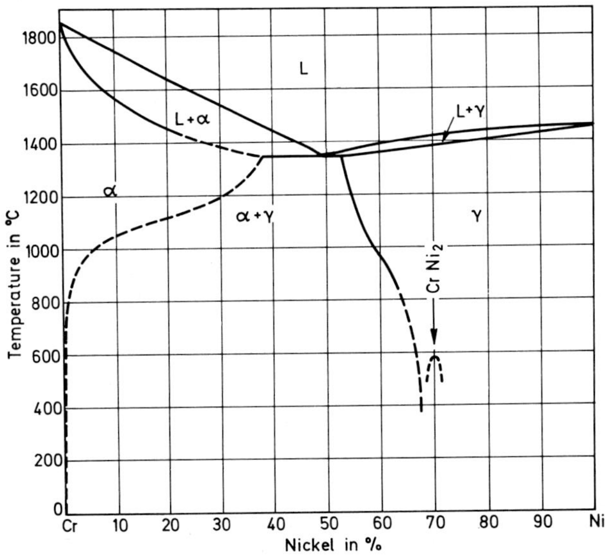

Fig. 3. Chromium-nickel constitution diagram to Hansen and Anderko [2] with supplements to Barin, Knacke and Kubaschewski [17]

The chromium-nickel constitution diagram according to Hansen and Anderko [2] with supplements according to Barin, Knacke and Kubaschewski [17] is shown in Fig. 3. It is eutectic at about 48% nickel. Gamma crystals are formed on the nickel side and alpha crystals on the chromium side. The transformation processes in the chromium-nickel constitution diagram excert no great influence on the properties of stainless steels, since the chromium and nickel contents of these grades, e.g. steel 18/8 (AISI 302), are relatively low. They are of importance, however, for nickel-based materials with nickel contents of more than 50% Ni.

1.2 The Ternary Iron-Chromium-Nickel Constitution Diagram

This ternary constitution diagram has been the subject of numerous investigations, but most of them deal only with certain areas. Because of the importance of this diagram for understanding the metallurgical phenomena which occur during the welding of stainless steels, a comprehensive survey of this ternary constitution diagram is given below, with special emphasis on the importance of specific phenomena for the welding of stainless steels.

Since the reading of ternary systems may sometimes present certain difficulties, we would like to refer the reader to the books of Rhines [6] and Schumann [7]. These books contain detailed descriptions of the ternary phase diagrams together with a discussion of their representation and interpretation.

1.2.1 Solidification and Crystallization Phenomena of Iron-Chromium-Nickel Steel Alloys

As mentioned previously in the discussion of binary iron-chromium and iron-nickel diagrams, chromium as well as nickel will inhibit γ-α transformation with the result that this secondary transformation is totally absent in a large number of stainless steels and weld metals. Consequently, the phenomena during primary crystallization from the melt exert a much stronger influence on the properties of the deposited weld metal than in the case of alloys with γ-α transformation. The type of crystallization, for example, greatly influences the hot cracking sensitivity of austenitic weld metals, including the heat affected zone, with low melting phases which precipitate at the grain boundaries of primary crystals playing a special role in some cases. The configurations of the liquidus area (start of crystallization from the melt) and solidus area (end of crystallization) are discussed in the works of Wever and Jellinghaus [10], Jenkins, Bucknall, Austin and Mellor [11], Schürmann and Brauckmann [12], Perteneder [13] and Perteneder and Jeglitsch [14]. Other works discuss the crystallization of austenitic weld metal. A thorough discussion of the primary crystallization phenomena in austenitic weld metal is found by Perteneder [15]. Schürmann and Brauckmann [12] provide the most exact data on the temperature of the beginning and end of crystallization in the iron-chromium-nickel system, since individual measuring points could be determined with great accuracy by long holding at temperature and extremely slow cooling. For welding operations, however, with their rather high cooling rates, this is of less significance. The investigations of Jenkins [11] include higher chromium contents of up to 50% which seems to be important for the understanding of the segregation phenomena. For this reason, the representation according to Jenkins et al. [11] has been chosen for illustrating the metallurgical reactions during welding.

Fig. 4 shows the isothermal profiles of the liquidus and solidus ranges with up to 50% chromium with plotted isotherms according to Jenkins et al. [11]. The liquidus range (upper diagram) is distinguished by a eutectic groove connecting the peritectic point in the iron-nickel system (Fig. 2) with the eutectic point in the chromium-nickel system (Fig. 3). To the left of this groove δ crystals (ferrite) and to the right γ crystals (austenite) are precipitated from the melt. The solidus range is

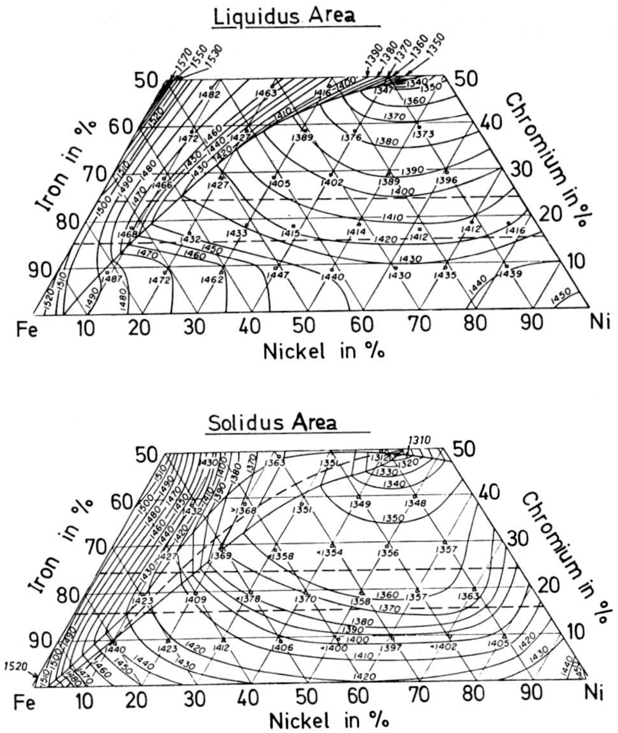

Fig. 4. Ternary iron-chromium-nickel constitution diagram to Jenkins, Bucknall, Austin and Mellor [11]

shown in the lower diagram of Fig. 4. The area of δ crystals is separated in the solidus range from that of γ crystals by 2 boundary lines. The area formed by the latter represents the lower periphery of the three-phase sector, in which δ and γ crystals are in equilibrium with the still remaining melt (L). The upper periphery of this three-phase sector ($\delta + \gamma + L$) is formed by the eutectic groove (see upper diagram). This groove is also plotted by a dashed line in the lower diagram (solidus area). The three-phase sector ($\delta + \gamma + L$) is of triangular cross section with a twist taking place from the peritectic of the iron-nickel system to the eutectic of the chromium-nickel system. The exact configuration of this three-phase sector in the iron corner can be studied in the work of Schürmann and Brauckmann [12].

For welding operations, the two areas of primary precipitated ferrite and austenite are of particular importance because of their strong influence on hot cracking sensitivity of austenitic weld metals. Also of importance is the configuration of the solidus area, particularly in view of a possible formation of segregated low-melting residual melts between the grain boundaries of primary precipitated crystals. As indicated by the isotherms in Fig. 4, the solidus area slopes rather steeply up to higher temperatures to the left of the three-phase sector ($\delta + \gamma + L$), i.e. where primary ferrite crystals are precipitated, whereas it drops to lower temperatures in the area of primary austenite crystal precipitation to the right of the three-phase sector, even down to as low as 1310° C. In addition, the slope of this area becomes flatter with increasing chromium-nickel contents. We are confronted here with two decisive differences in the temperature induced behaviour of the areas of primary ferrite and primary austenite crystallization which shall be discussed in more detail in section 2.7.

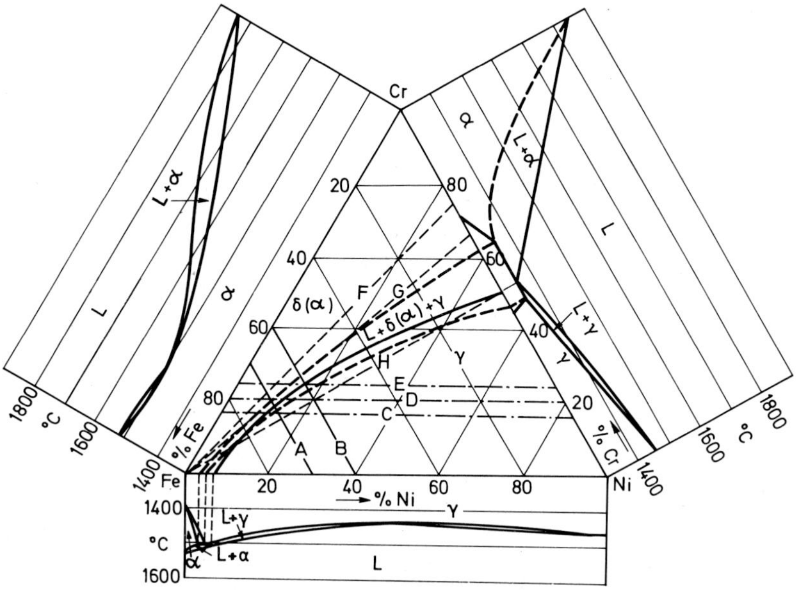

Fig. 5. Liquidus and solidus lines in binary iron-chromium, iron-nickel and chromium-nickel constitution diagrams and their correlation to liquidus and solidus areas in the ternary iron-chromium-nickel diagram. The lines *A*–*H* represent the concentration profiles of Figs. 6, 7 and 9

For a better understanding of the correlation between ternary and binary constitution diagrams, the configuration of the three-phase sector with its twist from the peritectic of the Fe-Ni system to the eutectic of the Cr-Ni system has been plotted in Fig. 5. It is apparent, how the peritectic line expands to a triangular sector in the ternary system which must turn with increasing nickel and chromium contents to meet with the eutectic point in the chromium-nickel-system (see Fig. 3).

As a supplement to the ternary iron-chromium-nickel diagram, Fig. 6 shows concentration profiles at 70% and 60% Fe and Fig. 7 at 16%, 20% and 24% chromium according to Schafmeister and Ergang [16] with additional data according to Jenkins et al. [11], Hoffmeister and Mundt [19], Lippold and Savage [20], Nicholson et al. [24], Westendorp [25], Gerlach [26], Bungardt, Dietrich and Arntz [172], and Mundt and Hoffmeister [149] which correlate to lines *A* through *E* in Fig. 5. We are able to distinguish the distribution of precipitation of primary δ and γ crystals as well as the three-phase sector ($\delta + \gamma + L$) which separates the areas of primary solidification to ferrite and austenite.

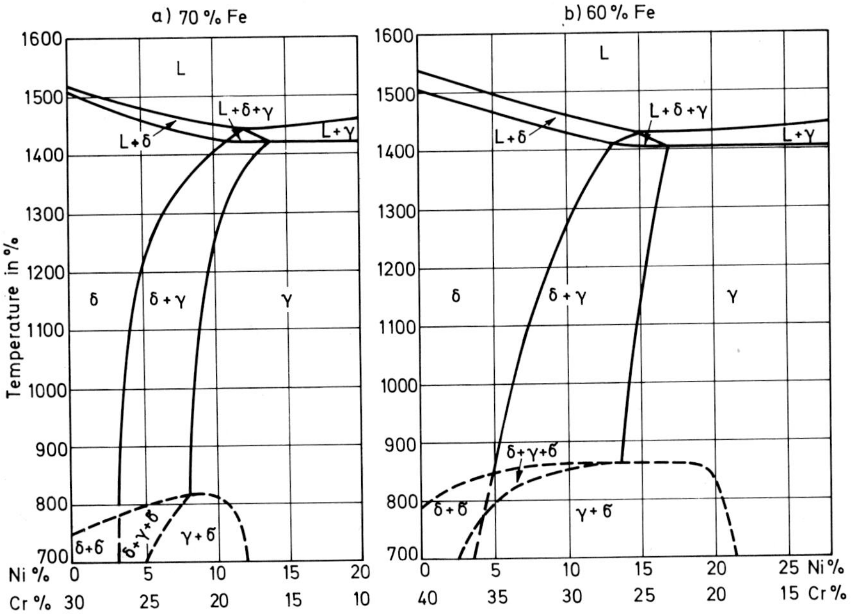

Fig. 6. Concentration profiles in the ternary iron-chromium-nickel constitution diagram at 70% and 60% iron according to Schafmeister and Ergang (16) with supplements. The profiles correlate to lines *A* and *B* in Fig. 5

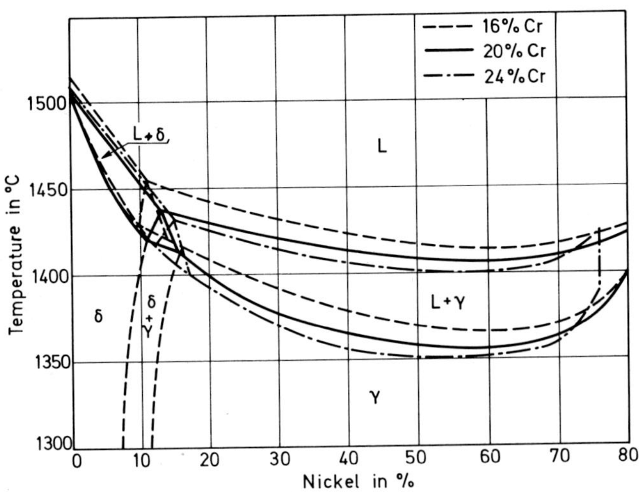

Fig. 7. Concentration profile in the ternary iron-chromium-nickel constitution diagram at 16%, 20% and 24% chromium. The profiles correlate to lines *C*, *D* and *E* in Fig. 5

1.2.2 Phenomena During Cooling from the End of Solidification to Ambient Temperature

Iron-chromium-nickel alloys can consist immediately after the end of solidification of the following phases:

- primary δ crystals (δ ferrite)[3],
- primary γ crystals (austenite),
- a mixture of $(\delta + \gamma)$ crystals which have been formed from the three-phase sector $(L + \delta + \gamma)$.

In the ternary iron-chromium-nickel constitution diagram, it is the configuration of the periphery of the $(\delta+\gamma)$ area which is of decisive influence for all further reactions during cooling after the end of solidification. The chromium-nickel constitution diagram (Fig. 3) shows that the $(\alpha+\gamma)$ area is expanded during further cooling after the end of solidification, in particular on the chromium side. The iron-nickel constitution diagram (Fig. 2) indicates that the area of primary δ crystals is limited to the peritectic transformation in the iron corner and disappears completely below 1400° C, with δ phase (ferrite) being transformed here into γ phase (austenite). In the iron-chromium system (Fig. 1) there are only δ (α) crystals below the solidus line. The phenomena of the ternary iron-chromium-nickel constitution diagram are shown in Fig. 8 in the form of two isothermal profiles at 1400° C and 1100° C according to Brewer, Chang and Shih-Ger [18]. The areas of primary precipitated δ and γ crystals can be gathered from the profile at 1400° C. In between we find in the iron corner the area of already solidified δ $(\alpha) + \gamma$ mixture which, with higher chromium and nickel contents, passes into the three-phase region δ $(\alpha) + \gamma + L$ and finally into the still liquid melt.

The isothermal profile at 1100° C (Fig. 8 right) shows that the $\delta(\alpha)+\gamma$ area has greatly expanded. It meets with the binary iron-chromium system at about 12% chromium, i.e. roughly at the point where we find in the binary constitution diagram (Fig. 1) the constricted $\delta(\alpha)+\gamma$ loop. In both profiles of Fig. 8, a ferrous alloy with 20% chromium, 10% nickel and 70% iron has been plotted as a specially marked

---

[3] δ (delta) ferrite and α (alpha) ferrite represent the same type of crystal, characterized by a body centered cubic crystal lattice. In welding technology literature, ferrite which is formed by primary precipitation (from the melt) is commonly called δ ferrite. In constitution diagrams, however, the term α ferrite is sometimes chosen to indicate that we are dealing with the same type of crystal. Ferrite which is formed by secondary precipitation from γ phase during the socalled γ-α transformation at temperatures below 911° C (A_3 point) is always called α ferrite in the literature.

Since the two kinds of ferrite show a different microscopic appearance, in particular with regard to the welding of stainless steels, and also excert a different influence on the properties of the weld metal, it seems to be proper to use a different designation for the two kinds of ferrite which indicates the nature of their formation. For this reason, we shall keep in the context of this book to the following designation:

- ferrite which is predominantly formed from the melt is designated as δ ferrite,
- ferrite which is formed at the γ-α transformation below 911° C is designated as α ferrite.

If another designation is found in a constitution diagram, it is put in parenthesis in the text, e.g. δ (α) ferrite.

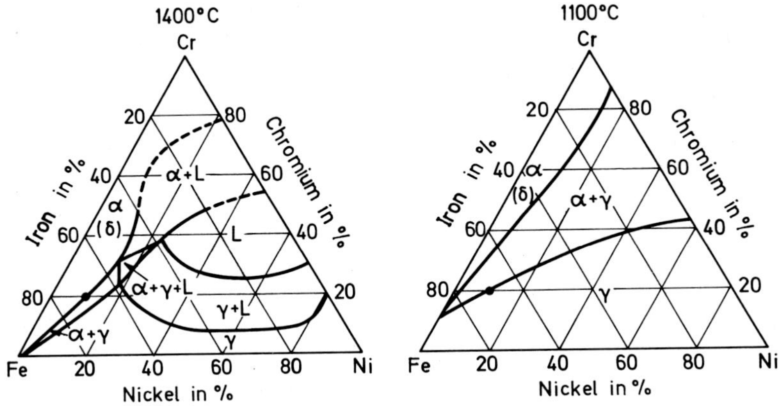

Fig. 8. Isothermal profiles of the ternary iron-chromium-nickel constitution diagram at 1400° C and 1100° C to Brewer, Chang and Shih-Ger [18]

dot. It is apparent that it solidifies from the melt at approx. 1400° C to primary δ phase which consist, however, at 1100° C of secondary formed γ phase. This means that primary precipitated δ ferrite is transformed during cooling from 1400° C to 1100° C into secondary γ austenite (see also Fig. 6). Depending upon the steel grade and cooling rate, this transformation can be partial or complete. In the case of the former, a content of residual ferrite remains in the structure which shall be discussed later (see chapter 3.2).

The concentration profiles for 70% and 60% iron (Fig. 6) also show quite clearly the solidification range of δ and γ crystals and the subsequent δ-γ transformation. In addition, Fig. 9 shows concentration profiles with constant chromium : nickel ratios

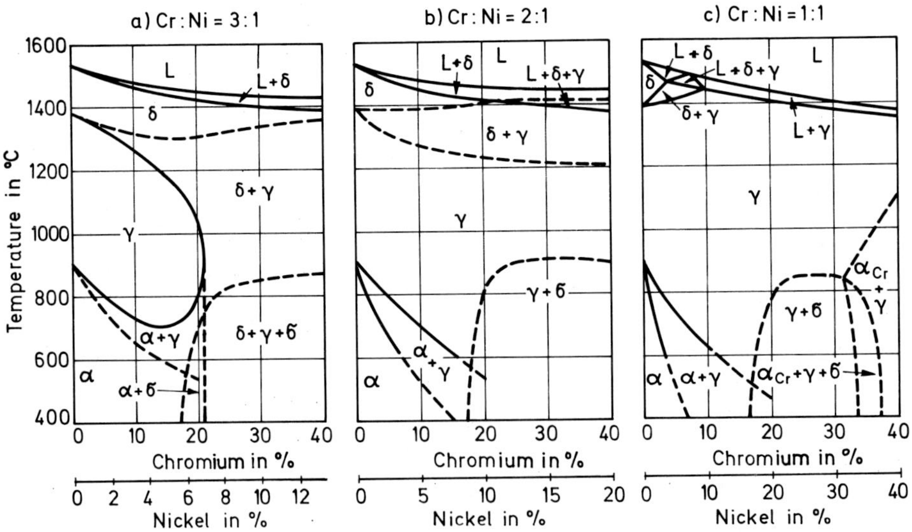

Fig. 9. Concentration profiles of the ternary iron-chromium-nickel constitution diagram for chromium : nickel ratios of 3 : 1, 2 : 1 and 1 : 1. The profiles correlate to lines *F, G* and *H* in Fig. 5

which correspond to lines *F*, *G* and *H* in Fig. 5. Low carbon martensitic and ferritic-austenitic weld metal alloys, such as 12/4, 21/7, 24/8 or 29/9.5 chromium-nickel filler metal grades can be alloted to Fig. 9a (Cr : Ni = 3 : 1). Fig. 9b illustrates the behaviour of most austenitic weld metals with a small amount of supercooled delta ferrite after cooling. Fig. 9c shows the phenomena occurring during solidification and cooling of fully austenitic weld metal alloys.

As indicated by the iron-chromium constitution diagram (Fig. 1, page 2), a transformation of primary δ crystals to γ crystals (austenite) takes place with chromium contents below approx 12% in the temperature range between approx. 1400° C and 1100° C. The γ crystals are transformed back into α iron below 900° C, often in combination with a grain refinement and the formation of a martensitic structure.

The iron-nickel system (Fig. 2, page 3) also shows a secondary δ-γ transformation with δ phase precipitating from the melt being completely transformed into γ crystals only up to about 5% Ni. With higher Ni contents, part of the primary precipitated γ phase may be maintained down to ambient temperature, with secondary γ-α transformation being only partially possible. With still higher nickel contents, the γ-α transformation is completely supercooled and the primary austenite is fully maintained down to ambient temperature (Fig. 9b and 9c).

In the ternary iron-chromium-nickel constitution diagram, a connection must exist between the γ-α transformation ranges of the binary iron-chromium system and the binary iron-nickel system, which is limited on the one side by about 12% chromium content and is shifted on the other with increasing Ni content to lower temperatures. There are numerous works about alloys with low carbon contents with compositions which correspond to steel grades used in practical applications. In this context, the phase boundaries in the iron corner of iron-chromium-nickel constitution diagrams according to Bungardt, Kunze and Horn [21] as well as Kunze

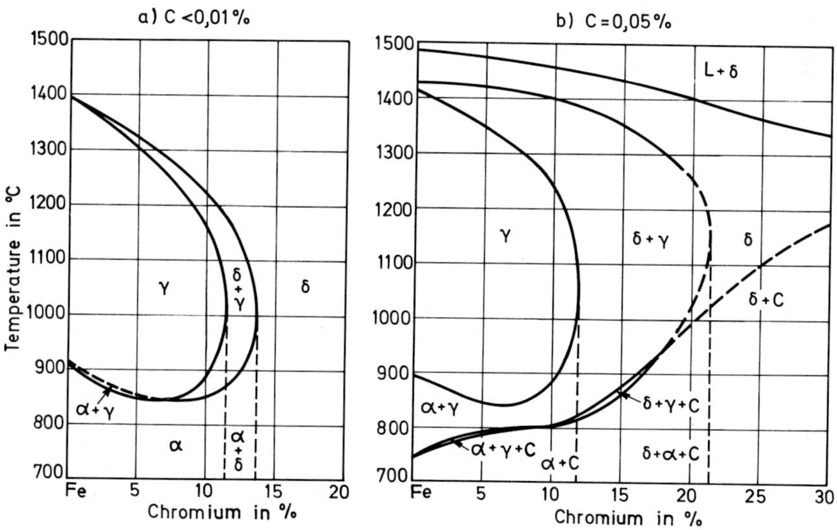

Fig. 10. Ranges of γ, δ and α phase in the iron-chromium constitution diagram with carbon contents of 0% (left) and 0.05% (right) to Bungardt, Kunze and Horn [21]

[22] have been investigated more closely. The following deliberations therefore refer to ternary iron-chromium-nickel steel alloys with a carbon content of 0.05%. As a supplement to Fig. 1, Fig. 10 (left hand diagram) shows the γ loop for a practically carbon-free iron-chromium alloy. The right hand diagram according to Bungardt, Kunze and Horn [21] shows the changes in the various ranges produced by 0.05% carbon. In particular the area of the two-phase range ($\delta + \gamma$) is shifted towards higher chromium contents with the possibility of carbide formation (designated with the letter C). Fig. 11 shows by way of concentration profiles for 0%, 1%, 4% and 6% nickel and 0.05% carbon, how the γ range in the iron-chromium system is altered by the addition of nickel according to data compiled by Kunze [22]. With increasing nickel contents, the γ range is expanded towards higher chromium contents and lower temperatures.

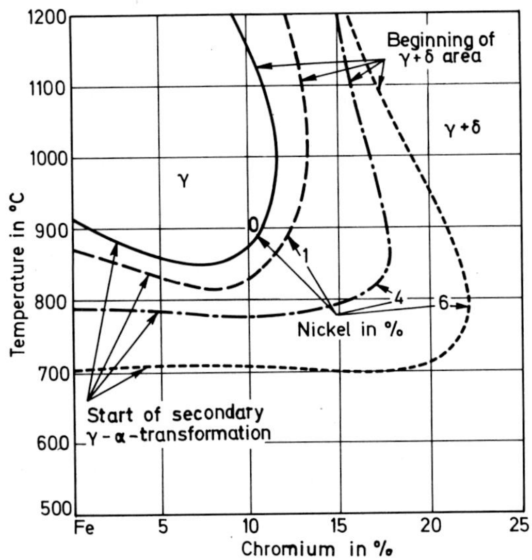

Fig. 11 Influence of nickel on the range of the γ phase (austenite) in the iron-chromium system to Kunze [22]

This nickel effect was of decisive importance for the development of the low carbon martensitic steels, where the austenite range is enlarged by the addition of nickel to such an extent that δ ferrite formation which often represents an undesirable structural constituent in low carbon martensitic stainless steels can be largely avoided. Gamma phase which is formed during cooling or heat treatment, is transformed during further cooling into secondary α crystals with a martensitic structure being formed in the case of rapid cooling. The A_3 point which in the case of pure iron is located at 911° C is shifted by chromium and nickel to lower temperatures which must be taken into account if heat treatment is being considered.

A further transformation in the ternary iron-chromium-nickel constitution diagram starts with higher chromium contents in the temperature range between 800 and 900° C. As shown in the binary iron-chromium constitution diagram (Fig. 1), the brittle sigma phase starts to precipitate below about 820° C with chromium contents of approx. 45%. The precipitation range is enlarged at lower temperatures towards lower and higher chromium contents. This sigma phase does not exist in the other binary constitution diagrams of iron-nickel and chromium-nickel which means that its range must be constricted in the ternary iron-chromium-nickel constitution diagram. Fig. 12 shows according to investigations by Nicholson, Samans and Shortleeve [24] and Westendorp [25] the areas of the beginning of sigma phase precipitation in the iron-chromium-nickel system by way of isothermal profiles at 700°, 800° and 900° C. It indicates from which alloying contents onwards, the formation of brittle sigma phase is to be expected. In addition, Fig. 13 shows according to the Gerlach [26] the exact ranges of the various phases in the form of an isothermal profile at 800° C, with additional plotting of phase boundaries between gamma phase and binary phase mixture ($\gamma+\sigma$) at temperatures of 650 and 550° C.

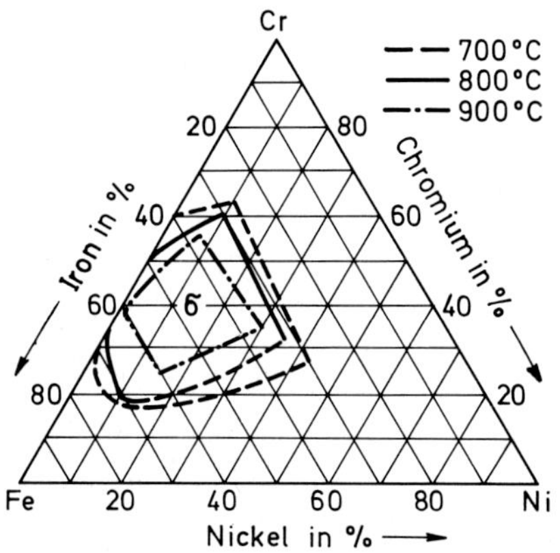

Fig. 12. Sigma phase range in the ternary iron-chromium-nickel constitution diagram to Nicholson, Samans and Shortleeve [24] and Westendorp [25]

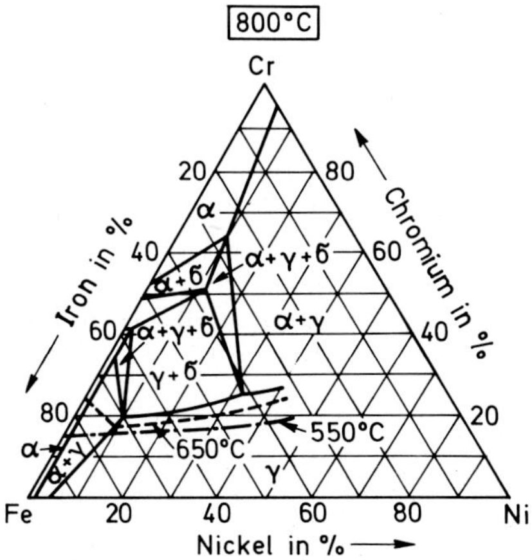

Fig. 13. Isothermal profile in the ternary iron-chromium-nickel constitution diagram at 800° C to Gerlach [26]

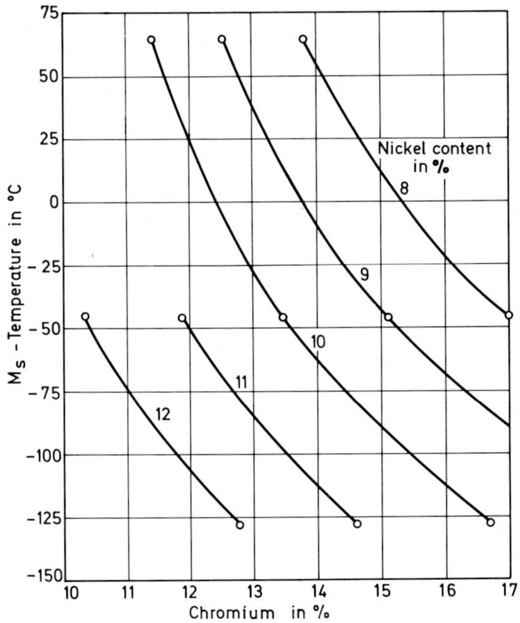

Fig. 14. Influence of chromium and nickel on the beginning of martensite transformation (M_s temperature) to Eichelmann and Hull [28]

From the concentration profiles at 60% and 70% iron shown in Fig. 6 (page 8), the distribution of various areas of sigma phase formation can also be gathered.

Since relatively small amounts of sigma phase are already capable of producing material embrittlement, it is mainly the precipitation of sigma phase at the grain boundaries which is of decisive importance for any deterioration of material properties. Due to the possibility, however, of supercooling sigma phase formation, embrittlement may often occur later during subsequent heat treatment or during long-time exposure to elevated temperatures. Annealing at temperatures above 1000° C produces a complete dissolution of sigma phase. In this way, embrittlement can be eliminated again if a renewed formation of sigma phase is suppressed by rapid cooling.

The influence of alloying elements on the temperature at the beginning of martensitic transformation (M_s temperature) is also of practical importance. It is substantially lowered by the presence of nickel. The simultaneous influence of chromium and nickel is shown in Fig. 14 according to Eichelmann and Hull [28].

1.3 Influence of Other Elements on the Iron-Chromium-Nickel Constitution Diagram

The incorporation of only one more element into the ternary Fe-Cr-Ni system would lead to quaternary systems which, because of the necessary 4th dimension, are no longer graphically representable. In order, however, to describe the often very important influence of additional elements, it is necessary to rely on quasi-binary and ternary diagrams in which one or two elements are kept constant. In view of the limited scope of this book, only the field of commercial stainless steel grades of the iron-chromium-nickel system shall be considered here.

1.3.1 Influence of Carbon

It is well known that carbon is the most important alloying element in steel. We would like to refer here to the widely used metastable iron-carbon constitution diagram [27].

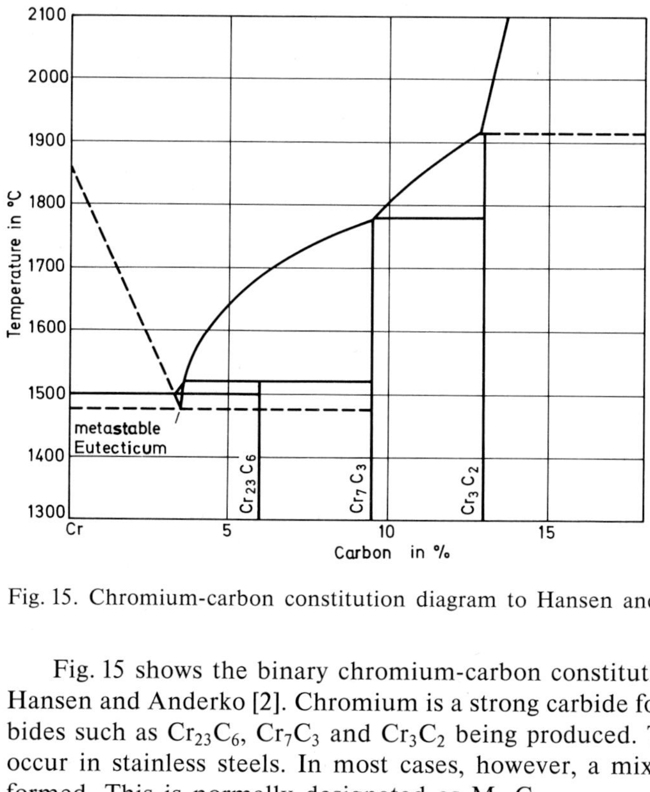

Fig. 15. Chromium-carbon constitution diagram to Hansen and Anderko [2]

Fig. 15 shows the binary chromium-carbon constitution diagram according to Hansen and Anderko [2]. Chromium is a strong carbide forming element, with carbides such as $Cr_{23}C_6$, Cr_7C_3 and Cr_3C_2 being produced. The former two may also occur in stainless steels. In most cases, however, a mixed carbide $(FeCr)_{23}C_6$ is formed. This is normally designated as $M_{23}C_6$.

Nickel does not form carbides. In the nickel-carbon system, the latter element always precipitates as graphite. In common stainless steels carbon is first of all bound to chromium and secondly to iron and practically never occurs as graphite.

The ternary iron-chromium-carbon system is shown in Fig. 16 by way of concentration profiles at 0.1% and 0.2% carbon according to Bungardt, Kunze and Horn [21]. A comparison with Fig. 10 shows that the γ and $(\delta + \gamma)$ range is increased more and more with rising carbon contents and is shifted towards higher chromium contents, showing that carbon has a strong austenitizing effect. In addition, the formation of mixed iron-chromium carbides is also increased with rising chromium content. Their iron content is according to Bungardt, Kunze and Horn [21] in case of

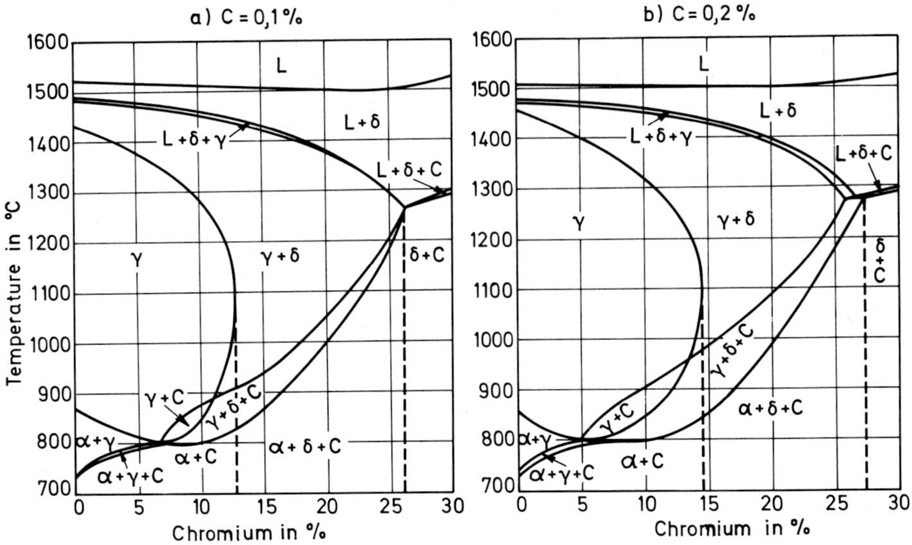

Fig. 16. Concentration profiles in the ternary iron-chromium-carbon constitution diagram at 0.1% and 0.2% carbon to Bungardt, Kunze and Horn [21] (carbide is designated by the letter C)

stainless low carbon chromium steels in the range of about 30—40%, their chromium content is 42—65% and their carbon content 0.4—1.1%. Within the above limits, the composition of mixed iron-chromium carbides is determined on the one hand by the alloy's chromium level and on the other by the annealing temperature during heat treatment. The ranges of δ and α ferrite have been separated in Fig. 16 to point out the differing grain structure. As commonly known, δ ferrite is formed as primary crystals from the melt and is often rather coarse. α ferrite is formed during secondary γ-α transformation and is normally rather fine grained.

Fig. 17 shows concentration profiles at 13% and 17% chromium according to Castro and Tricot [190]. It is apparent that the mixed carbide $M_{23}C_6$ (designated C_1 in Fig. 17) is already precipitated from α phase at temperatures below 700° C at very low carbon contents of around 0.01%, a fact which is of importance for the susceptibility of ferritic chromium steels to intergranular corrosion due to carbide precipitation.

Quasi ternary constitution diagrams of the quaternary iron-chromium-nickel-carbon system at 17% chromium and 4% nickel respectively are included in the works of Kunze [22] and Bungardt, Kunze and Horn [23].

The influence on the γ phase range by the presence of carbon in the ternary iron-chromium-nickel system is also of importance for stainless austenitic steels and weld metals with higher contents of chromium and nickel. In the case of pure iron, a considerable amount of carbon is dissolved, according to Staudinger [27], by γ crystals at elevated temperatures (2.1% at 1147° C and 0.80% at 723° C—see the metastable iron-carbon constitution diagram). Solubility decreases rapidly at γ-α transformation, since the α phase which is formed is only capable of dissolving 0.02% C. In other words, at 723° C and with pure iron, γ phase shows a carbon solubility which is 40 times that of α phase.

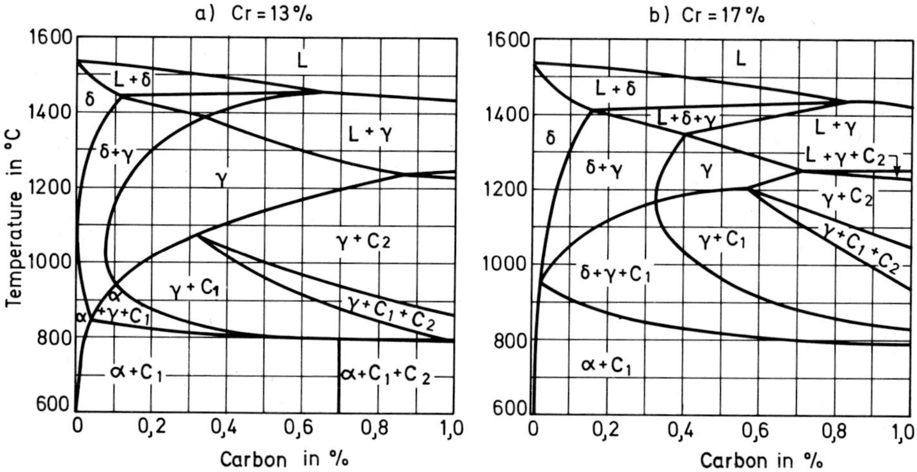

Fig. 17. Concentration profile in the ternary iron-chromium-carbon constitution diagram at 13% and 17% chromium to Castro and Tricot [190, Part 3]

Consequently, one ought to assume that the austenite crystals of stainless steel (e.g. grade 18/8) too posses good carbon solubility. This potential is counteracted, however, by the fact that chromium diminishes the activity of carbon within the austenitic phase because of the strong tendency towards mixed carbide $M_{23}C_6$ formation. This carbide already forms at very low carbon contents with the effect that in austenitic stainless steels the carbon solubility limit of austenite is greatly lowered.

Fig. 18 shows the carbon solubility limit of γ phase in stainless steel alloys with 18% chromium and nickel contents between 0–40% according to Tama, Vyklicky

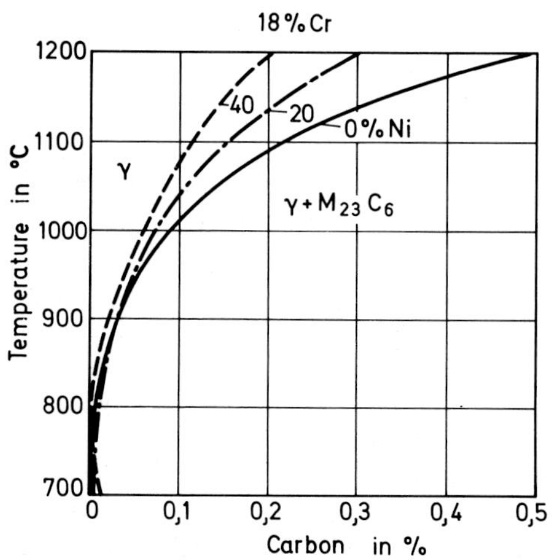

Fig. 18. Influence of nickel content on carbon solubility and precipitation boundary of mixed carbide $M_{23}C_6$ in ferrous alloys with 18% chromium to Tama, Vyklicky and Löbl [29] and Gerlach [30]

and Löbl [29] and Gerlach [30]. It shows how the carbon solubility of pure iron can be greatly reduced by the addition of 18% chromium and increasing nickel contents (e.g. at 723° C to approx. 0.01—0.02% C as compared to 0.80% C of pure iron). This solubility limit diminishes further as the temperature drops down to ambient temperature. This also means that austenitic chromium-nickel steels too may precipitate the mixed carbide $M_{23}C_6$ at correspondingly low carbon contents as is the case, for example, with ferritic chromium steels. There is a major difference, however, with regard to precipitation kinetics. In general, the rate of precipitation in α phase is much higher than in γ phase (see also section 4.1).

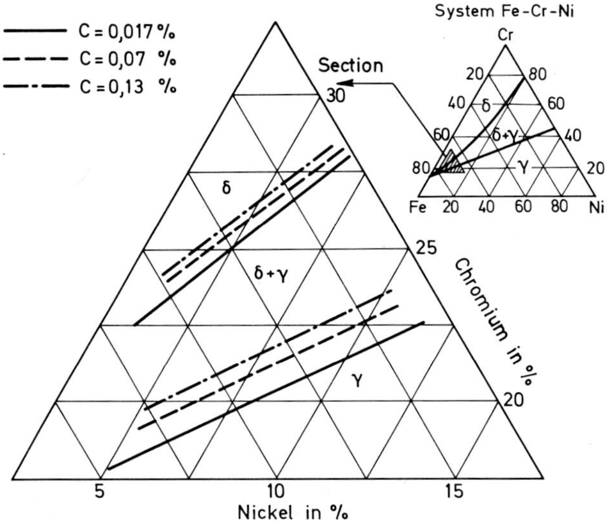

Fig. 19. Isothermal profile of ternary iron-chromium-nickel constitution diagram at 1200° C. Influence of carbon content on γ, $\delta + \gamma$ and δ phase ranges to Hoffmeister and Mundt [31]

In the case of austenitic steel grades, the transformation of primary δ crystals precipitating from the melt into γ crystals is also of major importance, in particular with regard to heat treatment processes (see also section 3.2). Fig. 19 shows the isothermal profile at 1200° C in the ternary iron-chromium-nickel constitution diagram at 0.017%, 0.07% and 0.13% carbon according to Hoffmeister and Mundt [31]. Rising carbon contents increase the γ range and constrict the δ range. At 1200° C, for example, an alloy with 20% chromium and 7% nickel is still in the $(\delta + \gamma)$ range at 0.017% C but already in the γ range at 0.07% C. To stay with 20% chromium, for example, in the γ range takes at least 8% nickel at 0.017% C but only 6.8% Ni at 0.07% C. In other words, carbon has a strong effect on the expansion of the γ range.

Fig. 20 shows according to Hoffmeister and Mundt [31] the reduction of the delta ferrite content caused by carbon in the temperature range of 800—1330° C by way of a concentration profile at approx. 25% chromium and 7% nickel. This diagram represents a major tool for the selection of proper temperatures for the heat treatment of carbon containing austenitic-ferritic chromium-nickel steels.

The formation of sigma phase is influenced by the addition of carbon. Rising carbon contents will produce increased carbide formation, with part of the

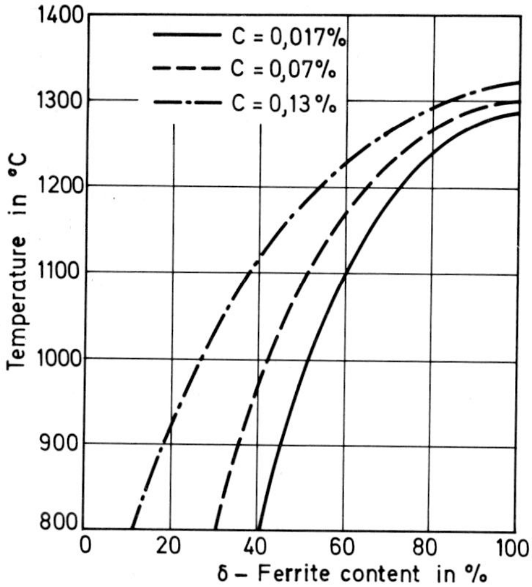

Fig. 20. Influence of carbon content on the content of delta ferrite in iron-chromium-nickel alloys with about 25% chromium and 7% nickel to Hoffmeister and Mundt [31]

chromium being bound to the high-chromium mixed carbide $M_{23}C_6$. Thus, the chromium content of the matrix is reduced and the precipitation of sigma phase is slowed down. With regard to the constitution diagram this means according to Ergang, Günther and Weik [32] a constriction of the range in which sigma phase precipitates.

In general, the effect of carbon on the formation of intermetallic phases can be described such that the solubility of carbon in the pertinent phase is the decisive criterion. If solubility is very low, e.g. in the sigma phase, carbon will inhibit its formation. If, however, carbon is arrested by additions of titanium or niobium, its inhibiting effect is reduced. According to Wiegand and Doruk [33] sigma phases can in such a case precipitate just as if there were only very little carbon present.

The influence of low carbon contents on the liquidus and solidus ranges as well as on the three-phase sector ($\delta + \gamma + L$) seems to be rather small in the iron-chromium-nickel system. The influence of higher carbon contents can be gathered from the work of Bungardt, Kunze and Horn [21] (see also Fig. 17).

1.3.2 Influence of Nitrogen

In low alloyed steels, nitrogen is often an undesirable element but in stainless steels it may be used as an alloying element. Even with conventional melting in electric arc furnaces, we may find in stainless steels relatively high nitrogen contents in the range of 0.03–0.05%. Due to the good nitrogen solubility that the alloying element chromium offers, these nitrogen contents are considerably higher than those in unalloyed or low alloyed steel grades. During electric arc welding nitrogen can be absorbed from the air by the molten weld droplet as it transfers across the arc particularly if a long arc length is used. According to Tösch and Schabereiter [34], nitrogen contents of 0.1% or more may be obtained in this way.

As an austenitizing element, nitrogen has an equally strong influence on the transition phenomena in the ternary iron-chromium-nickel constitution diagram as does carbon. But nitrogen possesses a much higher solubility in austenitic chromium-nickel stainless steel alloys than carbon does which results in a much lower susceptibility to form undesirable precipitations in austenitic steel grades.

Fig. 21 shows the nitrogen solubility of liquid iron-chromium-nickel alloys at 1600° C according to Humbert and Elliot [35]. A strong rise in solubility with increasing chromium contents is apparent. Kohl [36] reports on the technological

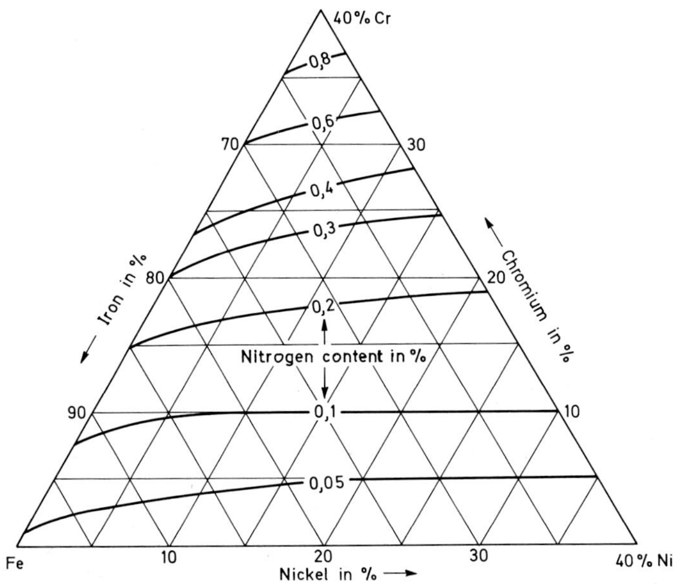

Fig. 21. Solubility of nitrogen in iron-chromium-nickel alloys at 1600° C to Humbert and Elliot [35]

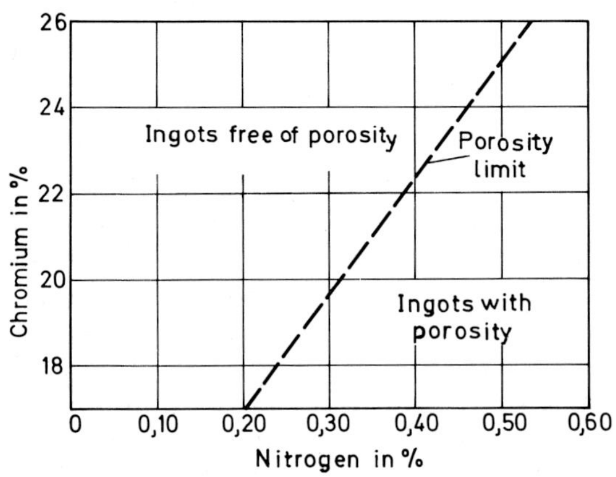

Fig. 22. Solubility of nitrogen in austenitic Cr-Ni steels at solidification, represented by the porosity free solidification limit of ingots to Kohl [36]

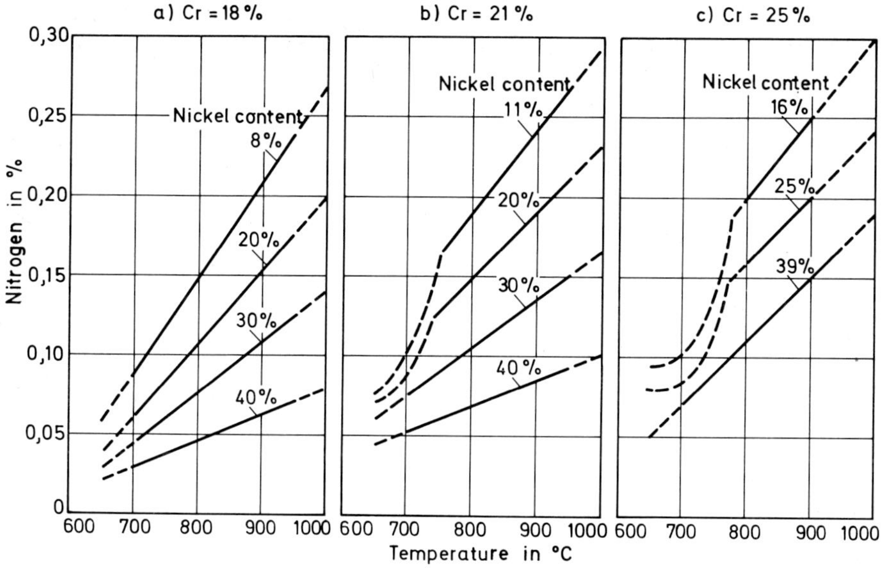

Fig. 23. Solubility of nitrogen in austenitic Cr-Ni steels with low carbon content in the temperature range between 600° and 1000° C to Bungardt, Laddach and Lennartz [37]

nitrogen solubility of austenitic Cr-Ni steels during solidification by way of the expansion limit for porosity-free solidification of ingots (Fig. 22). The maximum nitrogen content for porosity-free solidification rises from 0.20% at 17% chromium to 0.50% at 25% chromium in linear proportion with the chromium content. The interstitial nitrogen solubility in austenitic low carbon chromium-nickel steels with 18, 21 and 25% chromium as dependent on temperature and nickel content according to Bungardt, Laddach and Lennartz [37] is shown in Fig. 23. It is apparent that the γ crystals are capable of dissolving relatively high nitrogen contents in interstitial locations. In this respect, nitrogen differs considerably from carbon, the solubility of which is much lower, not only in ferrite but also in austenite of stainless steels at the same temperature (see also Fig. 18, page 17).

In the diagrams for 21% and 25% chromium (Fig. 23), the sigma phase range which possesses only a low nitrogen solubility can be detected by a small dip. If sigma phase precipitates, part of the chromium is retained and this part is then lacking in the matrix which leads to a reduction in nitrogen solubility of the latter.

While nitrogen solubility is high in austenite, it is much lower in ferrite. At temperatures around 900° C it is already in the range of only 0.01% and it is dropping further with lower temperatures. The solubility limits of nitrogen in iron-chromium alloys are shown in Fig. 24. It represents the concentration profiles at 18% and 26% chromium in the iron-chromium-nitrogen system according to Brewer, Chang and Shih-Ger [18]. Apart from the low nitrogen solubility in ferrite, these diagrams also show the strong austenitizing effect of nitrogen. Without nitrogen, such alloys solidify purely ferritic and even during further cooling there is no austenite formed. With increasing nitrogen additions, however, there is an increasing amount of austenite besides ferrite in the structure.

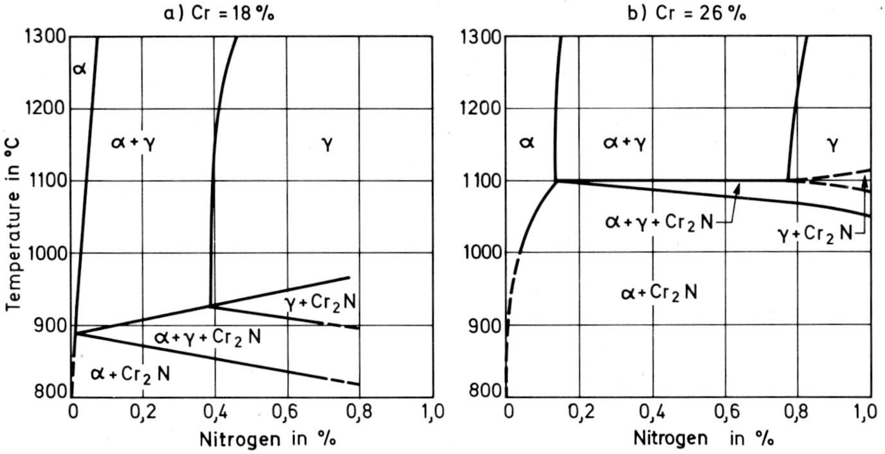

Fig. 24. Concentration profiles in the ternary iron-chromium-nitrogen diagram at 18% and 26% chromium to Brewer, Chang and Shih-Ger [18]

If the solubility limit for nitrogen is exceeded in stainless steels, nitrogen precipitates as chromium nitride Cr_2N. If carbon is also present, the carbide $M_{23}C_6$ is also precipitated simultaneously. The precipitation of all phases which are not able to dissolve nitrogen such as the carbide $M_{23}C_6$, the molybdenum containing χ phase and the Laves phase Fe_2Mo are shifted by nitrogen according to Thier [38] and Thier, Bäumel and Schmidtmann [41] to longer periods of time. However, the precipitation of the carbide M_6C which is able to dissolve nitrogen is encouraged. According to Gerlach and Schmidtmann [39], prolonged annealing will also produce the formation of complex nitrides of the type M_2N (*Z* phase) and the carbonitride $M_6(CN)$.

Fig. 25 shows the effect of nitrogen in the iron-chromium-nitrogen system in the form of concentration profiles at 0.1% and 0.3% nitrogen according to Brewer,

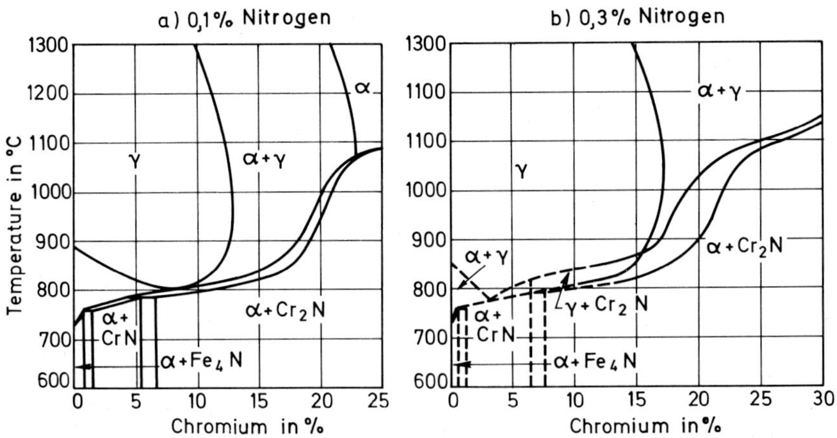

Fig. 25. Concentration profiles in the ternary iron-chromium-nitrogen constitution diagram at 0.1% and 0.3% nitrogen to Brewer, Chang and Shih-Ger [18]

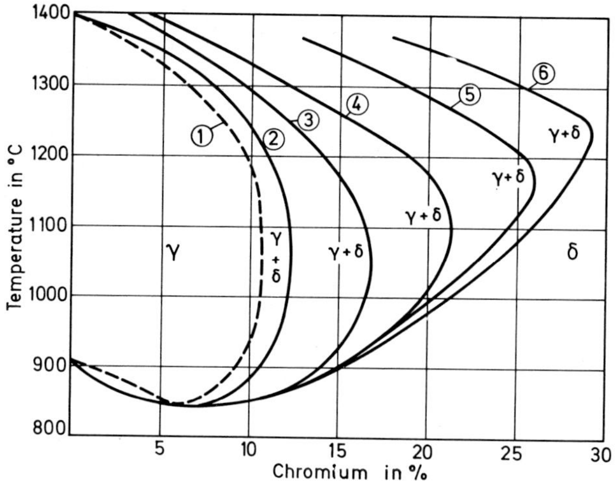

Fig. 26. Shifting of phase boundaries of γ, $\gamma + \delta$ and δ phase in the iron-chromium system caused by rising nitrogen and carbon contents to Baerlecken, Fischer and Lorenz [40]. Contents of carbon and nitrogen: 1 and 2: C = 0.004%, N = 0.002%; 3: C = 0.013%, N = 0.015%; 4: C = 0.040%, N = 0.030%; 5: C = 0.040%, N = 0.120%; 6: C = 0.050%, N = 0.250%

Chang and Shih-Ger [18]. It illustrates the strong austenitizing effect of nitrogen as well as the precipitation limits of nitrides at the γ-α transformation.

The effect of nitrogen and that of nitrogen and carbon on the boundaries of the ($\gamma + \delta$) range is shown in Fig. 26 according to Baerlecken, Fischer and Lorenz [40]. With the simultaneous presence of nitrogen and carbon, their effects are combined. The influence of nitrogen on the δ-γ transformation in the iron-chromium-nickel system is shown in Fig. 27 according to Hoffmeister and Mundt [31] by way of an

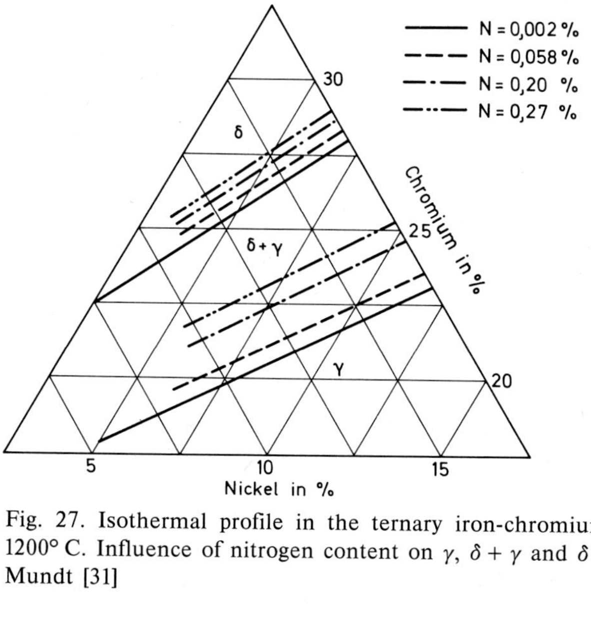

Fig. 27. Isothermal profile in the ternary iron-chromium-nickel constitution diagram at 1200° C. Influence of nitrogen content on γ, $\delta + \gamma$ and δ phase ranges to Hoffmeister and Mundt [31]

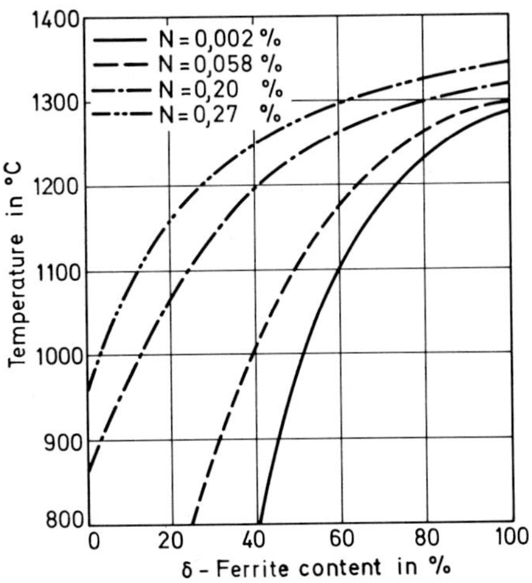

Fig. 28. Influence of nitrogen content on the content of delta ferrite in iron-chromium-nickel alloys with about 25% chromium and 7% nickel to Hoffmeister and Mundt [31]

isothermal profile at 1200° C. Like carbon, rising nitrogen contents expand the γ range to the detriment of the δ range (see also Fig. 19, page 18). In analogy to Fig. 20 (influence of carbon), Fig. 28 shows the influence of nitrogen on the reduction of delta ferrite content in the iron-chromium-nickel system according to Hoffmeister and Mundt [31] by way of a concentration profile at approx. 25% chromium and 7% nickel, with nitrogen producing similar effects as carbon.

With regard to the effect on the precipitation of sigma phase in ferrous iron-chromium-nickel alloys, nitrogen shows a similar behaviour as carbon. Both elements shift the precipitation limit of the sigma phase in the iron-chromium-nickel system towards higher chromium contents. The effect of nitrogen on other intermetallic phases is similar to that of carbon. Here too, the principle applies according to Thier, Bäumel and Schmidtmann [41] and Gerlach, Kautz and Schäffler [42] that the precipitation of phases not capable of dissolving any or only small amounts of nitrogen is impeded by nitrogen, with the effect that the precipitation boundaries in the iron-chromium-nickel system are shifted towards higher chromium contents by rising nitrogen contents.

1.3.3 Influence of Molybdenum

Molybdenum is besides chromium and nickel another important alloying element in stainless steels. Fig. 29 shows the iron-molybdenum constitution diagram according to Kubaschewski [5]. Like chromium, molybdenum also constricts the area of γ phase. This means, molybdenum also supports the formation of ferrite. In the case of chromium, it takes about 11.5% to produce a complete constriction (loop) of the pure γ range, in the case of molybdenum 2.7% is suffice (see detail in Fig. 29). Together with iron, molybdenum forms intermetallic phases. The most

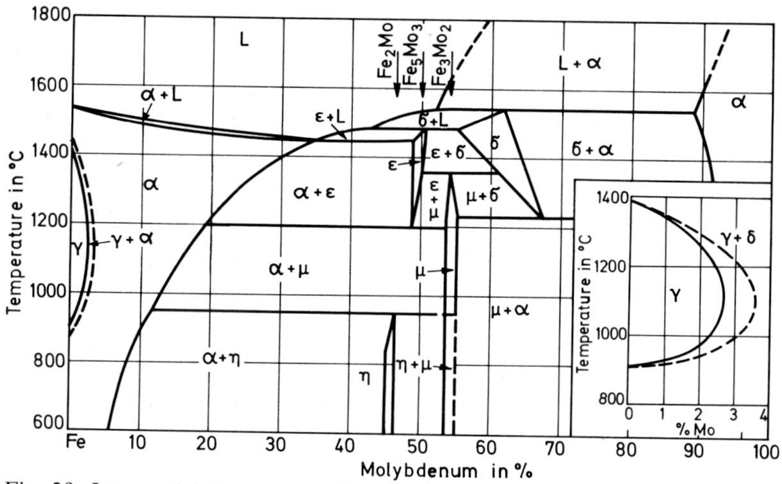

Fig. 29. Iron-molybdenum constitution diagram to Kubaschewski [5]

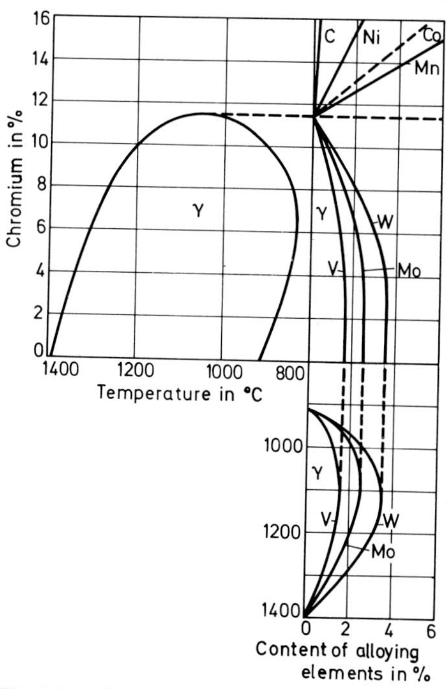

Fig. 30. Influence of different alloying elements on γ range size in the iron-chromium system to Bungardt, Kunze and Horn [48]

important one is the Laves phase Fe_2Mo with approx. 45% Mo[4]. It already starts to precipitate at relatively low molybdenum contents of around 5%. The effects of dif-

[4] Laves phases are intermetallic phases with the general formular A_2B. The major criterion for their formation is the tendency towards interstitial packing in the crystal lattice which reaches a maximum at a ratio of atomic radii of B : A = 1.1225. Examples of Laves phases in stainless steels are Fe_2Mo, Fe_2Ti or Fe_2Nb.

ferent elements on the constriction of the γ range can be illustrated very clearly by way of the changes produced by additions to iron-chromium alloys. Fig. 30 shows according to Bungardt, Kunze and Horn [48] the effects of carbon, nickel, manganese and cobalt (expansion of γ field) and molybdenum, tungsten and vanadium (constriction of γ field). An isothermal profile at about 1050° C through the ternary systems being formed, shows a shift of the γ field boundary towards higher (upper right) or lower chromium contents (lower right) depending on whether an austenite or ferrite forming element is being added. For example, with a content of 1% nickel, the γ field expands to about 13% chromium. But the γ field is constricted to 7% chromium by 1% vanadium and to about 9% chromium by 1% molybdenum.

The combined effect of chromium, molybdenum and carbon on the constriction of the γ range is shown in Fig. 31 on the basis of results compiled by Bungardt, Kunze and Horn [21, 48]. Both the strongly constricting effect of molybdenum and the expanding one of carbon which are excerted on the γ range in the iron-chro-

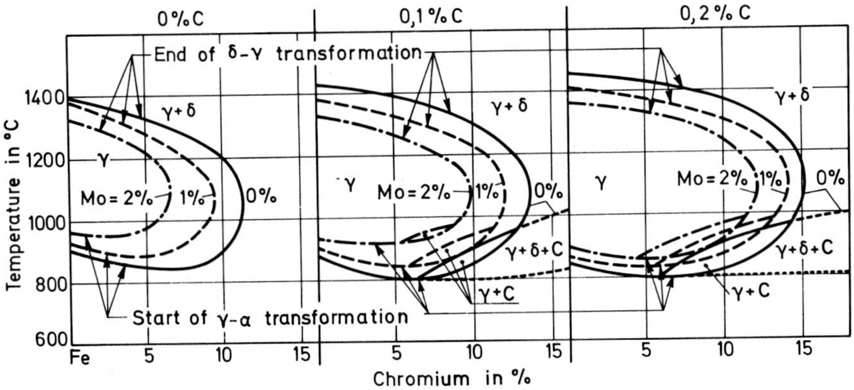

Fig. 31. Influence of carbon and molybdenum on the expansion of the γ field in the iron-chromium system compiled on the basis of results to Bungardt, Kunze and Horn [21, 48] (carbide designated by letter C)

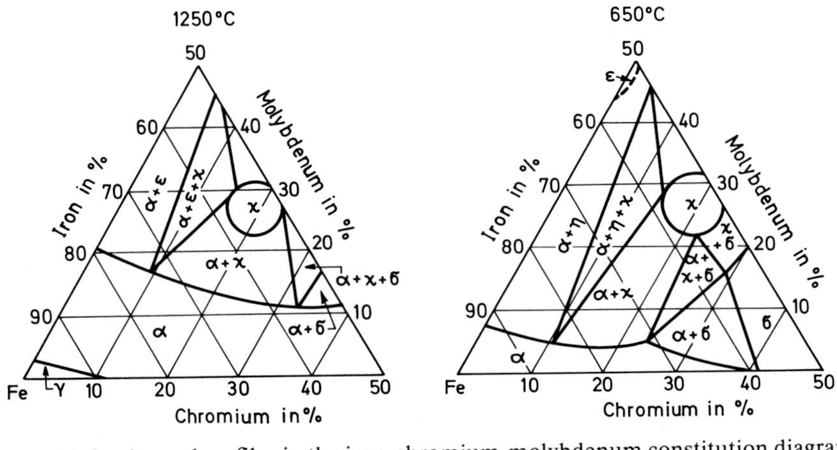

Fig. 32. Isothermal profiles in the iron-chromium-molybdenum constitution diagram at 1250° and 650° C to Brewer, Chang and Shih-Ger [18]

mium system are clearly visible. The boundaries of the γ range indicate which contents of chromium, molybdenum and carbon must be met to avoid or keep within certain limits the formation of δ ferrite, a structural constituent which is often unwanted in stainless ferritic chromium-molybdenum steels. The shifting of the boundary of the γ range towards higher temperatures due to the presence of molybdenum is of importance for heat treatment processes. Molybdenum containing stainless chromium steels must be transformed into the γ range at higher temperatures than those without molybdenum. If all carbides too are to be dissolved, e.g. in the case of hardening, a temperature must be chosen which is above the boundary line of the $(\gamma + C)$ range.

The ternary iron-chromium-molybdenum constitution diagram is shown in Fig. 32 according to Brewer, Chang and Shih-Ger [18] in the form of isothermal profiles at 1250° and 650° C. It appears that the γ range is limited at 1250° C to a very small area. Due to secondary γ-α transformation, the γ range has completely disappeared at 650° C. New phases appear such as the Laves phase Fe_2Mo (η range) and the chi phase (χ range), for which Bechtoldt and Vacher [43] have stated a formula $Fe_{36}Cr_{12}Mo_{10}$. In the isothermal profile at 650° C, the range of this phase passes partially over into the sigma phase range, the area of which is determined by the binary iron-chromium system. The three intermetallic phases, i.e. Laves, chi and sigma phase, may have a detrimental effect—in particular during long-time exposure to elevated temperatures and after heat treatment—on toughness and corrosion resistance of molybdenum containing stainless steels and weld metals.

Fig. 33 shows a concentration profile in the iron-chromium-molybdenum system at 80% iron according to Bechtoldt and Vacher [43]. The precipitation range of the chi phase is shifted by chromium towards lower molybdenum contents. At 17% chromium, this phase already starts to precipitate from about 3% molybdenum onwards. With higher molybdenum contents, Laves phase is also precipitated. The precipitation range of the chi phase extends to much higher temperatures than,

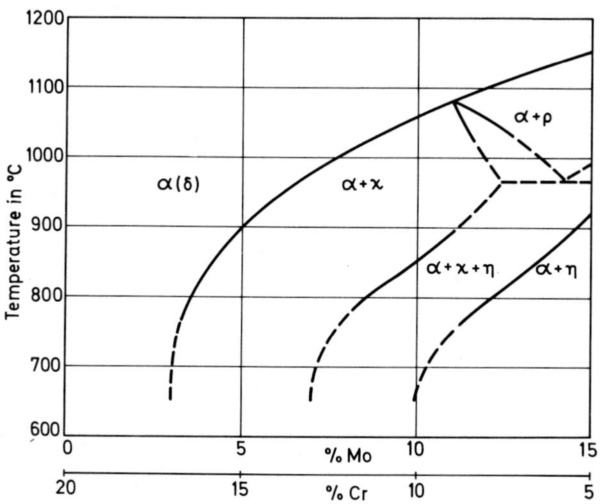

Fig. 33. Concentration profile in the ternary iron-chromium-molybdenum constitution diagram at 80% iron to Bechtoldt and Vacher [43]

for example, that of the sigma phase in the iron-chromium system. This fact also explains why molybdenum containing stainless steels must be solution annealed at higher temperatures than molybdenum-free grades.

The quaternary iron-chromium-molybdenum-nickel constitution diagram has been investigated by Bechtoldt and Vacher [43]. For reasons of representability it is based on alloys with a constant iron content of 70%. The precipitation phenomena of intermetallic phases in this system have been investigated by Kügler, Bäumel and Schmidtmann [44], Weiss and Stickler [45], by Kügler [46], Thier [38], Bungardt and Lennartz [47] and Bungardt, Kunze and Horn [48]. For further information on this topic, please refer to section 4.2.4 and the comprehensive list of supplementary literature at the end of this book. Fig. 34 shows a simplified version of isothermal profiles at 1093° and 816° C of the iron-chromium-molybdenum-nickel

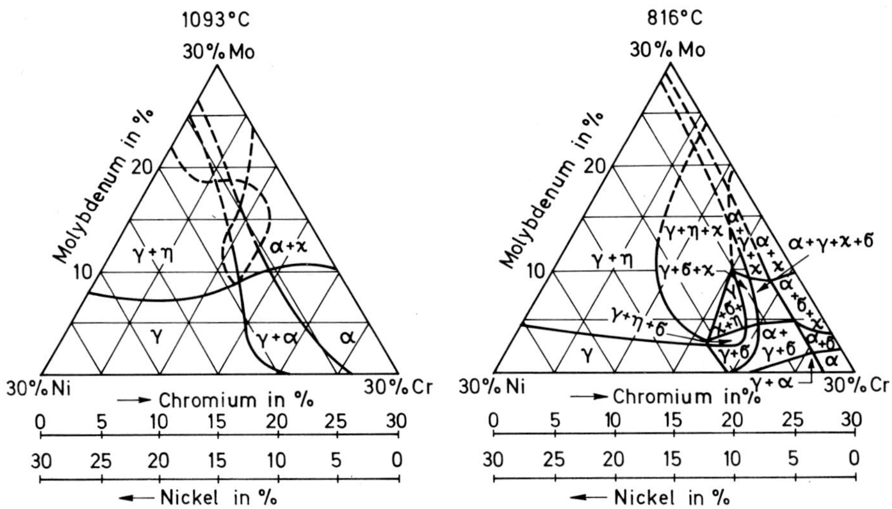

Fig. 34. Isothermal profiles in the iron-chromium-molybdenum-nickel system with constant iron content of 70% at 1093° and 816° C to Bechtold and Vacher [43]. σ = sigma phase, χ = chi phase, η = eta phase (Laves phase Fe_2Mo)

system with 70% iron according to Bechtold and Vacher [43]. It represents a survey of possible phase precipitations in the alloying range up to 30% chromium, molybdenum and nickel with an otherwise constant 70% iron. An investigation of alloys with 20% chromium and 10% nickel which starts at 0% Mo and moves up to higher molybdenum contents indicates that these alloys are at 1093° C and 0% Mo in the pure γ range. With molybdenum contents up to 3%, the $(\gamma + \alpha)$ range is shifted towards lower chromium and higher nickel contents which is due to the relatively strong ferritizing effect of molybdenum. With 1% molybdenum, a 20/10 chromium-nickel alloy is already located in the $(\gamma + \alpha)$ range. The precipitation ranges of the Laves and chi phases are located at higher molybdenum contents. These two phases still exist at 1093° C, i.e. at considerably higher temperatures than the sigma phase.

The isothermal profile at 816° C (Fig. 34) shows that the sigma phase range is shifted by molybdenum to lower chromium and higher nickel contents. The precipitation ranges of the Laves and chi phases start between 2% and 3% molybdenum. Starting with such a content, the occurrence of these phases must be expected, in particular during heat treatment processes and long-time exposure to elevated temperatures, with the decisive criterion being, however, the precipitation kinetics of these phases. These phenomena can no longer be represented in the form of constitution diagrams, since we are confronted in stainless steels with the simultaneous effect of six or more elements. They shall be discussed in detail, however, in section 4.2, (page 125) which deals with the precipitation kinetics of intermetallic phases.

1.3.4 Influence of Manganese

Manganese is like nickel an austenite former. As indicated by the iron-manganese constitution diagram according to Kubaschewski [5], shown in Fig. 35, the area of γ phase is expanded by manganese. Similar to nickel, the γ-α transformation is

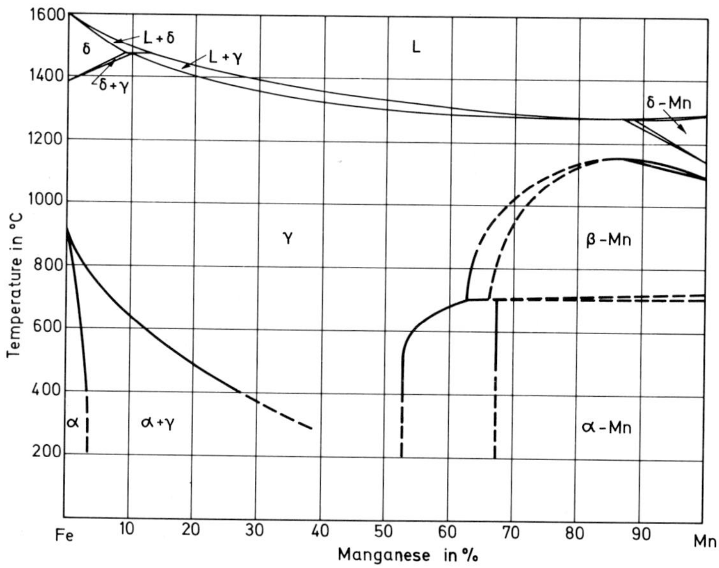

Fig. 35. Iron-manganese constitution diagram to Kubaschewski [5]

shifted again towards lower temperatures with the result that austenite becomes stable down to ambient temperature. But contrary to nickel, manganese forms mixed carbides of the type $(Fe\text{-}Mn)_3C$. For stainless steels, however, these carbides are of no great significance due to the fact that carbon is always linked to stronger carbide forming elements such as chromium or niobium.

If we look at the ternary iron-chromium-manganese system shown in Fig. 36 according to Krainer and Mirt [50] by way of a profile at 700° C and compare it to the iron-chromium-nickel system (Fig. 13, page 13), we see that the austenitizing effect of manganese is weaker than that of nickel. In Fig. 36, the area of γ phase is constricted at about 15% chromium by the precipitation of sigma phase. In the iron-chromium-nickel system (Fig. 13) the area of the γ phase extends from approx. 18%

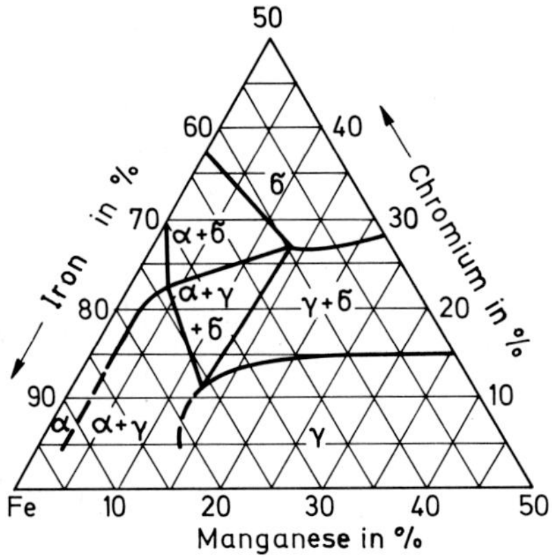

Fig. 36. Ternary iron-chromium-manganese constitution diagram at 700° C to Krainer and Mirt [50] and Schafmeister and Ergang [49]

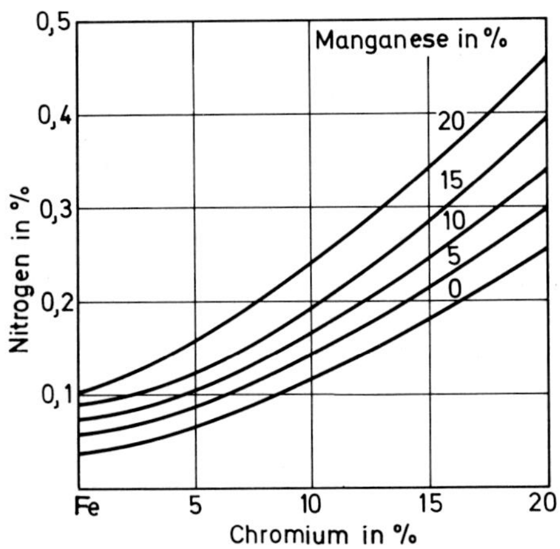

Fig. 37. Solubility of nitrogen at 1600° C in iron-chromium-manganese alloys to Wentrup and Reif [51]

chromium at 10% nickel to about 26% chromium at 40% nickel. Therefore, with regard to its effect on the formation of sigma phase, manganese is not a full but only a partial substitute for nickel. With the addition of nitrogen, the solubility of which is increased by manganese, it is possible to expand the range of pure austenite to about 17% chromium. However the addition of nitrogen does not have the same strong austenizing effect in Cr-Mn steels as it has in Cr-Ni steels. This is also one of the reasons, why corrosion resistant chromium-manganese steels did not prevail and manganese is only used as an alloying addition in stainless chromium-nickel steels. In this way, we profit mainly from two effects of manganese, i.e. the improvement of the hot cracking resistance of fully austenitic weld metal and an increase in nitrogen solubility. Fig. 37 shows the supportive effect of manganese on nitrogen solubility at 1600° C in iron-chromium alloys according to Wentrup and Reif [51].

1.3.5 Influence of Niobium

Niobium possesses a high affinity to carbon. This effect is used in stainless steels and weld metals in particular, to arrest carbon in a stable way in the form of niobium carbide, thereby improving the resistance to intergranular corrosion. We are talking then of niobium stabilized stainless steels and weld metals. Fig. 38 shows the iron-niobium constitution diagram according to Kubaschewski [5]. Niobium severely constricts the area of γ phase (see detail in Fig. 38). Initially, the liquidus temperature is reduced by niobium. At approx. 18% Nb, a eutectic is formed with a melting point of 1356° C. At approx. 45% niobium, the Laves (ε) phase (Fe_2Nb) is being formed, with the ($\alpha + \varepsilon$) and ($\gamma + \varepsilon$) ranges extending to relatively low niobium contents, i.e. 1% in the γ and 0.15% in the α range.

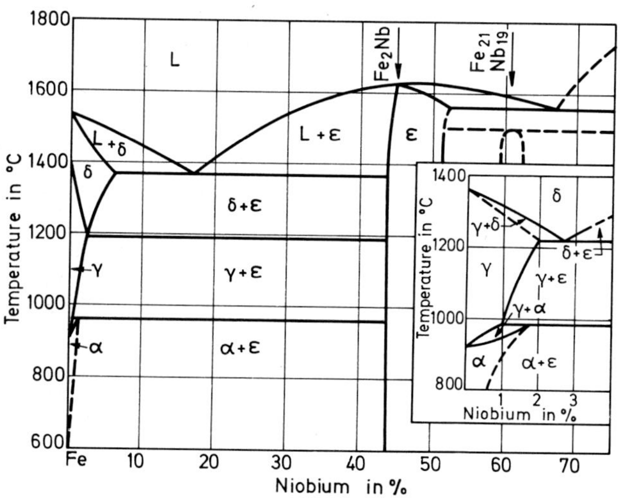

Fig. 38. Iron-niobium constitution diagram to Kubaschewski [5]

Together with carbon, niobium forms very stable carbides, represented by the formula NbC or—in older literature—Nb_4C_3. The affinity of niobium to carbon is that high that niobium is preferentially linked to carbon before it even enters the matrix. The full effect of niobium with regard to the arrestment of carbon is achieved by a socalled "stabilization" annealing which is executed, for example, with niobium containing austenitic 18/8 Cr-Ni steels at the same temperature as the solution annealing of niobium free grades, i.e. at 1050° C. At this temperature, carbon is "stably" linked to niobium on the one hand and residual amounts of ferrite and sigma phase are dissolved in austenite on the other. By subsequent rapid cooling (quenching) this state is maintained down to ambient temperature. To produce full arrestment of carbon, the niobium content must according to the formula NbC be stoichiometrically at least eight times as high as the carbon content. In other words, the minimum "stabilization ratio" of niobium to carbon to be met is $8 \times$ % C.

In case of the high cooling rates during welding, the time available during rapid cooling of the weld pool is often not long enough to achieve a full arrestment of carbon which corresponds to the metallurgical equilibrium. This is also the reason why weld metal which is not "stabilization" annealed, is sometimes alloyed higher with niobium than is required by the formula % Nb = 8 × % C. We are talking then of "over-stabilized" weld metal.

Another phenomenon which is of importance for weldments is a possible re-dissolution of already stably precipitated niobium carbide in the high temperature area of the heat affected zone both in the base material and in previously laid passes by deposition of further weld runs. During welding, this zone is heated up close to the solidus temperature of the stainless steels. With these high temperatures, it is possible that niobium carbide is dissolved again in the matrix closely below or besides the fusion line in the high heated part of the heat affected zone. The amount of carbon dissolved in this way can only partly precipitate again as niobium carbide during the rapid cooling after welding, with the residual carbon remaining in the matrix. Thus, the stabilizing effect can be partially lost. Fig. 39 shows the influence of temperature and holding time on the dissolution of niobium carbide in the matrix during the heat treatment of austenitic steels and during short-time heating up in the heat affected zone during the deposition of a new pass according to findings of Bungardt and Lennartz [47], Irvine, Murray and Pickering [53] and Kautz, Klärner and Schmidtmann [54]. The diagram shows that in case of steels 1–4 an annealing treatment which lasts between 30 minutes and 1 hour causes the amount of re-dissolved niobium carbide to quickly increase with rising temperature. During welding operations, the heat affected zone is only exposed to high temperatures for a very short time. A remarkable dissolution of niobium carbide is only experienced, however at temperatures above 1300° C. But the re-dissolution may

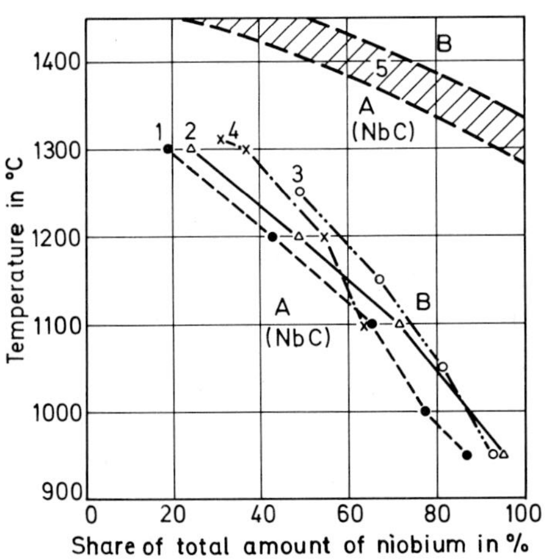

Fig. 39. Influence of temperature and holding time on re-dissolution of stably precipitated niobium carbide. *1–4* steel grades, *5* weld metal after short-time heating caused by deposition of a new pass. *A* – precipitated as niobium carbide, *B* – niobium carbide re-dissolved in the matrix; *1* – steel grade 16% Cr, 25% Ni, 2% Mo + Nb, holding time 1 hour [47]; *2* – steel grade 16% Cr, 16% Ni + Nb, holding time 1 hour [47]; *3* – steel grade 17% Cr, 13% Ni + Nb, holding time 1 hour [53]; *4* – steel grade 16% Cr, 13% Ni, 1% Mo + Nb, holding time 30 minutes [54]; *5* – weld metal grade AWS E 347, 20% Cr, 10% Ni + Nb, holding time approx. 5 seconds

reach considerable values at temperatures close to the solidus point (approx. 1430° C), i.e. immediately below or next to the fusion line of the new pass being deposited on top. This is shown in Fig. 39 by the hatched area 5 in the upper part of the diagram (see also section 9.2.3).

Another important criterion is the influence of niobium on the precipitation of undesirable intermetallic phases. We must remember here that niobium is primarily linked to carbon in form of niobium carbide. In the presence of nitrogen, however, to which niobium also shows a high affinity, other phases may precipitate as well which have been described in detail by Gerlach, Kautz and Schäffler [42]. Apart from niobium carbide and the mixed carbide Fe_3Nb_3C, the formation of carbonitride Nb(CN) and mixed nitride (CrNb)N also takes place. In the case of an excessive niobium addition, the Laves phase Fe_2Nb (in older literature also stated as Fe_3Nb_2) and a mixed carbonitride with the formula $(FeCr)_3Nb_3(CN)$ may also precipitate, in particular if there is more niobium present than is necessary for the arrestment of carbon and nitrogen.

The formation of intermetallic phases in niobium containing stainless steels and weld metals is normally a highly complex matter. Depending on the chemical composition of the grade in question, chromium can be replaced in the phases by iron, nickel or manganese and niobium by molybdenum. There is no generally applicable rule for possible precipitations. It must be noted, however, that apart from the major niobium containing phases such as niobium carbide NbC, niobium carbonitride Nb(CN), the mixed carbide Nb_3Fe_3C and the Laves phase Fe_2Nb, other complex phases will form only after extended exposure to the temperature range of 550–850° C. Precipitation mechanics are also notably influenced by the level of carbon content in comparison to niobium content, since both elements will precipitate stably as niobium carbide. The formation of intermetallic phases can thus be reduced or may even be suppressed completely. Also of importance is the influence of niobium on the precipitation of sigma phase in the iron-chromium-nickel system. According to Hull [55], the effect of niobium is similar to that of molybdenum, i.e. niobium primarily promotes the precipitation of sigma phase. Due to the fact that an excess of niobium will form other phases such as the Laves phase Fe_2Nb simultaneously with NbC, the effect of niobium on the precipitation of sigma phase can be diminished or even totally suppressed.

The iron-niobium constitution diagram according to Kubaschewski [5] (Fig. 38) shows that the γ phase range is severely constricted. Like chromium and molybdenum, niobium promotes the formation of ferrite to the detriment of austenite. As a consequence, niobium also reduces the γ range in the ternary iron-chromium-nickel system. There is, however, a major difference to the effect of molybdenum and chromium. Niobium shows such a high affinity to carbon that it will precipitate to a great extent as stable niobium carbide. Therefore, its effect on the gamma range in the ternary iron-chromium-nickel system is limited in the presence of carbon to the amount of niobium which is not being precipitated as niobium carbide. In other words, depending upon the niobium content, niobium loses more or less of its effect on the constriction of the γ range in carbon containing iron-chromium-nickel alloys.

Of importance with regard to welding operations is the influence of niobium on the formation of low melting phases which may be the cause of hot cracking in the weld metal and the heat affected zone. In the binary iron-niobium, chromium-niobium and nickel-niobium systems, eutectics are formed with melting points which are about 150—200° C lower than those of chromium, iron and nickel respectively. According to Dahl, Düren and Müsch [56], niobium forms low melting phosphides together with phosphorus, chromium and manganese and low melting sulphide-oxide inclusions together with silicon, manganese and chromium. On the other hand, the causes for hot cracking phenomena in the heat affected zone during the welding of fully austenitic niobium containing chromium-nickel steel castings with low phosphorus and sulphur content are related by Dahl, Düren and Müsch [57] to niobium concentrations at the grain boundaries in the coarse cast structure which lead to the formation of low melting niobium and nickel phases which solidify in part below 1160° C. Tamura and Watanabe [58] as well as Schüller [59] report that the formation of hot cracks in the heat affected zone of niobium containing austenitic steels can probably be attributed to a low melting eutectic of niobium carbide or niobium carbonitride with γ phase.

Comprehensive empirical data supports the assumption that niobium will only have a detrimental effect on hot cracking resistance if the solidification of austenitic weld metal is by way of the formation of primary γ crystals , i.e. if a fully austenitic weld metal is formed. If, however, the solidification of weld metal takes place by way of formation of primary δ ferrite crystals, with a ferrite containing weld metal at ambient temperature, the unfavourable influence of niobium on hot cracking resistance is largely compensated for (see also chapter 5.3).

1.3.6 Influence of Titanium

Titanium possesses a high affinity to carbon. It is used, therefore, like niobium, for the arrestment of carbon in the form of stable titanium carbide. Such steel grades are termed then as "titanium stabilized". In stainless filler metals, however, titanium can normally not be employed for stabilization since, unlike niobium, it is largely oxidized in the high temperature droplets during electric arc welding. Fig. 40 shows the iron-titanium constitution diagram according to Kubaschewski [5]. It resembles the iron-niobium diagram. The γ range is highly constricted. A eutectic is formed at 13.2% titanium and iron titanide with the formula Fe_2Ti (ε phase) at

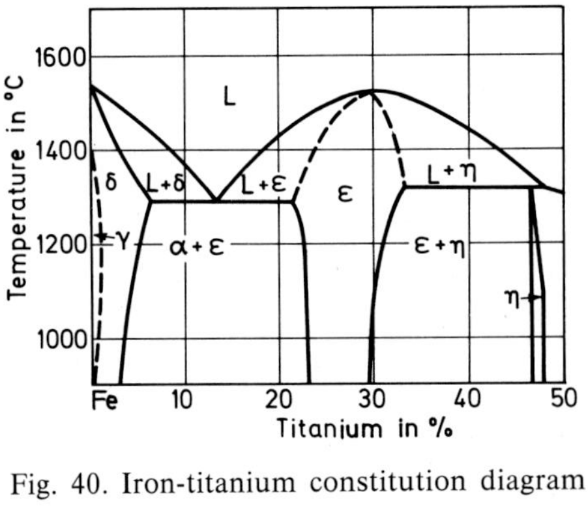

Fig. 40. Iron-titanium constitution diagram to Kubaschewski [5]

approx. 30% titanium. The $(\alpha + \varepsilon)$ range does, however, not extend to such low alloying contents as in the case of niobium.

Together with carbon, titanium forms the highly stable titanium carbide TiC. The theoretical stabilization ratio for a full arrestment of carbon by titanium is: % Ti = 4 × % C. In other words, stabilization with titanium requires only half the amount as compared with niobium stabilization. Because of the fact that titanium also shows a high affinity to nitrogen which is normally present in stainless steels in amounts of up to 0.05%, a stabilization ratio of % Ti = 5 × % C is used in practical applications. With nitrogen, titanium forms a nitride, TiN, and together with carbon a titanium carbo-nitride Ti(CN). The formation of these nitrides takes place parallel to the formation of titanium carbide. Together with nickel, titanium forms a compound with the formula Ni_3Ti with about 25% titanium, which is used for precipitation hardening purposes in the case of high nickel steels and nickel-base alloys.

According to its effect in the iron-titanium system, titanium constricts the γ range in the ternary iron-chromium-nickel constitution diagram. Normally, this effect appears only if titanium is present in excess, i.e. with a content above the stabilization ratio of % Ti = 5 × % C. Regarding its direct and indirect effects, the same considerations apply as with niobium, i.e. the formation of titanium carbide must take into account the arrestment of both titanium and carbon, the constricting or expanding effect of which on the γ range is cancelled the moment these elements are stably precipitated.

The effects of titanium up to contents of 2% on the location of the $(\delta + \gamma)$ range in the iron-chromium-nickel system and on the ranges of sigma phase precipitation have been investigated by Hattersley and Hume-Rothery [60] in low carbon steels with 17—34% chromium and 24—30% nickel in the temperature range of 650—1150° C. Like niobium, titanium promotes the formation of delta ferrite to the detriment of austenite. The effect on the $(\delta + \gamma)$ range is similar to that of molybdenum and niobium. In the ternary iron-chromium-nickel system, 1% titanium shifts the phase boundary between the $(\delta + \gamma)$ range and the pure γ range to about 1.2% lower chromium contents. The range in which the sigma phase precipitates in the iron-chromium-nickel system is shifted by titanium towards the iron corner. If, however, the phase Ni_3Ti appears which is only the case with relatively high nickel contents, the effect of titanium on sigma phase precipitation is diminished because part of the titanium has been absorbed. Please note, however, that in the case of carbon containing iron-chromium-nickel alloys this effect may occur only if the titanium content is greater than the stabilization ratio % Ti = 5 × % C.

The possible formation of low melting phases due to the presence of titanium have been investigated by Schüller [59] in 18/10 chromium-nickel steels. The first melting takes place at approx. 1340° C, with the predominant involvement of titanium carbo-nitrides. It can be assumed that we are confronted again, as with niobium, with a eutectic of titanium carbo-nitride and γ phase. The formation of liquid phases is less pronounced than in the case of niobium.

1.3.7 Influence of Silicon

In the last 20 years, silicon has gained in importance as an alloying element for stainless steels because additions of 4—5% greatly improve the corrosion resistance

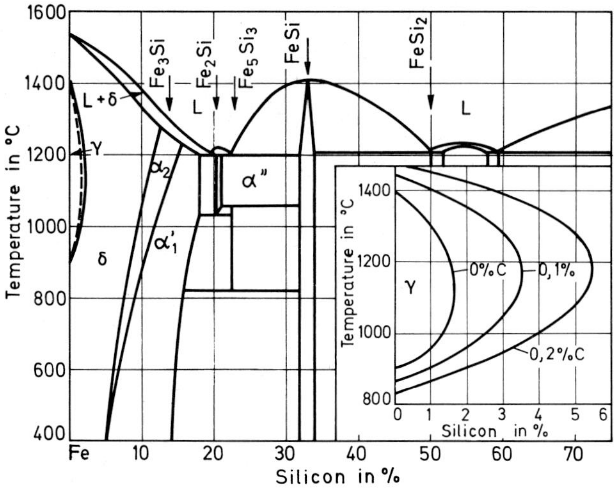

Fig. 41. Iron-silicon constitution diagram to Kubaschewski [5]

of austenitic chromium-nickel steels to strong nitric acid. In heat resistant chromium and chromium-nickel steels, silicon is used in contents of 1—3% to improve the scaling resistance. Fig. 41 shows the iron-silicon constitution diagram according to Kubaschewski [5]. Silicon severly constricts the range of γ phase, i.e. it promotes the formation of δ ferrite to the detriment of austenite. According to Houdremont [9], the γ range extends at 1150° C only to 1.75% Si if the carbon

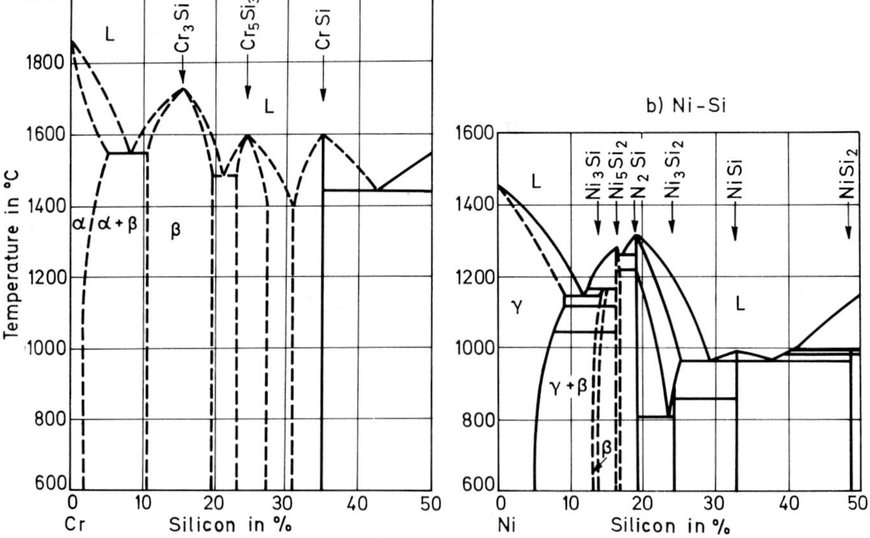

Fig. 42. Chromium-silicon constitution diagram to Elliott [3] and nickel-silicon diagram to Hansen and Anderko [2]

content is below 0.002%. The ($\gamma + \delta$) range extends at the same temperature to 2.15% silicon. As shown by earlier investigations according to Scheil [61], see detail in Fig. 41, the γ range is expanded towards higher silicon contents by the presence of carbon. With silicon contents above 5%, a number of brittle iron silicides appear (Fe_3Si, Fe_2Si, Fe_5Si_3, FeSi). Similar to the iron-silicon system, an intermetallic compound Cr_3Si (β phase) appears in the chromium-silicon system (Fig. 42 a) at approx. 15% Si, with the ($\alpha + \beta$) range extending to about 2% Si. In the nickel-silicon system (Fig. 42 b) too, we find a β phase Ni_3Si with about 14% Si, the range of which (γ Ni + β) extends to about 5% silicon. It is safe to assume, therefore, that in the ternary iron-chromium-nickel system too, intermetallic phases will already appear at relatively low silicon contents.

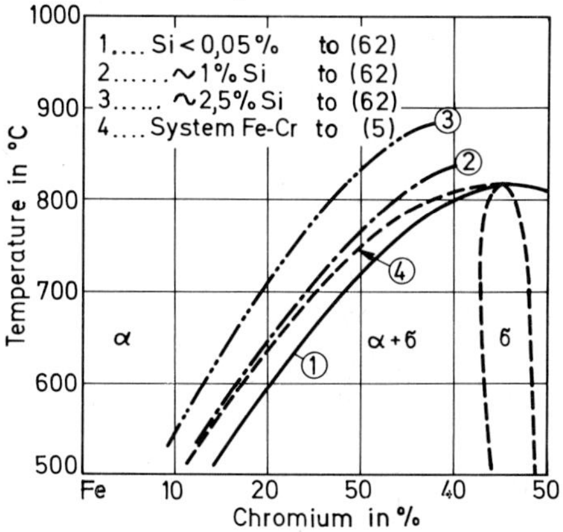

Fig. 43. Influence of silicon on the sigma phase range in the iron-chromium system. 1—3 Influence of silicon to Schüller [62], 4 equilibrium lines to Kubaschewski [5]

Of much greater importance is the influence of silicon on the expansion of the sigma phase range. This has been investigated in the binary iron-chromium system by Schüller [62]. The results of his work are supplemented with newer investigations on the iron-chromium system by Kubaschewski [5] and shown in simplified form in Fig. 43. This diagram includes both pure iron-chromium alloys and standard chromium steel grades with silicon contents up to 2.5%. Silicon greatly expands the sigma phase range towards lower chromium contents. At 2.5% silicon, the sigma phase range extends at approx. 550° C to about 10% chromium. The influence of silicon in the ternary iron-chromium-nickel system seems to be similar to that in the binary iron-chromium system. The presence of silicon not only shifts the sigma phase range towards lower chromium contents but also expands it with regard to precipitation temperatures. In austenitic chromium-nickel steels, precipitation of sigma phase starts normally only after 10 hours of annealing at temperatures above 620° C. With silicon contents above 2.5%, sigma phase precipitation already starts at temperatures around 550° C. The same applies to the upper temperature range of the sigma phase which is shifted by silicon additions above 2.5% by about 50—100° C towards higher temperatures. This means that such steel grades must be solution annealed at higher temperatures than low silicon grades in order to completely dissolve any sigma phase that may have formed.

Silicon also increases the activity of carbon. This means that the precipitation of the mixed carbide $M_{23}C_6$ is speeded up by silicon. According to Horn and Kügler [63], the mixed carbide $M_{23}C_6$ precipitates rather quickly in austenitic chromium-nickel steel in the case of higher silicon additions. If nitrogen is also added, however, a carbo-nitride (π phase) is also precipitated besides $M_{23}C_6$ from about 3% silicon onwards, for which Masumoto and Imai [64] state the formula $M_{11}(CN)_2$. With the appearance of this carbo-nitride, the precipitation of the mixed carbide $M_{23}C_6$ is slowed down since part of the carbon is required now for the formation of π phase. At 5% silicon, we find according to Horn and Kügler [63] apart from small amounts of $M_{23}C_6$ predominantly π phase and sigma phase and after prolonged annealing times also chromium silicide Cr_3Si (β phase). Additions of nitrogen have a positive effect on silicon containing austenitic chromium-nickel steels since the precipitation of the mixed carbide $M_{23}C_6$ and sigma phase is slowed down by the formation of the carbo-nitride $M_{11}(CN)_2$.

Silicon also influences the location of the ($\delta + \gamma$) range in the ternary iron-chromium-nickel system. Investigations in the form of constitution diagrams are not available, but the effect of silicon as a strong ferrite forming element ought to be similar to that of molybdenum, i.e. silicon additions in the ternary iron-chromium-nickel system will expand δ ferrite and ($\delta + \gamma$) range to the detriment of austenite. In other words, the formation of an austenitic structure in higher silicon containing austenitic chromium-nickel steels and weld metals requires higher nickel contents than in the corresponding low silicon grades.

The influence of silicon on the formation of low melting phases is also of importance, in particular with regard to the hot cracking phenomena in the weld metal. In the binary iron-silicon, chromium-silicon and nickel-silicon systems, eutectics are formed with melting points which are considerably below those of the individual elements. This applies especially for the nickel-silicon system (Fig. 42), where we find a eutectic with a melting point of 1152° C at 11.5% silicon and one with 964° C at 29% silicon. It is safe to assume that silicon promotes the formation of low melting phases. This applies in particular to steels with primary solidification to γ crystals. Dahl, Düren and Müsch [56] have found silicon concentrations at the grain boundaries of such steels which they consider to be the cause of hot cracking, these being due to the formation of liquid phases with high contents of silicon, nickel and niobium. Ogawa and Tsunetomi [65] arrived at similar results. Because of segregation phenomena, they assume that the formation of low melting phases of types NiSi (996° C), NiSi-Ni_3Si_2 (964° C) and Ni-Ni_3Si (1152° C) occur at the grain boundaries of primary precipitated austenite crystals. In weld metal, the detrimental effect of silicon shows up much stronger with primary solidification to γ crystals than with solidification through the formation of primary δ ferrite crystals.

1.3.8 Influence of Sulphur

Sulphur represents an undesirable impurity in stainless steels and stainless steel weld metals which gives rise to hot cracking problems during welding due to the formation of low melting sulphides which may precipitate during solidification as liquid film between the grain boundaries of newly formed crystals. They disturb the adhesion of grains and produce separations during shrinkage which in the course of the cooling process of the solidifying crystals appear in the form of hot cracks. Fig. 44a shows the iron-sulphur constitution diagram according to Kubaschewski [5]. Sulphur forms at 32% with iron a low melting eutectic with a melting point of 988° C. It shows a very poor solubility in γ iron which amounts to 0.01% at 988° C. This means that this eutectic will be maintained during cooling to this temperature as a liquid phase if the solubility limit of 0.01% sulphur is being exceeded.

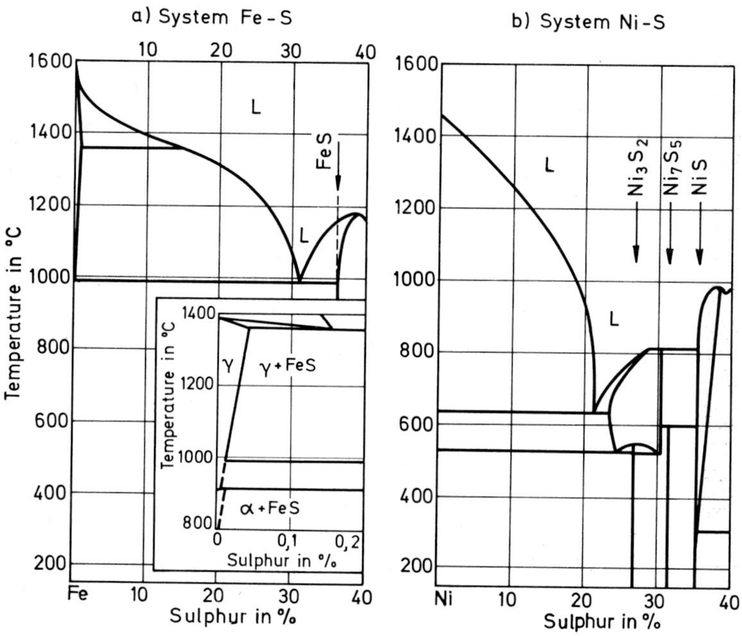

Fig. 44. Iron-sulphur constitution diagram to Kubaschewski [5], nickel-sulphur constitution diagram to Elliott [3]

Fig. 44b shows the nickel-sulphur constitution diagram according to Elliot [3]. At approx. 22% sulphur, a very low melting nickel-sulphur eutectic is formed with a melting point of 637° C. The situation is aggrevated by the fact that sulphur is almost completely insoluble in nickel. According to Elliott, there is a solubility limit of less than 0.006% sulphur, but sulphide phases at the grain boundaries have also been found in high purity nickel with sulphur contents as low as 0.0009%. Chromium and sulphur form a eutectic at 32.5% sulphur with a melting point of 1350° C. Like the nickel eutectic it is almost completely insoluble in chromium.

With regard to the influence of sulphur in the ternary iron-chromium-nickel system, we can gather from the binary systems that rising nickel contents will produce progressively lower melting sulphur eutectics the solubility of which in the matrix is diminished by rising contents of nickel as well as chromium. This means, the precipitation range of sulphur eutectics which in the case of pure iron begins at 0.01% sulphur is shifted to ever lower contents if more and more nickel and chromium are contained in the alloy. We know from practical application that in nickel-based alloys with more than 50% nickel we may encounter low melting sulphide eutectics as insoluble sulphide phases as far down as 0.005% sulphur.

It is common knowledge that weld metal which solidifies in the iron-chromium-nickel system from the melt in the form of δ ferrite crystals (left of the eutectic groove in Fig. 4 and 5) is less susceptible to hot cracking and the negative effect of sulphur is much less severe than in the case of weld metal solidification as primary γ crystals (right of the eutectic groove in Fig. 4 and 5). According to Sadowski [66], this is due, amongst other reasons, to the higher solubility potential of δ ferrite crystals for the elements silicon, niobium, sulphur and phosphorus. According to latest investigations, there must be other factors too which are responsible for the strong formation of hot cracks in stainless steel weld metals which solidify by way of primary γ crystals which shall be discussed in more detail in chapter 5.3. For fully austenitic niobium containing chromium-nickel-molybdenum-copper alloyed weld metal with a nickel content of 34%, Brown and Koch [67] recommend a maximum sulphur content of 0.005% to avoid the danger of hot cracking.

The effect that manganese has on steels for the transformation of the low melting iron sulphide into the higher melting manganese sulphide with a melting point of 1620° C has been known for a long time and is successfully employed, for example, to avoid hot cracking in forging and rolling operations. This effect of manganese has also been used since the mid-thirties in austenitic weld metals. In stainless steels manganese was used during the forties as a substitute for nickel where it aquired a bad reputation because of poor results of corrosion resistance. During the last 20 years, however, manganese has been added with great success to highly corrosion resistant fully austenitic chromium-nickel-molybdenum weld metal, to reduce sulphur induced hot cracking problems. The effect of manganese, however, is reduced by increasing nickel contents, with the formation of the very low melting nickel-sulphur eutectic becoming more and more apparent. Also of importance in this context is the fact that the two sulphides MnS and Ni_3S_2 are completely soluble in each other in the liquid state but show only very low solubility in the solid state. During the cooling of a mixture of MnS and Ni_3S_2, the higher melting manganese sulphide will precipitate first and the still liquid residual melt becomes enriched more and more with the very low melting nickel sulphide. For the welding of fully austenitic stainless steels with nickel contents between 7—25%, manganese still remains—especially for fully austenitic weld metals—a very valuable alloying element for the reduction of sulphur induced hot cracking phenomena, with strongest effects being achieved according to practical experience from about 4% manganese onwards.

According to Matsuda, Katayama, Nakagawa and Arata [68], the effect of manganese on fully austenitic 25/20 Cr-Ni weld metal can be supported by the addition of rare earth elements, in particular lanthanum. The positive effect of lanthanum is due on the one hand to the increase it produces in the melting point of manganese suphide and on the other hand to the formation of a high melting lanthanum oxysulphide La_2O_2S. In addition, lanthanum also produces a positive effect on the influence of phosphorus (see section 1.3.9). Optimum additions of lanthanum can be achieved if its content is determined by the formula $La = 4.5 \times \% P + 8.7 \times \% S$. Higher lanthanum contents, however, show according to Matsuda et al. [68] a negative effect on hot cracking resistance, due to the formation of a low melting lanthanum-austenite eutectic.

The low sulphur contents required in stainless steels for reasons of good hot forming properties and good weldability exert no noticeable influence on the ranges of δ and γ phases in the iron-chromium-nickel system or on the precipitation phenomena of intermetallic phases.

1.3.9 Influence of Phosphorus

The iron-phosphorus constitution diagram according to Kubaschewski [5] and the nickel-phosphorus diagram according to Hansen and Anderko [2] and Elliott [3] are shown in Fig. 45. Like sulphur, phosphorus forms low melting eutectics with the

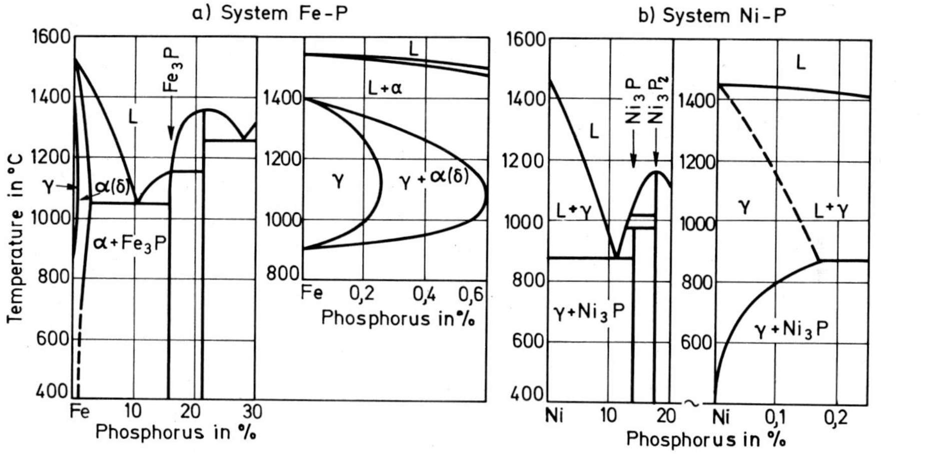

Fig. 45. Iron-phosphorus constitution diagram to Kubaschewski [5], nickel-phosphorus constitution diagram to Hansen and Anderko [2] and Elliott [3]

basic elements of stainless steels, i.e. iron, chromium and nickel. These are found in the iron-phosphorus system at 1050° C and 11% phosphorus and in the nickel-phosphorus system at 880° C and 12% phosphorus. Contrary to sulphur, however, both iron and nickel are capable of dissolving relatively large amounts of phosphorus at the eutectic's melting point, i.e. γ iron 0.25% phosphorus and nickel 0.17% phosphorus (detail in Fig. 45). In theoretical terms, this means that low melting phosphorus eutectics can form only if the above solubility limits are exceeded. If contents are below that which is always the case with stainless steels and their weld metals, phosphorus is dissolved in δ ferrite or in austenite and can no longer appear in the form of an individual liquid phase. Nevertheless, phosphorus has a negative effect on hot cracking behaviour in combination with other low melting phases such as sulphides, silicides or borides. Houdremont [9] states two reasons for this effect, i.e.:

— the great difference in the phosphorus content of the solidus and liquidus lines in the iron-phosphorus system (Fig. 45) which induces the formation of strong primary segregations during solidification of the melt (see also section 2.7);
— the relatively low diffusion rate of phosphorus in both γ and α phase which inhibits the equalization of existing segregations (see also Table 1, page 50).

Sadowsky [66] states that the solubility of phosphorus is much lower in fully austenitic weld metal which solidifies through the formation of primary γ crystals than in the case of δ ferrite crystals. He indicates a solubility of 0.25% phosphorus in γ crystals at 1150° C as compared to 2.8% in ferrite at 1050° C.

Brooks and Lambert [69] recommend that for the crack-free welding of stainless steels which show after welding only very small ferrite contents or no ferrite at all in the weld metal deposit the total amount of phosphorus and sulphur be kept below the value of 0.01%. Bernstein, Carlen and Rick [70] on the other hand, recommend limiting the phosphorus and sulphur content in stainless steels with low ferrite contents to 0.015% max. each. According to Matsuda, Nakagawa, Katayama and Arata [68], additions of rare earth elements—in particular lanthanum—to fully austenitic weld metals will not only have a positive effect on the influence of sulphur (see section 1.3.8) but also on phosphorus, by reducing the crack propagating effect of this element. This is related to the formation of lanthanum phosphide and of a complex chromium—lanthanum phosphide of the type M_3P. Since an excess of lanthanum may form, however, a low melting eutectic together with γ phase, the optimum lanthanum content for the arrestment of sulphur and phosphorus, determined by the formula $La = 4.5 \times \% P + 8.7 \times \% S$, should not be exceeded to any great extent.

Our own experience has shown that phosphorus contents of up to 0.025% are still rather harmless regarding the hot cracking resistance as long as the solidification of the weld metal takes place by way of primary δ ferrite crystals. If, however, primary solidification is by way of primary γ crystals which results in a fully austenitic weld metal, the effect of all crack propagating and crack inhibiting elements must be fully considered (for more details see section 5.3.3 and 5.3.4).

Phosphorus has a severe constricting effect on the γ range in the iron-phosphorus system (Fig. 45a). The same applies to the ternary iron-chromium-nickel system. Normally, the phosphorus content of stainless steels and weld metals is so low that this effect is negligable. The same applies to the effect of phosphorus with regard to a possible influence on the precipitation phenomena of intermetallic phases. A possible precipitation hardening effect by iron or nickel phosphide cannot be utilized since the required phosphorus contents in stainless steels and their weld metals would substantially exceed the limit beyond which this element has a damaging effect.

1.3.10 Influence of Copper

Copper is added to stainless steels in contents of up to 3.5%, to improve the corrosion resistance as well as increase the hardness and tensile strength through precipitation effects of this element. In the iron-copper constitution diagram according to Kubaschewski [5], copper is soluble in γ iron at 1100° C up to approx. 7.5% and at 835° C up to approx. 3%. In α iron, the solubility at 850° C is 2.0% and drops to approx. 0.4% at 650° C. The precipitation hardening of copper alloyed steel grades is based on this effect. Nickel and copper are completely soluble in each other. Chromium on the other hand shows practically no solubility for copper. According to Houdremont [9] copper shows an austenitizing effect in the ternary iron-chromium-nickel system which is weaker, however, than that of nickel.

The influence of copper on the hot cracking sensitivity of fully austenitic 17% Cr/19% Ni steel alloys has been investigated by Hull [71] by way of the cast pin tear test. According to this test, copper increases the hot cracking sensitivity, but its effect is smaller than that of silicon, niobium and titanium. With regard to stainless steel weld metal, however, this effect of copper has not yet been investigated but it seems to be of no great significance.

Copper shows a lesser affinity to oxygen than iron. During scaling, copper enrichement takes place beneath the layer of scale being formed which, induced by copper penetrating into the grain boundaries, may lead to the formation of a low melting iron-copper eutectic and to surface cracks. With higher contents of chromium, as present, for example, in austenitic chromium-nickel steels, this effect is inhibited by the rather thin and well adhering layer of scale which is formed to such an extent that the negative effect of copper is no longer felt.

1.3.11 Influence of Boron

Boron is sometimes added in small amounts to stainless steels during melting for crystal refinement during solidification, in order to achieve better hot forming properties. Small amounts of boron may also be "carried over", for example, with ferro-titanium in titanium alloyed stainless steels. Stainless steels with higher boron contents up to 2% are used in nuclear engineering because of boron's excellent neutron absorption properties. The iron-boron constitution diagram according to Kubaschewski [5] is shown in Fig. 46a. Boron substantially lowers the melting point of iron and forms a low melting eutectic at 1149° C and 3.8% boron. At this temperature, the solubility of boron in austenite is only 0.021% which means that the appearance of this low melting eutectic must be expected with higher boron contents at temperatures of approx. 1150° C. The solubility for boron decreases with falling temperature (see detail in Fig. 46a). It is only 0.005% at 1000° C and practically drops to zero at 700° C. In this process, the brittle iron boride Fe_2B is precipitated. The nickel-boron constitution diagram according to Hansen and Anderko [2] is shown in Fig. 46b. Nickel also forms a low melting eutectic with boron at 1140° C and 4% boron. After solidification, the solubility of boron in nickel practically drops to zero.

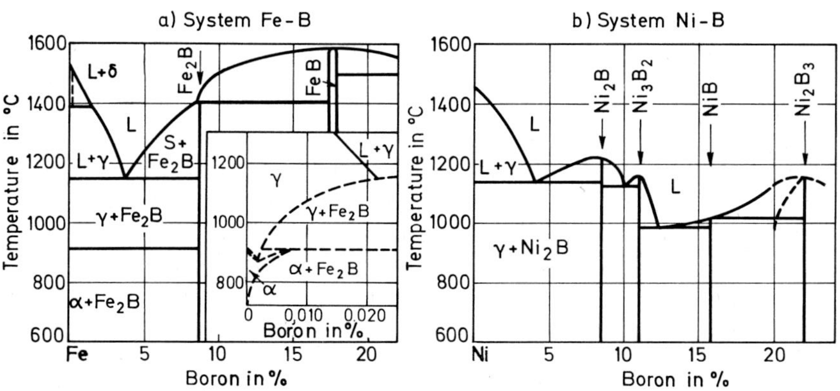

Fig. 46. Iron-boron constitution diagram to Kubaschewski [5], nickel-boron constitution diagram to Hansen and Anderko [2]

At approx. 8% boron, a nickel boride Ni_2B is formed with other borides such as Ni_3B_2, NiB and Ni_2B_3 being formed at higher boron contents.

According to Elliott [3], chromium and boron form a eutectic at approx. 1570° C and 3% boron. At 5% boron, a chromium boride Cr_4B is formed and further borides are formed with higher boron contents. Like nickel, the solubility of chromium for boron is practically zero after solidification.

In the ternary iron-chromium-nickel system, we can assume that the low solubility of iron for boron which amounts to 0.021% at 1149° C is further diminished by the addition of chromium and nickel. At the same time, we must assume that similarily to the binary iron-boron and nickel-boron systems, low melting eutectics will also appear in the ternary system which, due to the greatly reduced solubility for boron may show up as liquid phases even in the case of very low boron contents. The ternary iron-chromium-nickel system has been investigated by Robitsch [72] and Weingerl [73] for boron contents of up to 2%. Three areas of borides could be determined which contained predominantly iron or chromium or nickel, depending upon the pertinent range of individual elements. According to Swoboda and Malissa [74], boron which is practically insoluble in austenite shows up in 17/15 chromium-nickel steels as a mixed iron-chromium boride with about 9% boron, 88% (iron + chromium) and 1.3–2.3% (nickel + manganese). This mixed boride forms together with the austenitic matrix a eutectic of high hardness with a melting point around 1200° C. At 2% boron, the alloy is practically completely made up of this boride eutectic. Malissa, Grasserbauer, Hoke, Draxler and Weingerl [75] found that in chromium-nickel-molybdenum-copper steel grades with additions of up to 2% boron mainly chromium rich mixed borides are precipitated. According to Kügler [76], stainless 18/13 chromium-nickel steels with boron additions of between 0.2 and 1.8%, always represent a compromise between toughness and neutron absorption properties because increasing boron contents will promote the precipitation of brittle mixed borides which increasingly reduce the toughness of the austenitic matrix.

Because of the formation of low melting eutectics and the low solubility of boron in iron-chromium-nickel alloys, hot cracking susceptibility is a problem even with very low boron contents. Of 20 elements tested, Hull [71] classifies boron as the one that diminishes the hot cracking resistance of weld metal most severely. Donatti, Guttmann and Zacharie [77] detected in 18/10 chromium-nickel steels the low melting iron-chromium-boron eutectic $(FeCr)_2B$ with a melting point of 1180° C and recommended limiting the boron content to a maximum of 35 ppm (0.0035%), in order to avoid hot cracking. Gerlach and Schmidtmann [39] have investigated fully austenitic niobium-stabilized 16/6/2 chromium-nickel-molybdenum steels and found hot cracks only with boron contents above 0.006%, whereas Brooks [78] noted a marked improvement in hot cracking resistance if boron contents were lowered from 0.006 to 0.001%, the latter, however, in fully austenitic steels with 15% chromium, 25% nickel, 1.25% molybdenum, 0.26% vanadium and 2.2% titanium which, because of their alloying composition, must already be considered as highly susceptible to hot cracking. According to our own experience, the somewhat contradictory results as to the maximum admissible boron content in austenitic steels originate from the fact that any consideration of hot cracking susceptibility must take into account not only one single element but the combined effect of all crack propagating elements, together with the solidification processes which occur during welding. Nevertheless, boron is an element which may already have a negative influence on hot cracking resistance even if present in very low amounts.

Very little is known about the effects of boron on the area of γ and δ crystals in the ternary iron-chromium-nickel system. In low alloyed steels, for example, boron has an austenite stabilizing effect. This effect could also influence the austenite range in the iron-chromium-nickel system. Williams and Talks [79] found that for austenitic chromium-nickel-molybdenum steels boron contents as low as 60 ppm (0.006%) will already strongly inhibit the precipitation of sigma phase during long-time exposure to elevated temperatures. The same authors report the formation of a mixed carbo-boride of type $M_{23}(CB)_6$, particularly at the grain boundaries of austenite crystals, which also contributes to the slowdown of sigma phase formation in a similar way as the mixed carbide $M_{23}C_6$. The ratio of boron to carbon content is considered to be a major parameter in this context.

1.3.12 Influence of Oxygen

Oxygen is an undesirable tramp element in stainless steels and their weld metals if its content exceeds a certain limit. In the liquid state, the three base elements of stainless steels—iron, chromium and nickel—can dissolve around their respective melting points considerable amounts of oxygen in the range between 0.24 and 0.60%. With rising temperatures, the solution potential of iron, nickel and chromium for oxygen is considerably increased. When these elements solidify, however, solubility drops down to values in the range of one hundredth of a percent. In the ternary iron-chromium-nickel constitution diagram we find, similar to the binary systems, a relatively high oxygen solubility in the liquid metal. These processes are strongly influenced, however, by the addition of deoxidizing alloying elements such as carbon, manganese, silicon, aluminium or titanium which are always used for the treatment of stainless steels during melting, in order to lower the oxygen content. Thus, the oxygen content is already decisively influenced by these deoxidizing elements with iron still being liquid. Oxygen is retained in the form of oxides, i.e. transformed from its solution in iron into insoluble nonmetallic oxides.

Fig. 47a shows the reduction of the oxygen content dissolved in liquid iron at 1600° C through the addition of deoxidizing elements according to Knüppel [80]. Chromium shows the weakest and aluminium the strongest effect on the reduction of oxygen content. The effect of carbon can be increased very efficiently by the reduction of the carbon monoxide pressure, which is produced during deoxidation, to 0.01 bar. This is one of the major effects of the vacuum treatment of liquid steel. The other deoxidizing elements are changed into their oxides such as MnO, SiO_2, TiO_2, and Al_2O_3 which being lighter than liquid steel float on the surface, where they are absorbed by the slag. In this way, the oxygen content in liquid steel can be substantially reduced. The metallurgical reactions during deoxidizing are largely temperature dependent, with the temperature around 1600° C being of decisive importance for final oxygen content in conventional steel melting processes.

The weld metal situation is completely different. The process of melting a filler metal in an electric arc creates totally different conditions as compared to the situation in the steel melting furnace. In the steel plant, metallurgical processes

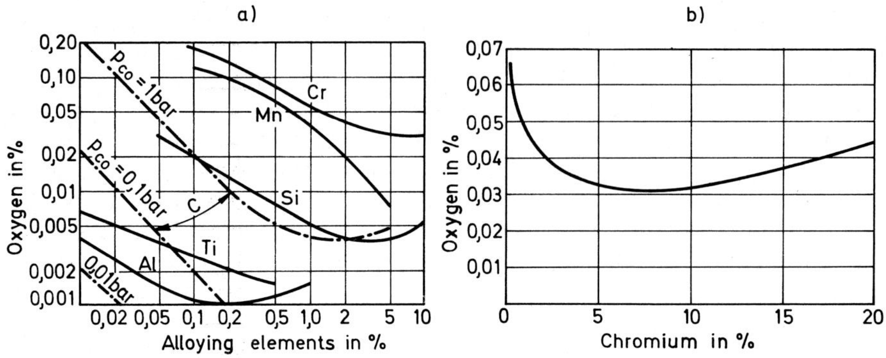

Fig. 47. *a* Influence of deoxidizing elements on the oxygen content of iron at 1600° C to Knüppel [80], *b* Influence of higher chromium contents on the oxygen content of iron at 1600° C to Kojima, Sakao and Sano [83]

normally last between 0.1 and 1 hour at temperatures of around 1600° C, with the possibility of separating nonmetallic deoxidation products and to apply a subsequent vacuum treatment. In electric arc welding, melting takes place in the course of seconds and at much higher temperatures in the range of 2000–2700° C in the liquid droplet. Oxidation and deoxidation processes are taking place simultaneously whereas in the steel furnace they happen in sequence. During welding, time spans are that short that the separation of oxidic deoxidation products is hardly possible. The predominant part remains in the weld metal in the form of nonmetallic inclusions. In electric arc welding, metallurgical reactions are "frozen" according to Rabensteiner [81] during the very rapid cooling process in the temperature range of approx. 2000–2100° C. This means that we must expect much higher final temperatures of these metallurgical reactions in temperature dependent deoxidation processes as compared to steel melting in the electric furnace.

Therefore, we find much higher oxygen contents in the weld metal which may differ by the power of ten as compared to the contents established in steel, because the oxygen content in the weld metal is determined or by the quantity of nonmetallic oxydic inclusions which are "frozen" in the weld metal. The fact that weld metals possess, inspite of this relatively high content of nonmetallic inclusions excellent properties which compare favourably to those of steel grades is a result of the extraordinary fineness of these inclusions. They strongly promote nucleation in all crystallization and transformation processes, exerting an additional grain refining effect on the micro-structure of the weld metal.

Fig. 47a shows that chromium as a major element in stainless steels has a rather weak deoxidizing effect. The influence of higher chromium contents of more than 10% is shown in Fig. 47b according to Kojima, Sakao and Sano [83]. Initially, chromium reduces the oxygen content. The lowest point is reached at 8% chromium. Then, the oxygen content rises again. With regard to the weld metal, the reactions at temperatures above 1600° C are of interest. Plessing [84] determined by way of calculation a strong temperature dependence of the chromium-oxygen equilibrium constant. This explains, why oxygen contents again rise with higher chromium contents.

In the melting of stainless steels, oxygen content can be limited to values around 0.005% and below by the application of strong deoxidizing agents in combination with advanced steel making processes such as vacuum technology. The situation regarding the weld metal is quite different, especially if welding is executed by the process most widely used, i. e. electric arc welding with coated electrodes. In this process, oxygen will enter the weld metal on the one hand through the electrode coating, e.g. the silicates contained therein, and on the other by contact with the surrounding air, e.g. when welding with a long arc. Vacuum treatment is only possible in electron beam welding but not in the commonly used electric arc welding process. Regarding the choice of deoxidizing agents, there are also limits in electric arc welding, e.g. in the case of silicon by a limitation of silicon content in the weld metal, in order to avoid the precipitation of sigma phase or an increased risk of hot cracking. Aluminium too, which is a very effective deoxidizing agent, has often a negative influence on the welding operation. Alloying elements such as magnesium or rare earth elements may be evaporated or oxidized by the very high temperatures in the electric arc which renders them ineffective. Higher contents of carbon, another strong deoxidizing agent is often not feasible for reasons of corrosion resistance. What remains as the most important deoxidizing agents in electric arc welding are the two elements manganese and titanium. These are employed to keep the oxygen content in stainless steel weld metals within admissible limits.

With pure manganese deoxidation, it is possible to set the oxygen content of low carbon stainless steel weld metal to values of around 0.06%. The type of coating (basic or acid) is of minor importance for stainless steel electrodes, since the oxygen content is mainly governed by the high chromium weld metal content in accordance with Fig. 47b. By an additional deoxidation with titanium, it is possible to lower the oxygen content of the weld metal to around 0.03—0.05%, a range which in view of the extraordinary fineness of nonmetallic inclusion that determine the oxygen content of the weld metal is sufficient for nearly all stainless steel welding applications. When using the GTAW process, where the oxygen supply is largely eliminated because of the use of uncoated welding wires and shielding by inert gas, it is possible to achieve the same low oxygen contents as in steel. In the GMAW process, however, which for welding technology reasons is normally used with an inert gas with 1% oxygen (e.g. Argon M 1), oxygen can enter the weld metal in a similar way as with coated electrodes. This oxygen supply can be governed by the level of deoxidizing agents in the welding wire. In submerged arc welding too, the situation is similar to coated electrode welding, i.e. oxygen can enter the weld metal through the flux mainly from the silicates which it contains.

The low oxygen contents found in high alloyed steels which are produced by advanced melting technology, do not exert any noticeable influence on the transformation and precipitation processes of stainless steels. The same applies to weld metal, even if the oxygen content of the latter is much higher. However, the nucleation effect of the extremely fine nonmetallic inclusion frozen in the weld metal must be emphasized.

The hot cracking resistance of stainless steel weld metal is normally not influenced negatively by the amount of oxygen it contains. The fine oxide inclusions of the main deoxidizing agents, manganese and titanium, as well as the major alloying element, chromium, have generally no decisive influence on low melting phases.

1.3.13 Influence of Hydrogen

Hydrogen may enter the weld metal through damp coatings in electric arc welding, through flux in submerged arc welding and through hydrogen containing or moist inert gas in inert gas arc welding. It may induce cracks in the weld metal (hydrogen induced cold cracking). The hydrogen content in steel is normally stated in milliliters per 100 gram of deposited metal = ml/100 g, with 10 ml/100 g being equivalent to the content of 9 ppm = 0.0009%. In steel, hydrogen is an undesirable element. It shows, however, a high solubility in most liquid metals and is taken on eagerly during welding by the high temperature droplets. The high solubility in the liquid state is greatly diminished during solidification and drops during further cooling to very low values. Hydrogen is the element with the smallest atomic diameter and thus possesses a very high diffusion potential. This fact permits hydrogen to diffuse rather well in steel at ambient temperature, giving it an extraordinary mobility within the crystal lattice. The literature on the effects of hydrogen in metals is very extensive. A good survey on the present state of research is contained in the works by Smialowski [85], Alefeld and Völkl [86], Ortner [87] and with regard to welding technology by Coe [88].

The solubility of hydrogen in iron, chromium and nickel is shown in Fig. 48 according to Rapatz [8], findings of Schenk and Wünsch [91] and Schwarz and Zitter [92]. During solidification, the hydrogen solubility of pure iron drops from 28 to 8 ml/100 g and that of pure nickel from 38 to 18 ml/100 g. At 911° C, the hydrogen solubility of γ iron is approx. 2 ml/100 g higher than that of α iron. In the upper temperature range, the hydrogen solubility of nickel is nearly twice that of iron. At

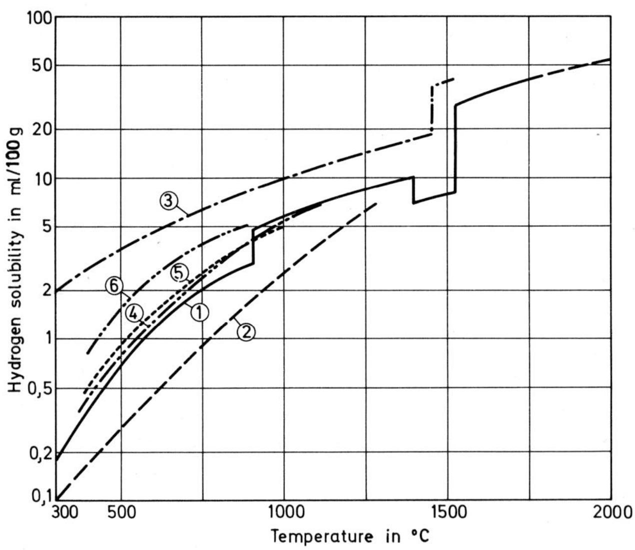

Fig. 48. Hydrogen solubility in different alloys as dependent on temperature at a pressure of 1 bar. 1 in iron to Rapatz [8], 2 in chromium to Luckemeyer and Schenk [90], 3 in nickel to Luckemeyer and Schenk [90], 4 in 20/10 chromium-nickel steel to Luckemeyer and Schenk [90], 5 in 12% chromium steel to Schwarz and Zitter [92], 6 in 4% manganese-iron alloy to Schwarz and Zitter [92]

lower temperatures, the difference is even greater. The hydrogen solubility of chromium is about half that of iron. With regard to hydrogen solubility at ambient temperature only values for iron are available. According to Geller and Sun [89], it is below 0.005 ml/100 g, i.e. less than 1/1000 of hydrogen solubility at 1000° C. The hydrogen solubility of three ferrous alloys have also been plotted. Graph 4 shows the behaviour of a ferrous alloy with 0.01% carbon, 20.3% chromium and 9.66% nickel and graph 5 that of a ferrous alloy with 0.03% carbon and 12.2% chromium. The difference to pure iron is rather small.

The influence of other elements on the hydrogen solubility of iron have been investigated by Schwarz and Zitter [92]. Carbon, silicon, chromium and nickel produce only a slight increase in the hydrogen solubility of iron. The strongest effect is produced by manganese. In the temperature range between 400—800° C, hydrogen solubility will increase 1.5 times with the addition of 2% manganese and almost double with the addition of 4% manganese (see graph 6 in Fig. 48).

The influence of alloying elements on the diffusion coefficient of hydrogen in pure iron has been investigated by Schwarz and Zitter [92]. At higher temperatures in the range of the α-γ transformation of iron, differences are mainly caused by microstructural changes, with diffusion being more rapid in cubic body centered ferrite lattice than in cubic face centered austenite lattice. At ambient temperature, the diffusion rate of hydrogen in iron diminishes with rising alloying contents of chromium, nickel and manganese. Silicon has only a minor influence with contents of up to 1.8%. Geller and Sun [89] have compared the diffusion coefficient of hydrogen in ferrous alloys of ferritic and austenitic structure, i.e. a ferritic iron-chromium alloy with 27.5% chromium and an austenitic iron-chromium-nickel alloy with 18% chromium and 8.9% nickel. In the iron-chromium alloy, the diffusion coefficient is diminished in comparison to pure α iron, but it is still in the same range as pure γ iron. In the austenitic chromium-nickel alloy, however, the diffusion coefficient is diminished more severely. At 400° C it is diminished according to Eckstein [112] in comparison to pure α iron by about twice the power of ten (see also Table 1).

The high diffusion potential of hydrogen in iron is also employed for the determination of the socalled "diffusible hydrogen content" in the weld metal which is the decisive criterion in the evaluation of the negative influence of this element in welding operations. This consists of quenching a welded specimen immediately after welding in water to force the hydrogen contained therein to stay in solution. Immediately afterwards, the specimen is either cooled to very low temperatures for the purpose of intermediate storage, or placed directly in the testing apparatus. In the latter, the specimen is left to effuse hydrogen at a specified temperature and time. The effused amount of hydrogen is measured. It represents the so called diffusible hydrogen content (see also ISO Standard 3690 and AWS Standard A 4.3—86: "Standard Methods for Determination of the Diffusible Hydrogen Content of Martensitic, Bainitic, and Ferritic Weld Metal Produced by Arc Welding" and DIN Standard 8572).

The forcibly dissolved hydrogen of a weld metal not only diffuses at the surface but also between individual passes and between weld metal and heat affected zone. The more rapidly a pass is cooled, the higher will be the amount of hydrogen which is forcibly dissolved and retained in the weld metal. With greater hydrogen supply, e.g. from the coating of an electrode, more hydrogen will be accumulated by the

Table 1. *Survey of diffusion constants of alloying elements in α (δ) and γ iron, including the values for self-diffusion of iron. Most of the data at 20°, 400°, 800° and 1100° C are calculated values to Eckstein* [112]. *They were supplemented by data from the work of Fridberg, Törndahl and Hillert* [142]

	Diffusion coefficient in cm²/s									
	α(δ)-Iron					γ-Iron				
	Temperature °C					Temperature °C				
Element	20	400	800	1100	1400	20	400	800	1100	1400
α(γ)-Fe	10^{-46}	10^{-19}	10^{-12}	10^{-9}	10^{-7}	10^{-53}	10^{-22}	10^{-14}	10^{-11}	10^{-9}
Al			10^{-11}	10^{-9}	10^{-7}	10^{-36}	10^{-15}	10^{-9}	10^{-8}	
B						10^{-20}	10^{-9}	10^{-6}	10^{-5}	
C	10^{-17}	10^{-8}	10^{-5}	10^{-4}		10^{-27}	10^{-13}	10^{-8}	10^{-6}	
Cr	10^{-42}	10^{-18}	10^{-11}	10^{-9}	10^{-7}	10^{-37}	10^{-18}	10^{-13}	10^{-11}	10^{-9}
Co	10^{-44}	10^{-19}	10^{-12}	10^{-10}		10^{-68}	10^{-26}	10^{-16}	10^{-12}	
Cu			10^{-11}	10^{-9}		10^{-49}	10^{-20}	10^{-12}	10^{-10}	
H	10^{-5}	10^{-3}	10^{-3}	10^{-3}		10^{-10}	10^{-5}	10^{-4}	10^{-3}	
Mn						10^{-53}	10^{-22}	10^{-14}	10^{-11}	10^{-9}
Mo	10^{-46}	10^{-18}	10^{-11}	10^{-9}	10^{-7}	10^{-49}	10^{-21}	10^{-13}	10^{-11}	10^{-9}
N	10^{-17}	10^{-8}	10^{-6}	10^{-5}		10^{-31}	10^{-13}	10^{-8}	10^{-7}	
Nb								10^{-13}	10^{-11}	10^{-9}
Ni	10^{-45}	10^{-18}	10^{-11}	10^{-9}		10^{-54}	10^{-23}	10^{-15}	10^{-11}	10^{-9}
P	10^{-34}	10^{-12}	10^{-8}	10^{-7}	10^{-6}			10^{-14}	10^{-9}	10^{-8}
S			10^{-10}	10^{-8}	10^{-6}	10^{-39}	10^{-16}	10^{-10}	10^{-9}	10^{-7}
Si	10^{-50}	10^{-21}	10^{-13}	10^{-11}	10^{-7}	10^{-48}	10^{-20}	10^{-13}	10^{-12}	10^{-9}
Ti	10^{-47}	10^{-19}	10^{-12}	10^{-9}	10^{-7}	10^{-49}	10^{-21}	10^{-13}	10^{-11}	10^{-9}
V			10^{-12}	10^{-9}	10^{-7}			10^{-14}	10^{-11}	10^{-9}
W	10^{-60}	10^{-26}	10^{-17}	10^{-10}		10^{-68}	10^{-30}	10^{-20}	10^{-16}	

high temperature droplet and more hydrogen will also be forcibly dissolved in the weld metal. According to Luckemeyer-Hasse and Schenk [90], this forcible dissolution produces internal pressures which may reach values of 1000 bar or more in locations where hydrogen accumulates in the microstructure.

Due to its microstructure, austenite is better capable of forcibly dissolving hydrogen than is ferrite. During γ-α transformation, i.e. during cooling from the welding heat, solubility is rapidly diminished. If this transformation takes place at lower temperatures, the hard martensitic structure with low ductility which is being formed, can dissolve only little hydrogen. This may lead to the formation of hydrogen induced cold cracks. Regarding the different theories about the formation of hydrogen induced cracks, we would like to refer to pertinent works and to items [85] through [92] of our list of supplementary literature.

In stainless steels and their weld metals, the danger of hydrogen induced crack formation is only present if the austenitic structure contains high delta ferrite contents (e.g. more than 40%) or if a γ-α transformation with martensite formation takes place in the weld metal or in the heat affected zone. This applies, for example, to 13% chromium steels (AISI 410) and also to low carbon martensitic 13/4 chromium-nickel steel (AISI 410 NiMo), in which the formation of a martensitic structure is desirable both in the steel and the weld metal. With such materials, the hydrogen supply during electric arc welding must be kept to a minimum to eliminate the risk of hydrogen induced crack formation. Brezina [93] states a probable threshold value of 3 ppm (3.3 ml/100 g) for diffusible hydrogen in these steel grades and recommends vacuum treatment of molten steel to guarantee that this value is met. Tösch, Perteneder and Rabensteiner [94] state values for the content of diffusible hydrogen in the weld metal of coated 13/4 chromium-nickel electrodes which—after re-baking of the electrodes at 300° C for 2 hours—average 2.2 ml/100 g.

In austenitic Cr-Ni weld metal we find practically no diffusible hydrogen at all, i.e. there is practically no hydrogen diffusion at ambient temperature. This is not only due to the much lower diffusibility of hydrogen in the austenitic structure but also to the better solubility of hydrogen in austenite. Hydrogen is able here to deposit interstitially in the cubic face centered crystal lattice. During welding of austenitic steels with austenitic filler metals, hydrogen rarely shows up in the weld metal in the form of hydrogen induced cracks. If it appears at all, it is by the formation of porosity. The formation of surface pores which often are accompanied by an uneven weld surface runs caused by the formation of gas cavities in the slag, are influenced here aside from the absolute content of hydrogen but by the gas permeability of the slag formed during electric arc welding using coated electrodes or during submerged arc welding.

The influence of hydrogen on the transformation and precipitation processes in stainless steels have been investigated by Narita, Altstetter and Birnbaum [95]. The effects of cathodic loading with hydrogen which were found, require very high hydrogen contents which would lead to the formation of pores during welding. Therefore, they are of minor importance for the metallurgical processes occurring in welding operations and have been omitted here.

2 Metallurgical Processes Occurring During Solidification and Cooling in Stainless Steel Weld Metal

In steel, the cast structure of ingots is largely modified and positively influenced by hot forming and heat treatment operations, with processes of recrystallization, homogenization of segregations and the elimination of harmful precipitations through solution annealing playing a decisive role. In weld metals, such possibilities are rather limited, since welded components can only in rare instances be hot formed, solution annealed or quenched and tempered. In the untreated condition, weld metal normally emerges with a solidification structure and with previously executed passes of multilayer welds being exposed to a short reheating cycle in the course of depositing subsequent weld passes. But due to the steep temperature gradient, the high cooling rate and the short time span this reheating is in no way comparable to the controlled heat treatment processes of steels. Furthermore, only part of a previously laid pass is influenced in this way and the rest is maintained as a solidification structure.

Weld metal must show the same properties as the hot formed and heat treated steel, although its structure is considerably less favourable and less homogenous than the base metal. Because of the absence of a controlled recrystallization of the solidification structure, of a thorough homogenization of segregations and in most cases a solution annealing or tempering treatment, the properties of the weld metal are largely influenced by the processes which take place during solidification and subsequent cooling of the welding pass.

A positive influence in the case of weld metal is produced by the fine-grained structure and in many cases by rapid cooling which may promote a supercooling of harmful phase precipitations and thus compensate to some extent the inhomogenous structure. Nevertheless, weld metal often demands a different structural configuration in comparison to steel, which is not completely the same as the microstructure of the steel but only similar. This is achieved by modifying the chemical composition of the weld metal in comparison to the base metal, taking into account the specific conditions of the welding process. In this way, it is possible to match the properties of the weld metal closely enough to those of the steel grade so that it will meet all the necessary requirements in practical applications.

There is a major difference between stainless steel weld metal and unalloyed weld metal. Whereas unalloyed and low alloyed weld metal will always experience a secondary transformation from γ to α crystals, this is only true in stainless steel weld metals for a proportion of the ferritic alloys which appear with such a chemical composition that they are still within the range of γ-α transformation.

Austenitic weld metals show no γ-α transformation at all, which means that their solidification structure is much more prominent. The same is true for ferritic weld metals which are located outside the γ range in the iron-chromium-nickel constitution diagram, i.e. which show high contents of delta ferrite after cooling to ambient temperature.

The processes which take place during solidification and cooling of stainless steel weld metals are shown quite vividly in Fig. 9 (page 10) which represents concentration profiles in the iron-chromium-nickel system with different chromium-nickel ratios. If the latter is 3 : 1 (Fig. 9 a), the liquid metal will solidify completely to δ crystals. Up to about 21% chromium, it is transformed during further cooling into γ crystals (δ-γ transformation) which are changed again at lower temperatures back into α crystals (γ-α transformation). Such alloys can therefore be hardened as well as hardened and tempered. Low carbon martensitic stainless steel grades and weld metals of the type 12% chromium and 4% nickel (AISI 410 NiMo) belong to this group. If the chromium content is higher than 21%, δ crystals which are formed from the melt are only partially transformed into γ phase. During further cooling, the γ crystals are maintained as an austenitic structure down to ambient temperature. The remaining part consists of δ ferrite with a solidification structure, which is also maintained without further transformation down to ambient temperature, partly by supercooling effects with the possibility of sigma phase being precipitated below 900° C. In this alloying group we find the austenitic-ferritic materials. Because of their two-phase $\delta + \gamma$ structure, they are also called "duplex steels". Their structure consists normally of secondary austenite crystals with more or less primary δ ferrite crystals.

With a chromium-nickel ratio of 2 : 1 (Fig. 9 b), primary delta ferrite crystals are initially formed from the melt. With higher chromium and nickel contents, γ crystals are also precipitated from the melt apart from δ crystals in the three phase ($L + \delta + \gamma$) sector. After solidification has been completed, transformation of δ ferrite into austenite (δ-γ transformation) begins, with the structure being completely transformed into austenite. The γ range is no longer constricted and extends because of the higher alloying content of nickel up to high chromium contents. The γ-α transformation which is still visible in the constitution diagram is shifted towards lower temperatures. Due to high cooling rates, the austenitic structure is maintained down to ambient temperature by supercooling (see also Fig. 14, page 14). With chromium contents above 16%, the formation of sigma phase is possible at temperatures below 900° C.

In this alloying range, we find the common stainless austenitic steels and weld metals such as 18% Cr and 8% Ni (AISI 302) steel grades and 20/10 chromium-nickel weld metals (AWS E 308 and E 308 L). In the case of steels, any remaining residual ferrite or precipitated sigma phase can be dissolved again by a solution annealing treatment, e.g. at 1050° C, with γ-α transformation as well as the renewed precipitation of sigma phase being inhibited by subsequent quenching, thus achieving a pure austenitic structure at ambient temperature. In general, weld metal cools very rapidly. This creates the possibility that a small amount of primary δ ferrite is maintained in the structure down to ambient temperature in the form of residual ferrite (see section 3.2). Below 20% chromium, both γ-α transformation and precipitation of sigma phase are suppressed due to the rapid cooling which takes place during electric arc welding, therefore a weld metal structure such as that of a 20/10 chromium-nickel filler metal will consist of austenite with a small amount of residual delta ferrite.

Fig. 9 c shows the transformation behaviour of alloys with chromium-nickel ratios of 1 : 1. The melt no longer solidifies to δ ferrite but to primary γ crystals which are maintained in the weld metal as solidification structure down to ambient temperature. In this alloying group, we find the so called fully austenitic steels and weld metals.

The process of the solidification of stainless steel alloys in the iron-chromium-nickel system to δ or γ crystals is illustrated quite well in Fig. 5 (page 7) by the lines *F, G* and *H*. In the iron-chromium-nickel system, all alloys on the iron-chromium side of the eutectic groove and the three-phase sector $(L + \delta + \gamma)$ will solidify to primary δ crystals and those on the nickel side to γ crystals. Line *F* is located in the range of solidification to primary δ crystals. Line *G* slightly intersects the three-phase sector $(L + \delta + \gamma)$. Line *H* first intersects the three-phase sector in the iron corner and then runs through the region of solidification to primary γ crystals.

In fully austenitic steels and weld metals, γ-α transformation is practically always suppressed because of the higher nickel contents in combination with rapid cooling from the γ range, and the structure being formed is fully austenitic.

With higher chromium contents, we must also expect—in particular during extended exposure to elevated temperatures—the precipitation of sigma phase. The spatial configuration of sigma phase in the iron-chromium-nickel system may be gathered from Figs. 12 and 13 (page 13). Figs. 6 and 9 (page 8 and 10) show the alloying and temperature limits at which the precipitation of sigma phase must be expected, e.g. during prolonged annealing times. In the practical behaviour of stainless steels and their weld metals, the influence of other elements such as carbon, nitrogen, molybdenum and silicon must also be considered (see also chapter 4).

2.1 Primary Crystallization of Weld Pools

Each phase transformation consists of two steps, i.e. nucleation of a crystal and subsequent crystal growth. A major characteristic of all weld pools produced by fusion welding processes is, according to Wittke [96], the continuous melting and subsequent solidification of a relatively small metal volume. According to Savage, Lundin and Aronson [97], nucleation caused by heterogeneous nuclei (inclusions, impurities) is of minor importance in fusion welding compared to crystal growth, because the crystals which precipitate from the liquid weld metal pool can grow directly on a solid base, i.e. the solid material which surrounds the liquified weld pool.

The welding process is comparable to a miniature steel making plant, in which two processes take place simultaneously—continuous steel melting and continuous solidification of molten steel. There is, however, a major difference compared to the solidification of steel in the ingot mould. During welding, the crystallization process is almost completely controlled by the dissipation of heat into the solid metal, with only a small amount of heat being dissipated into the surrounding atmosphere. This process has a decisive influence on the type, size and orientation of the crystals being formed.

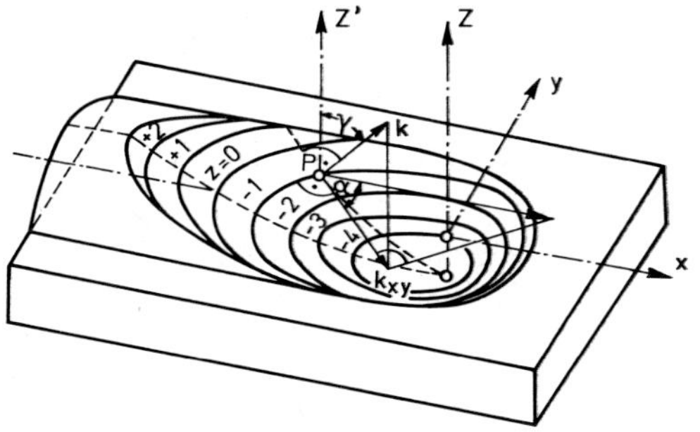

Fig. 49. Representation of a weld pool in relief with graphic determination of crystal orientation to Eichhorn and Engel [98]

Eichhorn and Engel [98] developed graphic models which illustrate quite well the crystallization processes in a weld pool. Fig. 49 which shows the spatial configuration of a weld pool is taken from their work. It facilitates the determination of crystal orientation for any given point by drawing an orthogonal line to the interface liquid-solid interface at this point. Regarding the determination of the rate of crystallization, please refer to the original work [98] which includes a comprehensive list of supplementary literature.

2.2 Primary Crystallization of Binary and Ternary Systems

The crystallization of binary and ternary systems is discribed very clearly by Leitner, Plöckinger and Straube [99]. Because of its great importance for understanding the crystallization processes in weld metals, we shall present here an excerpt of their work.

The solidification of a pure metal takes place at its solidification point, i.e. at a unique and exactly defined temperature, and the crystals which are produced show a uniform composition over their full cross section. The solidification from a homogeneous liquid solution in binary and ternary systems on the other hand—with the exception of purely eutectic compounds—does not take place at a single temperature but stretches over a certain temperature range. In binary systems, the extent of this solidification range is determined by the distance between liquidus and solidus lines and in ternary systems by the distance of liquidus and solidus areas.

If an alloy *L* is cooled down in a binary system from the melt (line *l*–*l'* in Fig. 50 a), crystallization starts when the liquidus line is reached in point 1 by the precipitation of crystals on the solidus line with deviating composition 1'. If the temperature drops further, the precipitated crystals change their composition along the solidus line from 1' over 2' to 3, while the composition of the melt is changed along the liquidus line from 1 over 2" to 3", i.e. being enriched by element *B*. At temperature *T*3, the melt has completely solidified.

The crystallization process in ternary systems can be gathered from Fig. 50 b. The solidification of a ternary alloy *D* (line *d*–*d'*) starts at intersection 1 on the liquidus area with the precipitation of crystals on the solidus area of composition 1'. As the temperature drops further the composition of crystals change along

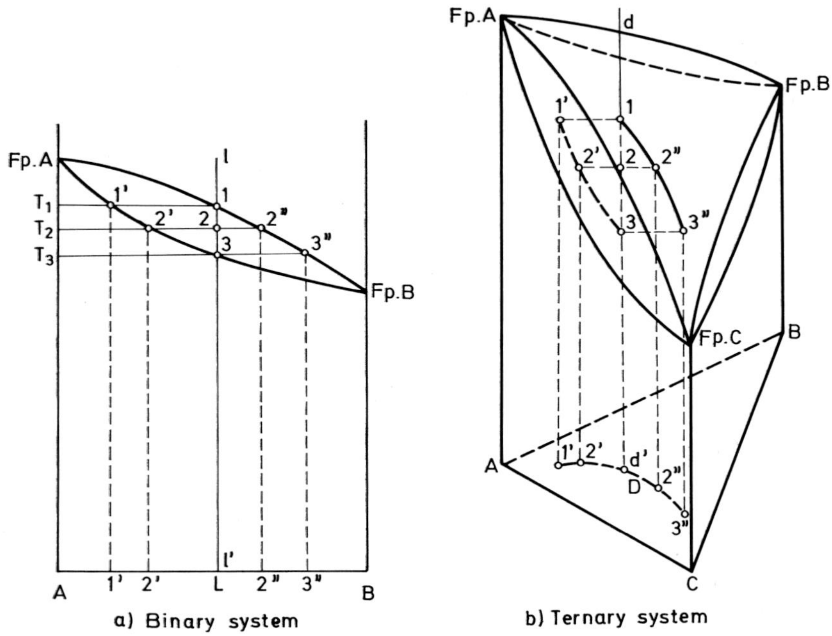

Fig. 50. Process of solidification in binary and ternary systems to Mitsche and Niessner [139], (Fp=Fusion point)

the curve 1′, 2′, 3 on the solidus area. The melt changes along curve 1, 2″, 3″ on the liquidus area and is enriched mainly by compound *C*. The respective composition of precipitated crystals and the remaining residual melt can be determined from the projection of precipitation curves on the base area.

In both binary and ternary systems, a change of alloy composition takes place in the crystals which are being precipitated during solidification, as compared to the original composition of the alloy. This can be detected in Fig. 50, for example, by the differing composition of crystals 1′ and 2′ in comparison to alloys *L* and *D* respectively. In this way, a state of nonequilibrium is created after solidification which tries to return to equilibrium by way of diffusion. The crystals 1′ and 2′ as well as the residual melts 2″ and 3″ try to approach by diffusion the original composition of the alloys *L* and *D*. If complete equalization by diffusion is possible because of high diffusion rates and very slow cooling, the segregations of the precipitated crystals and residual melts will be readjusted to the alloy. The crystals which are then present at the end of solidification completely match the alloys *L* and *D* in their chemical composition and are fully homogeneous.

In the case of normal cooling rates, in particular the quite rapid cooling of weld metals, the diffusion rate in the precipitated crystals is not suffice to readjust its composition over the full cross section of the crystal, in the amount of time available, to the original composition of the alloy. It is not a homogeneous crystal which is formed but a so called layer crystal consisting of layers with different chemical compositions. Its composition at the crystal nucleus, i.e. at the beginning of its precipitation, corresponds, for example, to point 1′ in Fig. 50 a. During cooling this crystal will grow, with layers of another composition being formed on the outside which may, for example, be in the range of composition $2'$. Meanwhile, the melt is enriched with element B and the outer zone of the crystal which solidifies from this residual melt is also enriched with element B. In this way, the crystal which is precipitated becomes inhomogeneous. This process is called microsegregation. The difference in composition between core and outer layer of a crystal will increase with increasing distance between liquidus and solidus lines, with decreasing inclination of the temperature gradient of these two lines (areas), with decreasing diffusion rate and with diminishing time span available for equalization by diffusion.

The segregation phenomena which occur during the crystallization of binary and ternary alloys in the case of insufficient equalization of alloy concentration are influencing the configuration of liquidus and solidus areas. Fig. 51 a shows the behaviour of a binary mixture, representing the course of solidification at empirical cooling rates and with impeded (delayed) equalization of concentration within the crystals. Solidification starts at temperature T_1 with the precipitation of a crystal $1'$. Because of impeded equalization of concentration, the composition of precipitated crystals does not change as the temperature drops to $2'$ but to $\overline{2}'$ and that of the melt to $\overline{2}''$. Further cooling takes place along the dotted line until the point where, at temperature T_4, the precipitated crystals show the original composition of alloy L. The residual melt follows the dotted line from 1 over $\overline{2}''$, $\overline{3}''$ to $\overline{4}''$. It is then greatly enriched with element B. Because of impeded equalization of the alloy concentration, this enrichement can only be equalized to a limited extent with the result that the remaining residual melt will show a rather high enrichement in element B between the crystals already formed. Compared to the state of equilibrium, the process of solidification is accompanied in general by a drop in solidus temperatures. In addition, there is an increase in segregations with the composition of individual

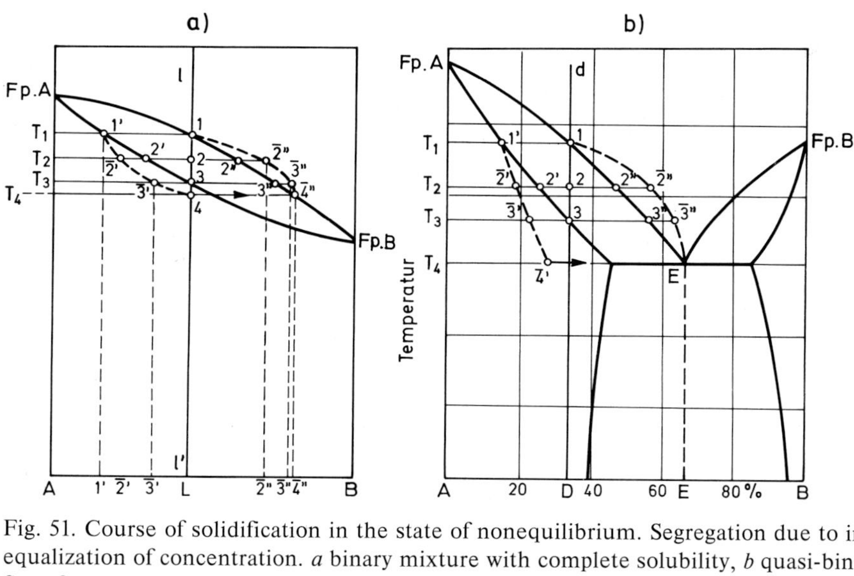

Fig. 51. Course of solidification in the state of nonequilibrium. Segregation due to impeded equalization of concentration. *a* binary mixture with complete solubility, *b* quasi-binary profile of a ternary system with a eutectic E to Plöckinger and Straube [99]

layers being shifted during crystal growth towards the dotted line from 1′ to 4 (Fig. 51 a). Due to the impeded equalization of concentration, their composition can only approach to a limited extent by way of diffusion that of alloy *L*. At the end of solidification, grain boundary films of residual melt $\bar{4}''$ will form between the crystals. They solidify with a composition close to that of residual melt $\bar{4}''$. In the case of rapid cooling, this process is often further intensified by a more severely impeded equalization of concentration.

Fig. 51 b shows the solidification of a multinary mixture with a eutectic by way of a quasi-binary profile according to Plöckinger and Straube [99]. Crystallization takes place similar to binary mixtures, with impeded equalization of concentration leading, however, to a substantial change in concentration of both crystals and melt. For this reason, the eutectic *E* is precipitated during solidification of alloy *D*, although the former should be unable to exist according to the constitution diagram in equilibrium.

In the solidification of commercial steels and alloys, we must always expect a partial segregation and the formation of layer crystals. This applies in particular also to weld metal where, due to the often quite rapid cooling, the possibilities for equalization of concentration through diffusion are limited.

2.3 Phenomena at the Liquid-Solid Interface (Solidification Front)

The liquid-solid interface, also called solidification front, represents the boundary between melt and precipitated crystal, i.e. the boundary line between liquid and solid phase during solidification. As shown in Fig. 51, there is a substantial difference in concentration between precipitated crystal and melt, e.g. in the contents of element *A* and element *B*. In case of complete equalization of concentration, this difference is equalized by diffusion, i.e. it exists only in the limited area of the solidification front. Fig. 52 a shows this phenomenon, with the points of different concentration during solidification (see also Fig. 51 a, page 57).

Fig. 52 a shows the ideal state with full equalization of concentration. At temperature T_1, a crystal 1′ is precipitated which adapts itself immediately by way of diffusion to the original alloy S with 50% element *A* and 50% element *B*. This process repeats itself in the course of solidification with the result that only homogeneous crystals with composition *S* will be present at the end.

Fig. 52 b–e shows a schematic and simplified representation of the concentration gradient which is formed in the case of impeded equalization of concentration at the liquid-solid interface and its immediate vicinity. It has been assumed that equalization of concentration is impeded more in the precipitated crystals than in the melt which corresponds to the reality of weld pools. Due to strong weld pool agitation, equalization of concentration can take place much easier than in precipitated crystals.

Fig. 52 b shows the phenomena occurring at the liquid-solid interface at the start of solidification. Crystal nuclei which are enriched with element *A* will precipitate from the melt L_1 with the concentration 1′. Their composition cannot, however, adapt itself completely by way of diffusion and a crystal S_1 which is enriched with *A* is formed. Due to the enrichement of precipitated crystals by element A, the melt suffers simultaneously a decrease in element A and an enrichment in element B. At temperature T_2, a crystal $\bar{2}'$ will precipitate from the melt L_2 with a composition changed by diffusion towards S_2 with a decreased A content compared with S_1. By way of temperature T_3 which is not shown in Fig. 52 temperature T_4 is reached. Here, the melt L_4 shows considerably less A than L_2. It precipitates a crystal $\bar{4}'$, the concentration of which barely reaches the composition of L. By diffusion, the composition of the crystal changes from $\bar{4}'$ to S_4 which is already below the original composition of the alloy with 50% element A and 50% element B. Towards the end of solidification, the content of the melt has dropped from the original composition L to the value L_5 with a substantially reduced content in element A, with crystals of

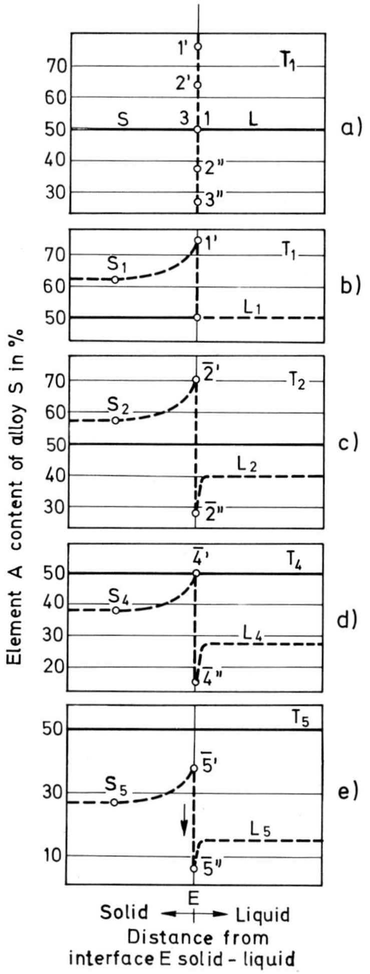

Fig. 52. Formation of liquid-solid interface during solidification (schematically). *a* full equalization of concentration (ideal state), *b–e* impeded equalization of concentration (real state), *b* start of solidification at temperature T_1, *c, d* during solidification at temperatures T_2 and T_4, *e* state at the end of solidification at temperature T_5

composition S_5 being formed. The last remainder of the melt which forms, for example, in the grain boundary corners between three crystals or between dendrite branches of crystals shows the composition $\bar{5}''$ with a high enrichment of element B. Since solidification of such residual melts which often form as grain boundary films between the grain boundaries of precipitated crystals take place rather rapidly, an equalization of concentration by diffusion is possible to a limited extent only, with the result that the final composition of the grain boundary film is very close to the composition of residual melt $\bar{5}''$ as indicated by the arrow in Fig. 52 e.

The above phenomenon is also illustrated in Fig. 53 according to Kurz and Lux [100]. If two crystal grains grow towards each other, with each of them pushing a concentration peak at its liquid-solid interface ahead of it, the residual melt at the grain boundary will show a high enrichment of, e.g., element B and a depletion in element A, as the gap between crystals and volume of residual melt is diminishing towards zero. Transportation of elements by diffusion is only possible to a limited extent and the residual melt solidifies to crystals of nearly equal composition. This process can often be superimposed, however, by the fact that the concentration of a melt reaches the precipitation range of a new phase, as shown in Fig. 51 b in form of the eutectic E. This phenomena is indicated in Fig. 53 by the precipitation boundaries at concentration C_E. In this case, the concentration of residual melt no longer rises and the eutectic E is precipitated at the grain boundaries of primary crystals, a phenomenon which ought not occur according to equilibrium conditions.

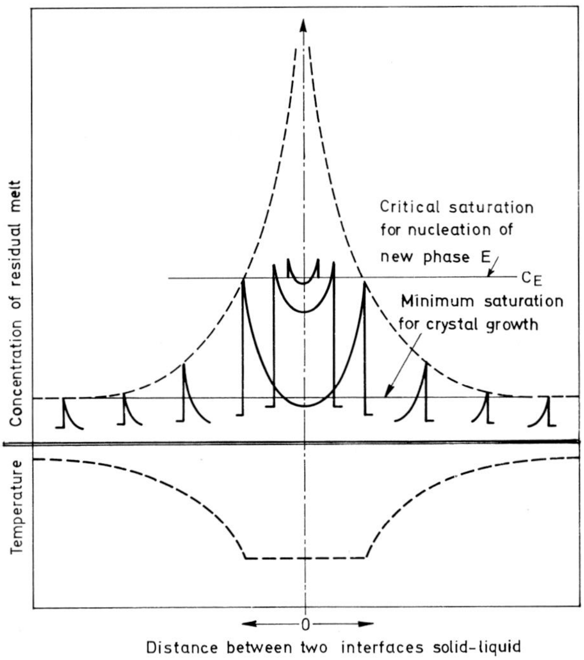

Fig. 53. Schematic representation of concentration and temperature gradients of two approaching interfaces solid-liquid, with consideration of the precipitation of a phase E at concentration C_E to Kurz and Lux [100]

For details on the mechanisms that take place during crystallization and the nature and properties of grain boundaries, please refer to the pertinent literature [101–110 and 487].

2.4 Crystal Growth During Solidification and Crystal Configurations

The formation of different crystal configurations (dendrites, globular crystals, cellular crystals) is determined by the temperature situation at the liquid-solid interface. The processes which take place have been described by Tiller [111] by way of a crystallization model. The adaptation to the processes which occur during welding have been carried out by Savage, Lundin and Aronson [97], Wittke [113], Matsuda, Hashimoto and Senda [114], and Arata, Matsuda and Nakata [115]. A comprehensive representation is given by Savage [116]. Kulmburg [117] deals with the problems arising with the description and representation of solidification phenomena. Fig. 52 shows segregation processes and enrichment in concentration at the liquid-solid interface together with a description of changes in concentration of precipitated crystals and the remaining residual melt. If we look now at the temperature gradient of the melt at the liquidus line, resulting from the element enrichment which takes place on the liquid side, we find according to Tiller [111] a dependency of crystal configuration on the occurring differences between local liquidus temperatures and the respective real temperatures of the melt, produced by external cooling parameters. A negative temperature peak of the melt is produced at the liquid-solid interface, i.e. a supercooling effect occurs compared to the temperature gradient established by the cooling rate which is limited to the liquid-solid interface, which is also called "constitutional supercooling". For more details on this model representation, please refer to the original work [111] or to the comprehensive representation thereof according to Eckstein [112].

Fig. 54 shows the influence of cooling rate and direction of base material crystallization on crystal configuration, crystal direction and constitutional supercooling according to Savage (116). High cooling rates promote a cellular solidification, medium cooling rates a dendritic solidification. In stainless weld metals, we can often find a cellular solidification structure at the bottom of a bead and dendritic structures in the middle of the same bead. In the case of very high cooling rates, there is the possibility of the formation of globular structures. In stainless steel weld metals, however, the latter are rather rare because of the fact that the base metal on which the weld metal is being deposited will often predetermine a preferred crystal orientation for newly formed weld metal crystals. An illustrative representation of the way in which various crystal configurations will grow from the liquid pool is given by Flemings in a comprehensive survey [125] on the state of knowledge in 1977. This work also communicates some idea of how crystallization in a weld pool is actually taking place. Direction of growth is strongly determined by the direction of greatest dissipation of heat. When applying solidification phenomena to welding processes, a number of special features must be considered. Constitutional supercooling may vary considerably within a weld pool. Fig. 55 shows according to Wittke [113] a theoretical model of the distribution of constitutional supercooling at the

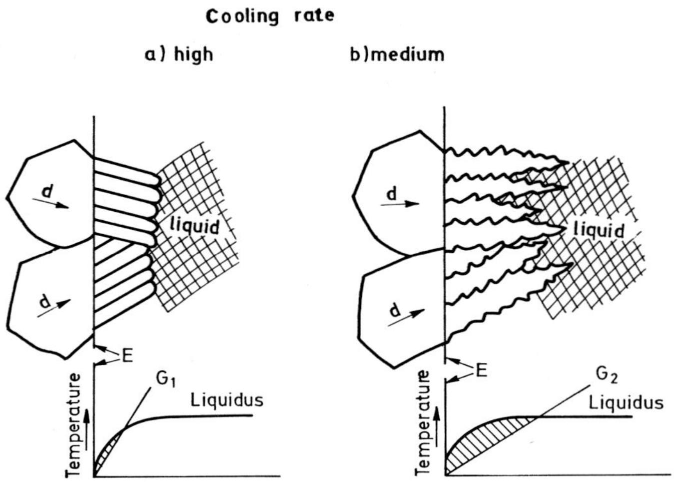

Fig. 54. Crystal configurations of weld metal at differing constitutional supercooling at the liquid-solid interface "E" caused by different temperature gradients G_1 and G_2 at varying cooling rates to Savage [116]. d = direction of crystallisation of base material crystals

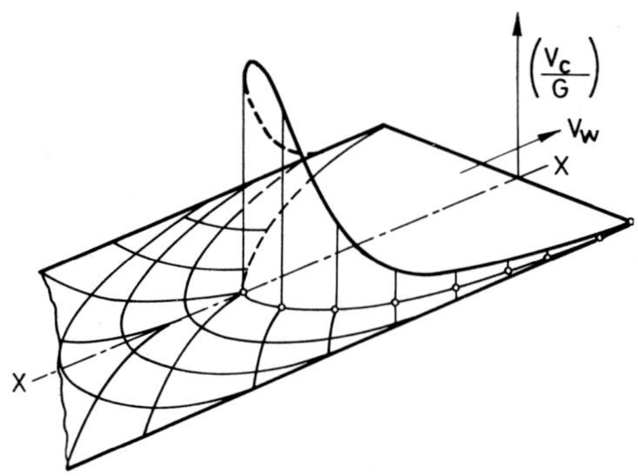

Fig. 55. Distribution of crystallization parameter "constitutional supercooling" at the liquid-solid interface of a weld pool to Wittke [113]

liquid-solid interface at the surface of a weld pool. The highest values appear at the bead center line. At the bottom of the weld pool, we also must expect high cooling rates when welding thick material, due to the larger material masses lying beneath, which promote the formation of cellular crystal growth.

Due to the varying solidification parameters within a weld bead, different crystal configurations are produced in a spatial sense over the whole bead cross section. Fig. 56a shows a schematic representation of the weld metal crystal structure according to Matsuda, Hashimoto and Senda [114]. Starting from the interface solid-liquid, directed cellular crystals are formed first which change over into directed dendrites. In the middle of the weld pool surface, nondirected dendrites are formed. Figs. 56b and c show weld pools as they are produced in welding operations as seen from above together with the resulting crystal configurations and their direction both for high and low, welding speeds.

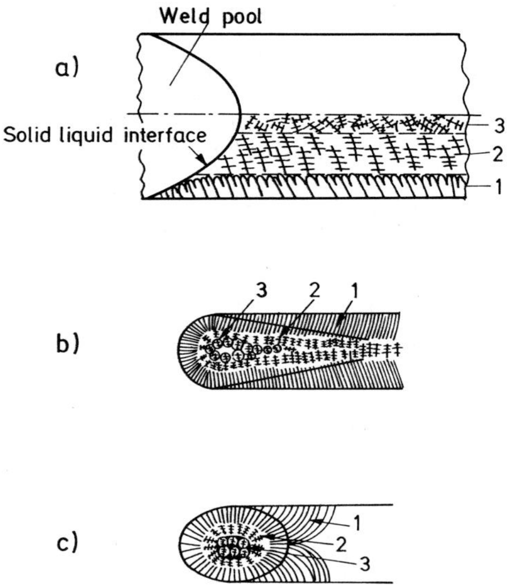

Fig. 56. Solidification structures of welding beads. *a* schematic. *b* real crystallization during welding with high welding speed to Matsuda, Hashimoto and Senda [114]. *c* same as *b*, but with lower welding speed. *1* directed cells, *2* directed dendrites, *3* nondirected dendrites

2.5 Primary Crystallization of Stainless Steel Weld Metals

Stainless steel weld metals can solidify from the liquid state either to primary δ ferrite or primary γ crystals. If an alloy is located in the ternary constitution diagram in the area of the eutectic groove and touches during solidification the three phase sector ($L + \delta + \gamma$), a mixture of primary δ and γ crystals may also be formed. If solidification takes place to primary δ crystals a large part of such alloys will experience a secondary transformation (e. g. δ-γ and/orγ-α transformation) during further cooling as is described in chapter 3. If solidification takes place to primary γ crystals, the austenitic structure will in most cases be maintained down to ambient temperature. In the upper temperature range, i.e. relatively close below the solidus area, the primary γ crystals formed from the melt are normally altered by recrystallization, i.e. new austenite grain boundaries are formed.

Similar phenomena take place during solidification to primary δ crystals which do not experience secondary transformation. Here, the weld metal when cooled to ambient temperature no longer shows the original primary crystals but new grain boundaries, with primary solidification structures produced by segregations being retained in a distinct way. In the case of steel, the transformation of the primary structure is much greater than with weld metals, due to hot forming and heat treatment processes.

Normal etching compounds react in particular to secondary structures. If they are employed, the primary structure is only barely visible or not at all. To make primary structure also visible requires the use of socalled primary etchants which react to crystal segregations of primary solidified crystals.

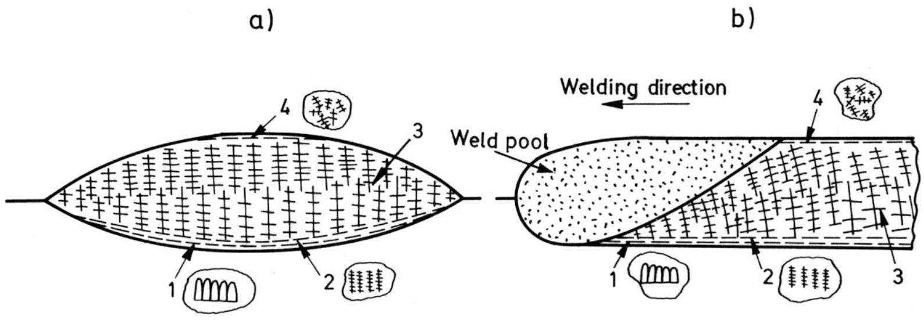

Fig. 57. Crystal configurations during solidification of austenitic stainless steel weld metal (schematically). *a* section transverse to welding direction, *b* longitudinal section. *1* zone of directed cells, *2* zone of moderately branched directed dendrites, *3* zone of medium to highly branched directed dendrites, *4* zone of nondirected dendrites

In case of unalloyed and low alloyed steels, primary etching is normally carried out with the Oberhoffer etchant [118]. It reacts particularly to phosphorus segregations but it cannot be employed for stainless steels and weld metals. For the latter, colour etching is normally used today which makes both primary and secondary structures visible. Since etching procedures used in this process have only been developed and improved in recent years, we would like to refer to the publications of Beraha [119, 123], Lichtenegger und Blöch [120], Perteneder and Lichtenegger [121], Beraha and Shpigler [122] and Weck and Leistner [124]. Procedures for etching and compositions of etchants for the practical application of colour etching are stated in the respective parts of the following literature [119, 120, 122, 123 and 124]. Illustrative examples of the possibilities which colour etching offers for making both primary and specific secondary structures visible are contained in the works of Perteneder [13] and Perteneder and Jeglitsch [14].

At normal cooling rates, zones with varying crystal configurations will be formed during the solidification of a welding pass from stainless steel weld deposits which are determined in size and extent by locally dominating constitutional supercooling (see also Fig. 54, page 62). Fig. 57 shows the crystal structures formed in a weld bead of stainless steel weld metal at medium cooling rates by way of a trans-

---

Fig. 58. Cellular solidification of 20/10 chromium-nickel weld metal (AWS E 308) at the bottom of a weld bead. Zone *1* according to Fig. 57. *a* micrograph transverse to direction of welding (see sketch), *b* micrograph parallel to surface (see sketch). Chemical composition in %: C: 0.034, Si: 0.61, Mn: 1.56, Cr: 20.10, Ni: 10.00, N: 0.050. Colour etching to Lichtenegger [120]

Fig. 59. Dendritic solidification of a 20/10 chromium-nickel weld metal (AWS E 308) in the bead center (zone 3 of Fig. 57). *a* micrograph transverse to direction of welding, *b* micrograph parallel to surface. Chemical composition: same as in Fig. 58. Colour etching to Lichtenegger [120]

Fig. 60. Nondirected dendrite crystals in the upper part of a bead of 20/10 chromium-nickel weld metal (zone 4 of Fig. 57). Chemical composition: same as in Fig. 58. Colour etching to Lichtenegger [120]

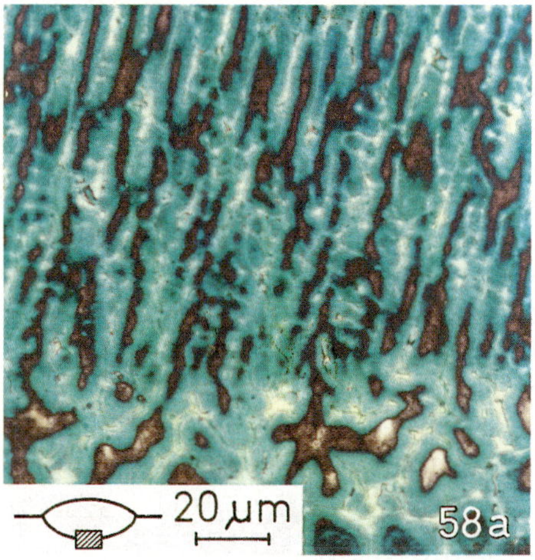
20 µm
58a

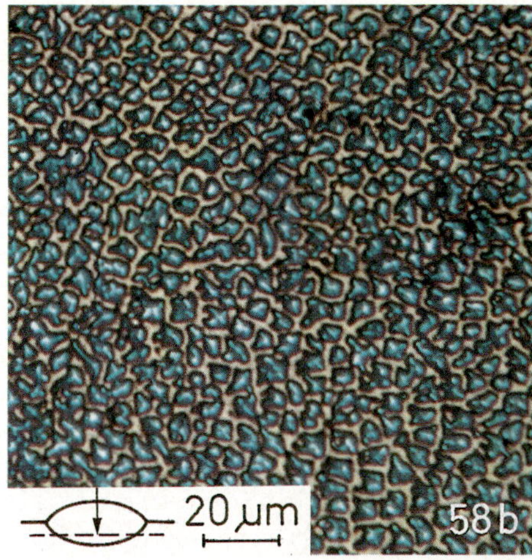
20 µm
58b

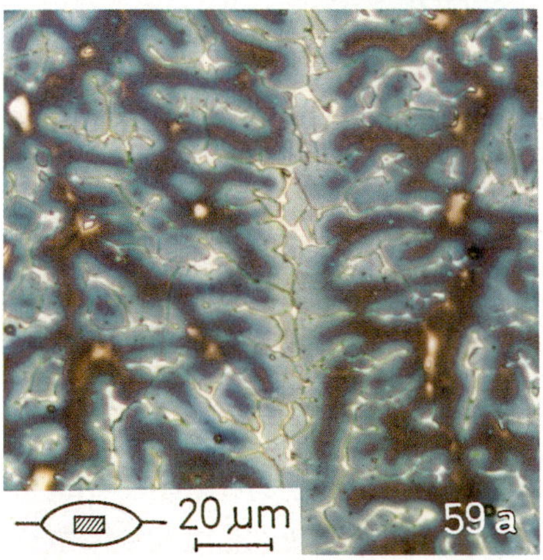
20 µm
59a

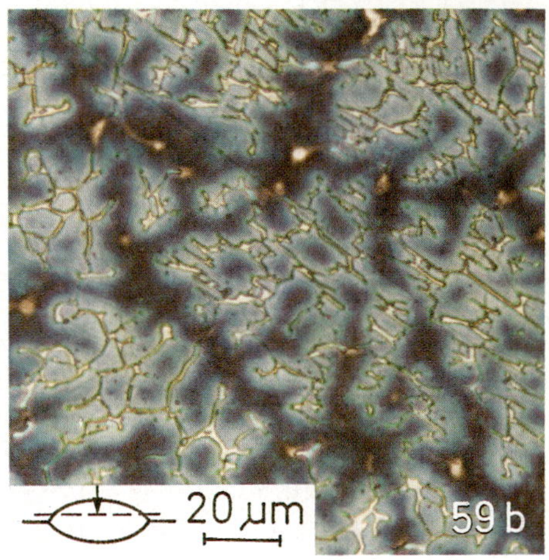
20 µm
59b

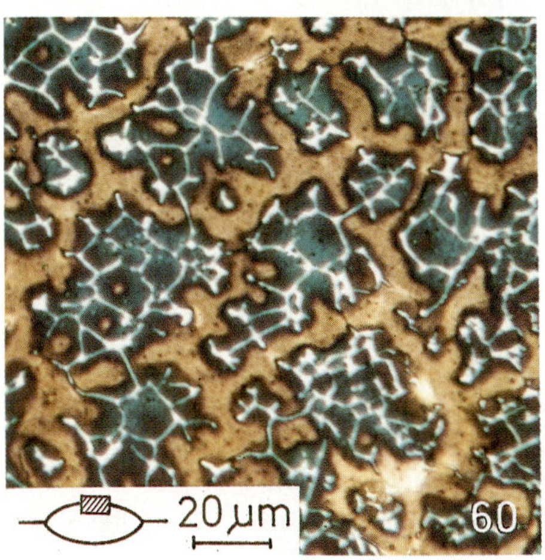
20 µm
60

verse and longitudinal section. At the points of highest cooling rate, i. e. at the bottom of the weld pool, cellular crystals will grow at first on the solid base material. Their direction of growth follows the orientation of base material crystals. With stainless weld metal, this zone is rather thin. There follows a narrow zone of moderately branched directed dendrites and a rather wide zone of distinct, directed dendrites which change at the pool surface into a small zone of nondirected dendrites. However, there are no distinct boundaries between these zones. Thus it is possible to find two different crystal configurations next to each other.

The following micrographs show characteristic primary solidification structures as they appear in different zones of a weld bead of austenitic stainless steel weld metal with normal cooling in air. The beads were deposited with 5 mm coated electrodes on previously laid weld metal of similar compositon and micrographs were taken thereof. The primary solidification structure has been made visible by way of colour etching according to Lichtenegger [120].

Figs. 58–60 show the solidification structures of 20/10 chromium-nickel weld metal (AWS E 308) which has solidified to primary δ crystals (see also profiles of ternary iron-chromium-nickel system shown in Fig. 6 and 9, page 8 and 10). Fig. 58 shows the microstructure (original in $\times$ 500 magnification) of zone 1 of Fig. 57 with cellular crystal structure. This zone is connected immediately to the base material which consists of previously laid weld metal of similar composition. Fig. 58a shows the transition from the base material to the deposited test bead in the form of a transverse micrograph according to Fig. 57a. The upper half shows the cellular crystals which grow in an upwards direction from the coarser base material of the previously laid weld metal. If these cellular crystals are cut transverse to the direction of growth which is equivalent to a micrograph parallel to the bead surface, the result is a structure as shown in Fig. 58b.

The colour etching has been set in such a way that it illustrates the situation during solidification of weld metal when about $^3/_4$ of the weld metal volume has solidified and $^1/_4$ is still present as liquid residual melt. The blue areas of the structure represent the already solidified cellular crystals consisting of primary delta ferrite. In between is the residual melt which appears as a brownish tinted structural constituent. The blue coloured cellular crystals show a slightly wavy surface, indicating the beginning of dendritic crystal formation. The already solidified crystals are still largely surrounded by liquid residual melt as illustrated by Fig. 58b. A bright structural constituent is also visible in the primary precipitated cellular delta ferrite crystals. This represents the residual delta ferrite which remains after δ-γ transformation from the primary precipitated delta ferrite crystals at ambient temperature and which is identified in conventional etchings as delta ferrite.

Fig. 59 shows the solidification structure of the 20/10 chromium-nickel weld metal (AWS E 308) in zone 3 of Fig. 57 with very distinct dendrites. The blue coloured structure represents the precipitated dendrite crystals and the brown structure the residual melt which is left between the dendrite branches. The dendrite configuration in Fig. 59a should be visualized such that the micrograph represents a longitudinal section through the dendrite trees. In the center of Fig. 59a appears the cut trunk of a dendrite tree. The structures to each side also represent sections through the branches of neighbouring dendrites, without the micrograph cutting into the main trunk of these dendrites. The dendrites appear in an upward direction in accordance with the flow of heat, i.e. they grow from the bottom to the top of the bead from the cellular crystals of zone *1* beneath. In accordance with the cooling rate which becomes slower at the bead surface, the structure is coarser than the more quickly cooled bottom zone *1* with a cellular structure. Fig. 59b shows a cross section through dendrites with brown coloured residual melt appearing between dendrite branches. Here too, we must visualize the structure in its spatial arrangement. In this section, the dendrite trunk is cut transversely with either the cut trunk or the cut branches appearing. In the primary precipitated delta ferrite dendrites we can see once again as a bright structural component the residual delta ferrite at ambient temperature which is left after δ-γ transformation.

Fig. 60 shows a section through the upper zone *4* of the bead, with any directionally formed dendrites lacking since heat is dissipated upwards, downwards and sideways. In this zone, there is no distinct flow of heat in any one direction and dendrites do not show any preferential direction either.

Figs. 61–63 show the primary structure of fully austenitic 20/14 chromium-nickel weld metal which has solidified to primary γ crystals, again in the form of zones *1, 3* and *4* of Fig. 57. Fig. 61a shows the cellular structure of zone *1* of the weld bead which connects to the dendritic structure of the bead below. This micrograph also shows that small areas in the heat affected zone of the base material have been liquefied during the execution of the newly deposited weld bead which is made visible by the bright brownish tinted colourations of diluted spots in the residual melt located between base material dendrites, close to the boundaries of the cellular crystals which grew on top of it. Such liquefied zones immediately below the boundary between base material and weld bead are predominantly found with primary solidification to γ crystals, i.e. with fully austenitic weld metal. They may be the origin for the formation of so called liquation cracks which are dealt with in detail in chapter 5. Fig. 61b shows the microsection parallel to the bead surface. Grain boundaries of coarse grains formed by recrystallization after solidification can be identified which overlap many primary precipitated small crystal cells.

Figs. 62a and 62b show the structure of zone *3* in the center of the bead with rather distinctly branched dendrites. On both sides of Fig. 62a, two cut dendrite trunks are visible whereas the center of the micrograph shows only cut branches. Fig. 63 shows the transition from directed to nondirected dendrites at the upper side of the bead.

If stainless steel weld metal solidifies in the immediate vicinity of the eutectic groove in the ternary iron-chromium-nickel constitution diagram, crystallization takes place preferably in the three phase sector $(L + \delta + \gamma)$. In this process, both primary δ and γ crystals are precipitated. The structure shows neighbouring areas of both types of crystals. If, for example, a crystal nucleus is formed of δ ferrite, the crystal will grow along the chromium richer boundary of the three phase sector, bordering the area of primary δ ferrite precipitation (see Fig. 6, page 8). Because of the precipitation of chromium rich crystals, the area of the melt located next to the precipitated crystals loses chromium and is simultaneously enriched with nickel. This promotes the precipitation of primary γ crystals next to the zone of the precipitated δ crystals. This process is repeated alternately until the melt has completely solidified (alternating crystallization of δ and γ crystal zones).

Fig. 64a (page 69) shows the structure in the center of the bead of a 20/12

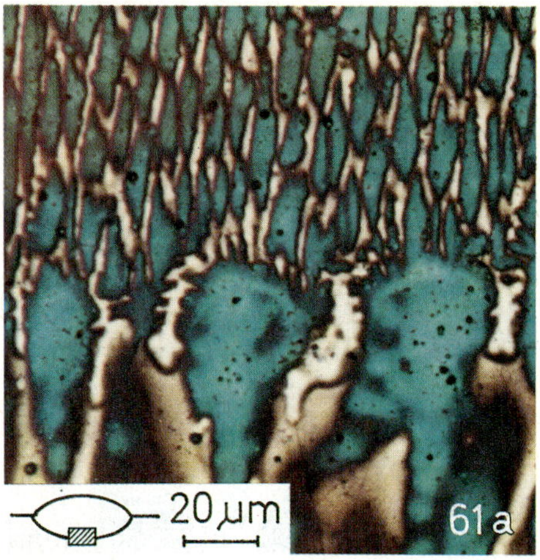
20 µm
61a

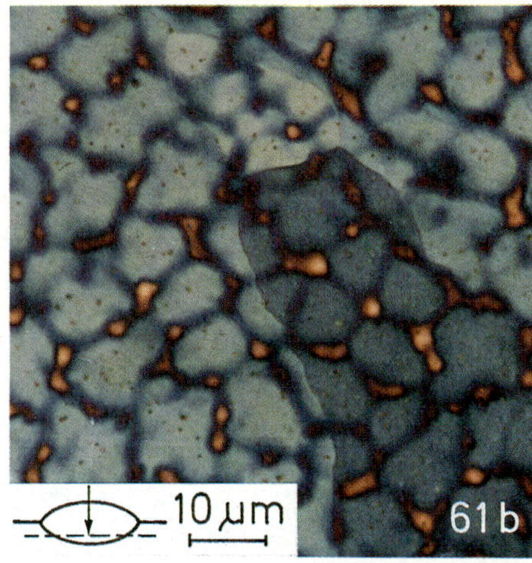
10 µm
61b

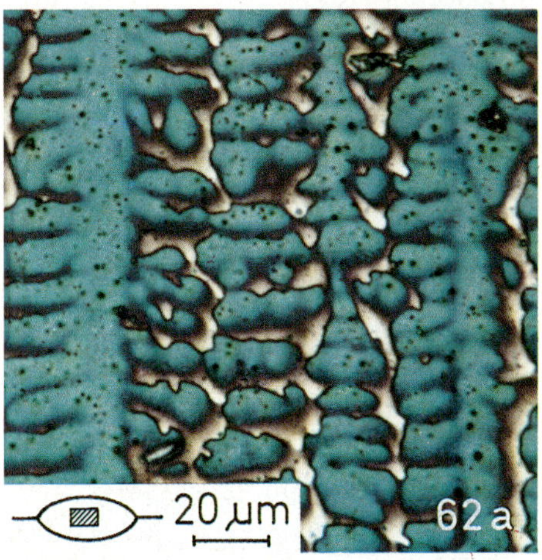
20 µm
62a

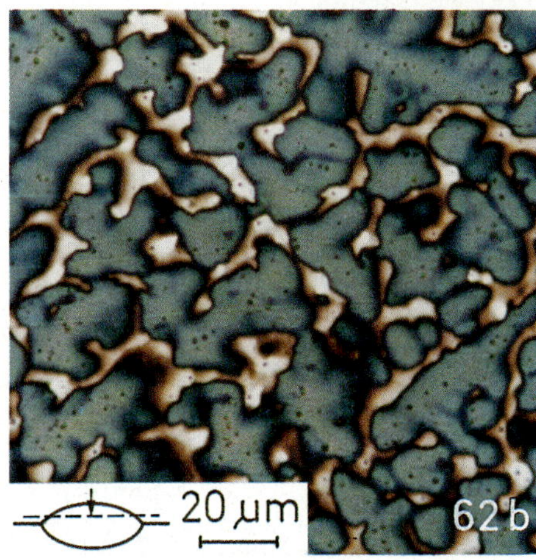
20 µm
62b

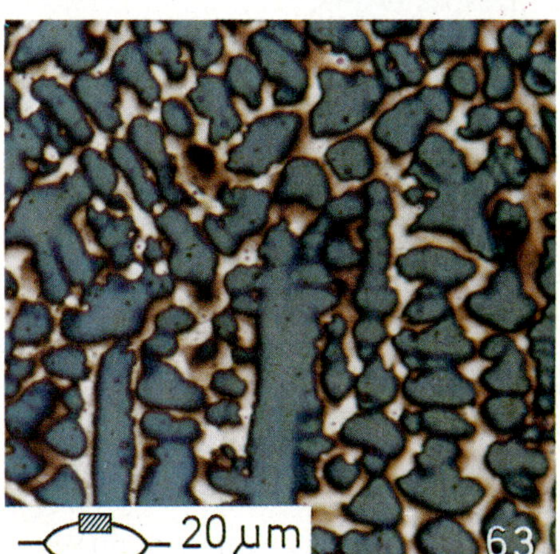
20 µm
63

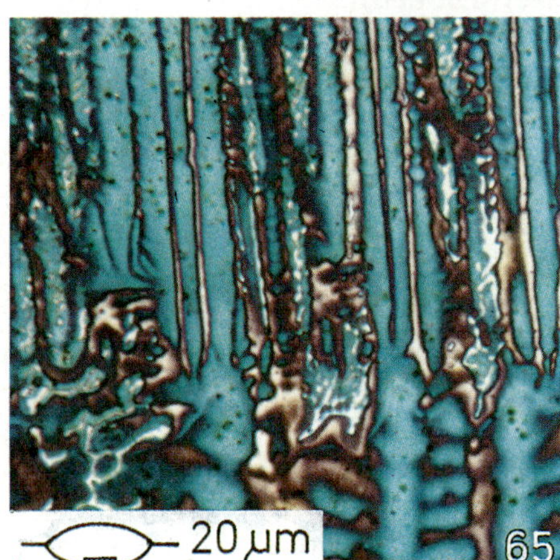
20 µm
65

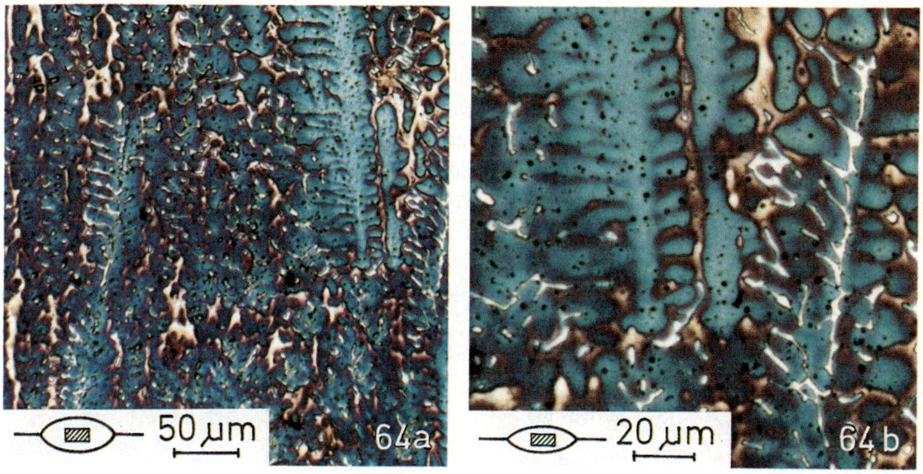

Fig. 64. Dendritic solidification of 20/12 chromium-nickel weld metal in the bead center with neighbouring zones of primary δ and γ crystals. *a* ×200 magnification (original), *b* ×500 magnification (original). Chemical composition in %: C: 0.034, Si: 0.56, Mn: 1.55, Cr: 19.90, Ni: 12.20, N: 0.045. Colour etching to Lichtenegger [20]

chromium-nickel weld metal with lower magnification. This alloy solidifies preferably in the three phase sector ($L + \delta + \gamma$). The right hand side of Fig. 64a shows a zone of primary blue coloured γ crystals, bordered to the right and left by zones of primary δ crystals. Fig. 64b shows a section of the same structure with stronger magnification. Left of the center, a large γ dendrite appears which is located in Fig. 64a on the righthand side. To the right of it, we can see a large δ dendrite which is recognizable by the bright residual δ ferrite in the dendrite trunk. In between, there are areas of residual melt (brown colouration) which partially also include residual δ ferrite.

The structure of the transition from bottom bead to the bead deposited on top of it (zone 1) when solidification takes place in the three phase sector is shown in Fig. 65 (page 68). The cellular crystals growing on top of the base material dendrites

Fig. 61. Cellular solidification of 20/14 chromium-nickel weld metal at bottom of a weld bead (zone 1 according to Fig. 57). *a* microsection transverse to direction of welding, *b* microsection parallel to surface. In this micrograph secondary grain boundaries formed by recrystallization after solidification can be identified (see e.g. dark blue coloured coarse grain on the right side of the micrograph). Chemical composition in %: C: 0.033, Si: 0.58, Mn: 1.53, Cr: 19.80, Ni: 14.00, N: 0.051. Colour etching to Lichtenegger [120]

Fig. 62. Dendritic solidification of 20/14 chromium-nickel weld metal in the center of the weld bead (zone 3 according to Fig. 57). *a* microsection transverse to direction of welding, *b* microsection parallel to bead surface. Chemical composition: same as in Fig. 61. Colour etching to Lichtenegger [120]

Fig. 63. Transition from directed to nondirected solidification at the upper part of the weld bead of 20/14 chromium-nickel weld metal (transition from zone 3 to zone 4 in Fig. 57). Chemical composition: same as in Fig. 61. Colour etching to Lichtenegger [120]

Fig. 65. Boundary line between two beads from 20/12 chromium-nickel weld metal, bottom dendritic solidification, top new bead with cellular solidification and neighbouring zones of primary δ and γ crystals. Chemical composition: same as in Fig. 64. Colour etching to Lichtenegger [120] solidify similarily to the mode shown in Fig. 64 to zones of primary γ and δ crystal cells which are both blue coloured and arranged side by side. Here too, residual ferrite may be found in some primary solidified δ ferrite crystals and partially also in some brown coloured zones of the residual melt.

The thickness of the zone with directed cellular growth is determined on the one hand by the cooling rate and on the other by the configuration of the liquidus area in ternary systems. The impeded equalization of concentration can also excert an influence since it changes the liquidus area (see Fig. 51, page 57). All these phenomena are highly complex and a theoretical forecast is often not posible. Empirical tests must be employed in individual cases to provide exact information on primary structure. It is also possible, for example, that a cellular structure disappears completely, with dendrite crystals growing directly from the base material. The crystal structures and growth orientations produced in that way have been investigated by Baikie and Yapp (126) on multilayered austenitic weldments, carried out in different welding positions. Again, the strong influence of heat flow on crystal orientation could be established.

Solidification parameters during welding will also influence the type of crystallization. The eutectic groove in the ternary iron-chromium-nickel system which represents the upper boundary of the three phase sector ($L + \delta + \gamma$), separates the sectors of primary solidification to delta and gamma crystals. According to Suutala [127] the location of this groove is influenced during welding by the rate of crystal growth which in turn is dependent on the rate of deposition. High deposition rates will promote, for example, the formation of primary gamma crystals. This means that with high welding speeds the sector of gamma crystals may become slightly enlarged to the detriment of the sector of primary delta ferrite crystals and the eutectic groove may be shifted a little towards lower nickel contents.

2.6 Cooling Rate and Thermal Supercooling During Solidification of Stainless Steel Weld Metal

The cooling rate of weld metal influences both constitutional and thermal supercooling of the weld pool. Constitutional supercooling is limited to a very small range at the liquid-solid interface and is caused by deviations in temperature and concentration of liquidus and solidus lines (areas) of alloys which are determined by the constitution diagram (thus–the term "constitutional", see also section 2.4). The cause of delayed crystallization by thermal supercooling results from the fact that in liquid pools both number of nuclei and solidification rate are rather small at the beginning of solidification. Only when the temperature of the weld pool drops somewhat below the theoretical solidification temperature, the solidification rate will assume noticable values and the number of nucleus will rise. Then, crystallization will start and with increasing supercooling effect both solidification rate and number of nuclei will quickly rise. In other words, thermal supercooling represents the difference between the theoretical temperature at the beginning of solidification with extraordinary slow cooling rates and that temperature at which crystallization will actually start with real cooling rates. In general, thermal supercooling will rise with the increase in cooling rate.

In principle, the cooling rate of weld metal is determined on the one hand by the amount of heat supplied per unit length of weld bead (heat input) and on the other by the rate of heat flow which occurs predominantly through the base metal which surrounds the bead. Regarding general correlations in welding operations, please refer to the comprehensive body of supplementary literature and available welding handbooks.

At ambient temperature stainless steels show a much lower thermal conductivity than unalloyed steel, i.e. at ambient temperature, welding heat will be dissipated slower into the base material. With rising temperature, however, thermal conductivity of unalloyed steel will decrease much stronger than that of stainless steel. The difference gets smaller and smaller until they meet in the case of unalloyed and austenitic 18/8 chromium-nickel steel according to Zitter [128] at about 1200° C and approx. 33 $J \cdot cm^{-1} \cdot s^{-1} \cdot {}^{\circ}C^{-1}$. Investigations on cooling phenomena in the heat affected zone during electric arc welding of stainless steels have been published by Zitter [128], Nippes [129] and Klug [289]. From these works as well as from our own investigations it appears that the time in which stainless steel weld metal solidifies is just as short as that of low alloyed weld metal. When executing a bead with low heat input (5 kJ/cm) on steel plate of 20 mm thickness, the temperature range between 1450° C and 1350° C will be covered in about 0.2 seconds, without consideration of released crystallization heat. If beads are deposited with high heat input of 30 kJ/cm on 10 mm plate, e.g. vertical weldments, the temperature range mentioned above is covered in about 2 seconds. If the released crystallization heat is also considered, these times will slightly increase.

The crystal growth rate of weld metal has been investigated in GTAW spot welds by Suutala, Sivonen and Moisio [130]. They found a rate of crystal growth in the range of 3—8 mm per second, with theses values being primarily influenced by the cooling rate of the weld pool. These results support the assumption that both thermal supercooling and impeded equalization of concentration must occur in electric arc welding of stainless steel during the solidification of weld metal. Similar conclusions were drawn by Lippold and Savage [20] who have assumed that equalization of concentration through diffusion on the solid side of the liquid-solid interface will be rather limited in the case of the solidification of austenitic stainless steel weld metal.

At a cooling rate of 0.1° C/s David, Goodwin and Braski [131] determined a liquidus temperature of 1445° C and solidus temperature of 1335 ± 2° C for a 20/10 chromium-nickel steel (AISI 308). For a 26/21 chromium-nickel steel (AISI 310 S), they found 1395° C and 1304° C respectively. If compared to the equilibrium temperatures in the ternary system according to Schürmann and Brauckmann [12], these values correspond to a thermal supercooling of the liquidus temperatures of about 30—60° C and of the solidus temperatures of about 80—120° C. Since cooling rates in electric arc welding are still much higher than those of the tests mentioned above, thermal supercooling must be even greater. Employing a process in which liquid weld metal is produced with the aid of an electron beam, Velkov and Schelev [132] determined a liquidus temperature of 1360° C and a solidus temperature of 1275° C for a 25/20 chromium-nickel steel (AISI 310 S). The effect of thermal supercooling is limited, however, by the fact that the solid base material on which the weld metal crystals are growing excerts a strong nucleating effect, with the result that the rate of crystal growth increases considerably with increasing supercooling.

2.7 Segregations in Stainless Steel Weld Metal During Solidification

Segregations may have a detrimental effect on material properties. Bäumel [133] and Weingerl, Straube and Blöch [134] report on the influence of segregations on corrosion resistance. Blanc and Tricot [135] deal with the influence of segregations on the process during primary crystallization of austenitic steels and their influence on the liquidus and solidus areas in the iron-chromium-nickel constitution diagram as well as the effect of homogenizing (annealing) on the reduction of segregations. Kulmburg [136] and Eckstein [112] describe the influence of segregations on γ–α transformation of steel and Hoffmeister [137] on δ–γ transformation of stainless steel weld metal. Masumoto, Tamaki and Kutsuna [138] report on the influence of segregations on the hot cracking behaviour of austenitic weld metal and Brezina [93] on the cold cracking behaviour and toughness properties of low carbon martensitic stainless chromium-nickel steels.

In stainless steel weld metal segregations which are produced during solidification and further cooling generally cannot—contrary to steel—be eliminated or reduced by hot forming operations, diffusion annealing or any kind of solution annealing. They are present, therefore, in the finished weldment or welded component. A positive effect results from the fact that the very fine grained weld metal structure limits these segregations to some extent and the pertinent literature is often more concerned about segregation damage in the steel rather than in the weld metal. Nevertheless, it seems to be of importance to investigate the segregation behaviour of weld metal more closely, in order to consider its effects on the practical welding of stainless steels.

Figs. 50 b and 51 b (page 56 and 57) show schematic representations of the segregation behaviour of ternary alloys. A specific composition of the melt in the liquidus area corresponds to a crystal being precipitated in the solidus area at the same temperature. Contrary to binary systems, where the composition of crystals is clearly defined by isothermal lines (conodes), this is not true for ternary alloys since the direction of isotherms from the liquidus to the solidus area which determine the composition of the precipitated crystals is according to Mitsche and Niessner [139] undefined. Therefore to exactly determine the composition of precipitated crystals in ternary systems, it is necessary to empirically determine the direction of segregation.

For the ternary iron-chromium-nickel constitution diagram, respective tests were carried out by Schürmann and Brauckmann [12], Blanc and Tricot [135], Kulmburg, Blöch and Swoboda [136], Hoffmeister [137], Kulmburg [117], and Siegel and Günzel [141]. Schürmann and Voss [140] have shown the configuration of the conodes which determine the direction of segregations. Fig. 66 represents a survey of the most important findings contained in the works mentioned above. The direction of segregation in the area of primary precipitation of δ crystals differs from those of primary γ crystals. In the area of solidification to δ ferrite, crystals are precipitated which are enriched by chromium and depleted by nickel, whereas the melt is enriched by nickel and depleted by chromium (direction of arrows in Fig. 66 to the left of the three-phase sector ($L + \delta + \gamma$).

In other words, there is a negative correlation of the segregation of chromium and nickel in the precipitation range of primary δ ferrite. In the precipitation range of pri-

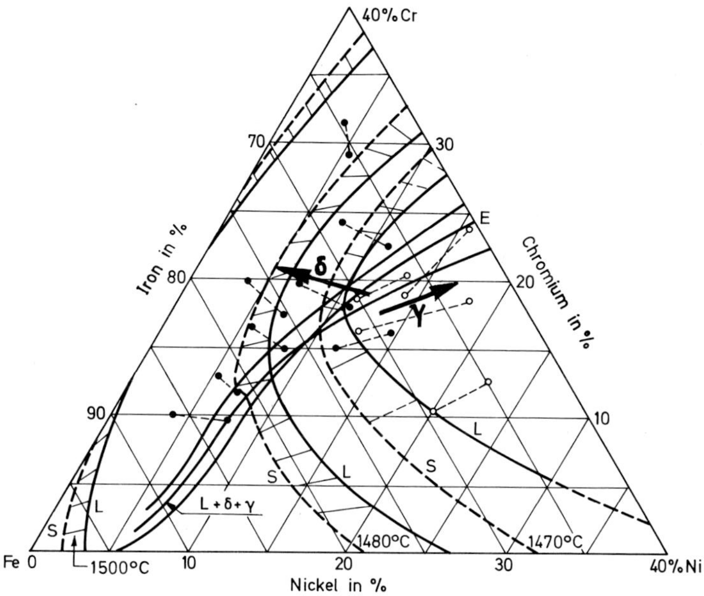

Fig. 66. Representation of conode configuration in the iron-chromium-nickel constitution diagram by way of isothermal profiles at 1470° C, 1480° C and 1500° C from investigations by Schürmann and Voss [140], supplemented with results from Kulmburg, Blöch and Swoboda [136] and Hoffmeister [137]. $L =$ isothermal profile at liquidus area, $S =$ isothermal profile at solidus area, ← main direction of segregation during primary solidification of δ or γ crystals for the range of austenitic chromium-nickel steels

mary austenite crystals, in Fig. 66 to the right of the three-phase sector ($L + \delta + \gamma$), crystals are precipitated which are depleted of both chromium and nickel, with the melt being enriched with both elements. This means that there is a positive correlation in the direction of segregation for chromium and nickel in this range.

To enable us to make correct predictions about possible changes in the concentration of precipitated crystals and the residual melt, as in the case of impeded equalization of concentration, Fig. 67 shows concentration profiles in the iron-chromium-nickel system which were not plotted with constant iron contents but with a direction of segregation determined empirically according to Fig. 66. These profiles were taken from a 20/10 chromium-nickel alloy (AISI 308) with primary precipitation of δ crystals (section A-A) and a 20/16 chromium-nickel alloy with primary precipitation of γ crystals (section B-B). The representations of solidus and liquidus profiles relating to the location of liquidus and solidus areas in the iron-chromium-nickel system (see Fig. 4, page 6) are based on the findings of Jenkins, Bucknall, Austin and Mellor [11]. They cover a relatively large area of this system which permits the expansion of concentration profiles to contain relatively high chromium and nickel contents. As a supplement, Fig. 67 also shows by way of a dotted line (which runs approx. 30° C above the values given by Jenkins et al. [11]) the profiles in the ternary iron-chromium-nickel system according to the results of Schürmann and Brauckmann [12].

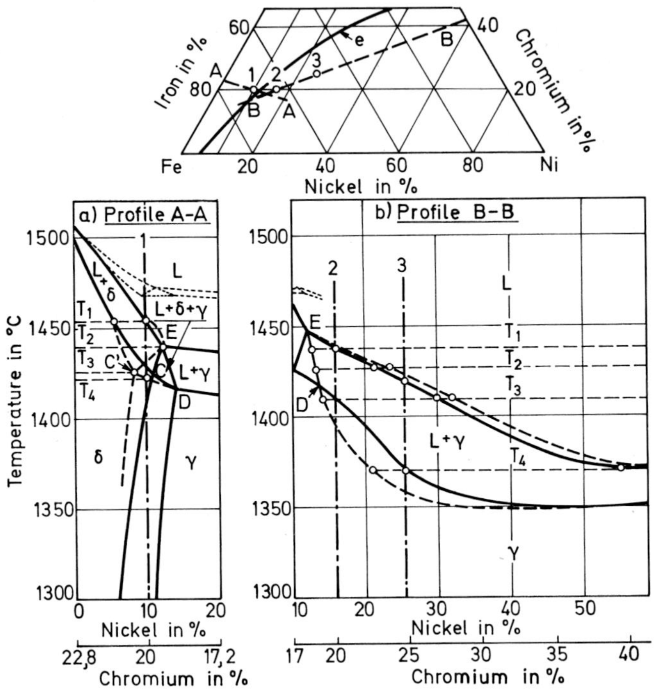

Fig. 67. Schematic representation of concentration profiles in accordance with the direction of segregation in the iron-chromium-nickel system as shown in Fig. 66. Solid line: Fe-Cr-Ni system according to results of Jenkins et al. [11]. Dotted line: Fe-Cr-Ni system according to results of Schürmann and Brauckmann [12]. Section A-A: concentration profile through 20/10 chromium-nickel alloy with primary precipitation of δ crystals. Section B-B: concentration profile through 20/16 chromium-nickel alloy with precipitation of γ crystals. The changes of the liquidus and solidus area with impeded equalization of concentration are plotted by the dashed lines in sections A-A and B-B

Fig. 67a shows the concentration profile A-A. The 20/10 chromium-nickel alloy [1] is shown by a vertical dash-dot line. Thus it solidifies completely as δ ferrite crystals. If we now assume that an impeded equalization of concentration exists which is a highly realistic assumption for the crystals being precipitated at the high cooling rates of the weld metal, a shift of the solidus and liquidus lines must appear which is similar to that shown in Fig. 51b (page 57). This is shown schematically in Fig. 67 by the dashed line. The left corner point C of the three-phase sector ($L + \delta + \gamma$) is shifted to a higher chromium and a lower nickel content at C′.

Fig. 67b shows the concentration profile B-B which represents in schematic form the solidification to primary γ crystals with impeded equalization of concentration (dashed line). The 20/16 chromium-nickel alloy [2], through which the concentration profile has been taken in the direction of segregation, is indicated by a dashed line. In addition, a 25/26 chromium-nickel alloy [3] has also been plotted which is located in the same sectional plane. Starting from the point of intersection (E) of the eutectic groove, the temperature of the solidus profile drops in the case of impeded equalization of concentration. The situation is aggrevated by the fact that an eventual equalization of concentration by diffusion (homogenization) is more difficult in γ crystals than in δ ferrite, because the diffusion coefficient of chromium and nickel at 1400° C is lower in austenite than in δ (α) iron (see Table 1, page 50).

The direction and size of crystal segregations occurring in an austenitic weld metal with different types of primary solidification structure have been investigated by Cieslak, Ritter and Savage [143] covering a number of different austenitic steel alloys, all remelted by the GTAW process. The authors found four different basic possibilities of primary solidification, leading to different types of segregations:

a) complete solidification to δ ferrite.

b) complete solidification to austenite.

c) partial solidification to austenite and partial solidification in the three-phase sector ($L + \delta + \gamma$) with the precipitation of δ ferrite in the area of residual melt.

d) partial solidification to δ ferrite and partial solidification in the three-phase sector with the precipitation of austenite from the residual melt.

The segregations which are produced in this way over the whole cross section of a crystal, are shown schematically in Fig. 68 according to Cieslack, Ritter and Savage [143]. The complete solidification to δ ferrite as shown in Fig. 68a, produces chromium enriched and nickel depleted zones at the crystal core. The dendrite boundaries at the end of solidification approach the composition of the chromium depleted and nickel enriched residual melt. Because of the positive correlation of the segregations, the complete solidification to austenite will lead—according to Fig. 68b—to a chromium and nickel depleted core with a boundary enriched by both

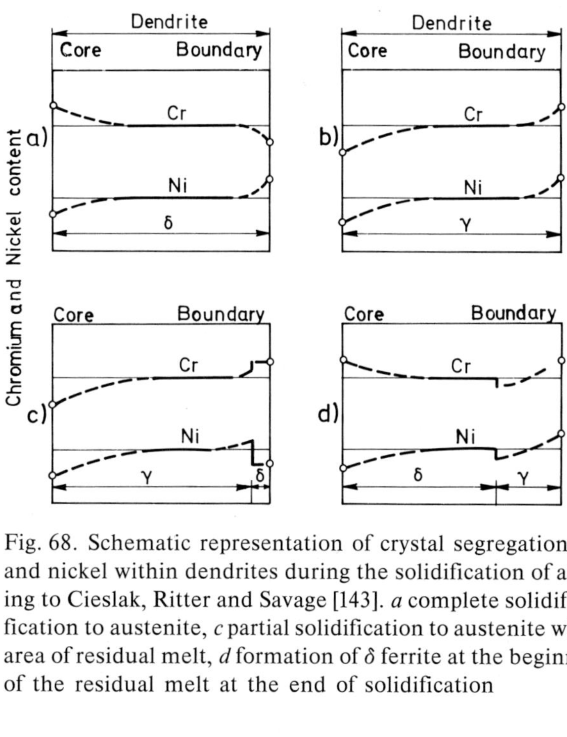

Fig. 68. Schematic representation of crystal segregations and the distribution of chromium and nickel within dendrites during the solidification of an austenitic weld metal alloy according to Cieslak, Ritter and Savage [143]. *a* complete solidification to δ ferrite, *b* complete solidification to austenite, *c* partial solidification to austenite with precipitation of delta ferrite in the area of residual melt, *d* formation of δ ferrite at the beginning of solidification and of austenite of the residual melt at the end of solidification

elements. If—according to Fig. 68c—solidification is mainly to austenite, with the residual melt precipitating delta ferrite in the three-phase sector, there is a jump in both size and direction of segregations. A similar effect, which differs only with regard to the direction of segregation, is produced if—according to Fig. 68d—primary δ ferrite is precipitated first and then due to the nickel enrichment of the melt—austenite is precipitated later on. This kind of solidification has also been referred to by David, Goodwin and Braski [131], with primary solidified ferrite dendrites being encased by austenite at the end of solidification. The segregations of chromium, nickel and molybdenum which can be determined by way of micro analysis, indicate a strongly impeded homogenization in the precipitated crystals which occur during welding.

Since homogenization by diffusion plays an important role in all transformation and precipitation processes, Table 1 (page 50) lists the diffusion coefficients of different elements in austenite and ferrite together with the coefficients of the self-diffusion of iron according to Eckstein [112] and to Fridberg, Törndahl and Hillert [142].

If the severe thermal supercooling caused by the high cooling rates in electric arc welding is also considered in the representation of segregation phenomena in Fig. 67, the liquidus and solidus profiles are shifted to lower temperatures. To enable us to imagine the solidification processes which take place during electric arc welding more easily, the probable configuration of the concentration profiles A-A and B-B with possible supercooling effects are shown in Fig. 69. Just above the

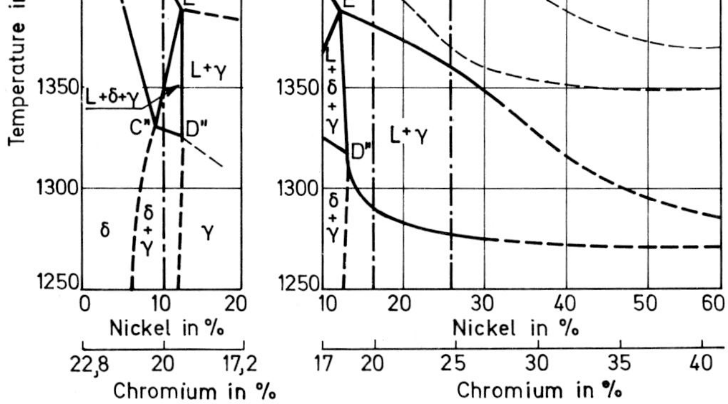

Fig. 69. Schematic representation of the probable configuration of concentration profiles in the direction of segregation on the liquidus and solidus area of the iron-chromium-nickel system, with consideration to the actual thermal supercooling and impeded equalization of alloy concentration during solidification of austenitic weld metal, under the cooling rates found in electric arc welding. The configuration under equilibrium conditions is shown by the dashed line above

assumed phase boundaries that occur during welding, the equilibrium phase boundaries of the ternary iron-chromium-nickel system according to Jenkins et al. [11] have also been plotted (dashed line). The tendency towards segregations is particularly high in the area of primary solidification to γ crystals because of the dependent marked differences in alloy concentration that exists between the precipitated crystals and the residual melt. It is also interesting to note that fully austenitic weld metal which solidifies primarily to austenite shows a higher tendency towards the formation of low melting phases than weld metal which solidifies primarily to delta ferrite. The noticeably stronger segregations found in primary precipitated austenite crystals as compared to primary precipitated delta ferrite crystals are confirmed by Blanc and Tricot [135], Hoffmeister [137] and Héritier and Lévy [144] for standard iron-chromium-nickel alloys and by Takalo, Suutala and Moisio [145] for austenitic stainless steel weld metal. The possibility of the formation of low melting metallic phases due to segregations is pointed out by Savage, Nippes and Miller [146].

It must also be noted here that the solubility of austenite for elements such as sulphur, phosphorus, silicon and boron is lower than that of δ ferrite and that it is further diminished by the segregation-induced nickel enrichment of the residual melt as in the case of the primary precipitation of γ crystals. This phenomena produces an increase of these undesirable elements in the residual melt.

The correlation of different factors favouring the formation of low melting phases is discussed in detail in Chapter 5. It is safe to assume, however, that the foregoing "constitutional" influence on the formation of low melting phases which results from the actual configuration of the liquidus and solidus areas in the ternary iron-chromium-nickel system in combination with a severely impeded equalization of concentration, particularly in austenite, and the strong thermal supercooling effect caused by the high cooling rates produced by electric arc welding, represents a major factor which must be taken into account in the evaluation of the hot cracking sensitivity of stainless steel weld metals.

3 Metallurgical Phenomena in Secondary Crystallization of Stainless Steels and Weld Metals

3.1 Recrystallization of Weld Metal

After the solidification of stainless steel weld metal to primary crystals, mainly of a dendritic structure, during further cooling immediately recrystallization processes occur. New grain boundaries are formed which are no longer dendritic but are mainly globular or cellular in structure. Recrystallization is most pronounced in weld metals that do not undergo further transformation, i.e. weld metals which experience neither δ-γ nor γ-α transformation. These alloying ranges can be seen in Figs. 6 and 9 (pages 8, 10). They cover alloys with high chromium and low nickel contents which possess a high degree of delta ferrite. They are classified as ferritic or semi-ferritic stainless steel weld metals. The second range covers alloys with primary solidification to austenite which are classified as fully austenitic stainless steel weld metals.

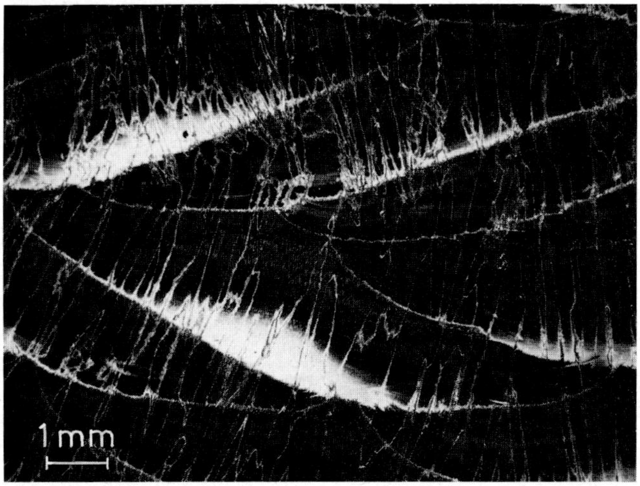

Fig. 70. Formation of coarse crystal grains due to recrystallization during GTA-welding in a multi-layer weld using weld metal with an average delta ferrite content of 55%. The newly formed crystal cells show epitaxial growth through the individual layers. The light coloured structural constituent locally contains up to 82% delta ferrite. Chemical composition of weld metal in %: C = 0.012, Si = 0.89, Mn = 0.75, Cr = 22.12, Mo = 3.06, Ni = 9.09, N = 0.061. Color etching according to Lichtenegger [120] shown in black and white

In general, during welding recrystallization produces coarser grains which normally overlap a higher number of primary precipitated dendritic or cellular crystals. Fig. 61 b (page 68) shows a micrograph of such coarse secondary grains, that were produced by recrystallization, which cover many primary formed crystal cells. They represent the lower area of a weld deposited using a fully austenitic 20/14 chromium-nickel weld metal. Fig. 70 shows a section of the coarse secondary crystals, formed by recrystallization, taken from a multi-layer weldment of stainless steel weld metal with a high delta ferrite content. The newly formed crystals grow beyond individual layers and may reach a length of 3—6 mm. Their growth is neither impeded by the rim of pure ferrite, formed at the boundaries between the individual weld layers (light structural constituents), nor by the considerably finer primary solidification structure. The recrystallization processes found in the weld metal not only take place during cooling but also in multi-layer welds during the deposition of further weld layers as illustrated by Fig. 70. Because of the reheating of the weld metal by subsequent passes, the recrystallization of previously deposited layers is further intensified. This phenomenon is particularly apparent when welding with high interpass temperature and high heat input.

Coarse grain formation caused by recrystallization during welding is largely dependent upon the nucleating conditions in the weld metal. New crystal grains will always start to grow from a crystal nucleus. If more crystal nuclei are present, more new crystals will be formed and the recrystallization structure will be finer. In other words, the nucleating condition has a direct influence on a possible coarse grain formation.

With stainless steel weld metal, the temperature range from which recrystallization normally starts is around 500—550° C and it ends at the melting point of the alloy. With lower temperatures, however, recrystallization takes more time which normally only happens with prolonged exposure to these low temperatures. From temperatures of around 700° C and upwards, the recrystallization processes of stainless steel weld metals take place during all heat treatment processes. In the case of the rapid reheating of the heat affected zone during electric arc welding, this temperature limit rises to above 1000° C. Due to the higher self-diffusion coefficient of δ (α) iron (see Table 1, page 50), ferritic weld metals will recrystallize easier than austenitic ones. If the weld metal experiences a transformation (e.g. δ-γ or γ-α transformation), this process will blanket a possible recrystallization and often considerably slow down the latter.

Besides the normal crystal nuclei such as grain boundaries, crystal lattice defects, or fine nonmetallic inclusions, the presence of any residual delta ferrite in austenitic weld metal also has a strong nucleating effect which explains why austenitic weld metals with, for example, 5—10% residual delta ferrite will show a lesser tendency towards coarse grain formation during recrystallization. This is because the coarse grain formation is impeded by the relatively fine grained formation of the new austenite crystals during δ-γ transformation. The danger of increased coarse grain formation will only arise with delta ferrite contents of about 30% and above. This can be compensated for, however, by influencing the nucleating condition of the material in question.

In stainless steel weld metal, the recrystallization process will have a disturbing influence in specific cases only. According to Tamura and Watanabe [58] it is possible at temperatures slightly below the melting point of the pertinent material that low melting phases are pushed ahead of the newly formed crystals and enrichment of these phases may occur at the newly formed grain boundaries. These may be the cause of liquation cracks in fully austenitic weld metal if other parameters for crack formation are present. In the case of extended exposure to elevated temperatures too, creep rupture strength can be influenced by recrystallization processes, with the latter often being accompanied by the precipitation of intermetallic phases. In the welding of chromium steels with high delta ferrite content, however, recrystallization may due to strong coarse grain formation produce a considerable embrittlement of the heat affected zones or even the whole welded cross section.

3.2 δ-γ Transformation of Stainless Steel Weld Metal

In the ternary iron-chromium-nickel system, the two-phase sector $\delta + \gamma$ is adjacent to the three-phase sector $(L + \delta + \gamma)$ which is located between the areas of primary solidification to δ ferrite and austenite. The configuration of phase boundaries can be seen from the concentration profiles which are shown in Figs. 6 and 9 (page 8 and 10). As a supplement, Fig. 71 shows that part of the ternary iron-chromium-nickel system which is rich in iron with the phase boundaries of the δ and γ sector which indicate the spatial position of the different phase sectors. The lines *c* and *d* represent the lower boundaries of the three-phase sector $(L + \delta + \gamma)$ according to Jenkins et al. [11] which are identical with the upper boundaries of the two-phase sector $(\delta + \gamma)$. These lines start in the iron corner of the binary iron-nickel system (see Fig. 2, page 3) in the temperature range of 1500—1510° C and drop at 50% iron to the range of approx. 1370—1390° C (see Fig. 4, page 6). The $(\delta + \gamma)$ sector, located below, is shown in Fig, 71 by the isothermal phase boundaries at 1300 and 1200° C according to Mundt and Hoffmeister [149]. In the binary iron-nickel system (Fig. 2, page 3),

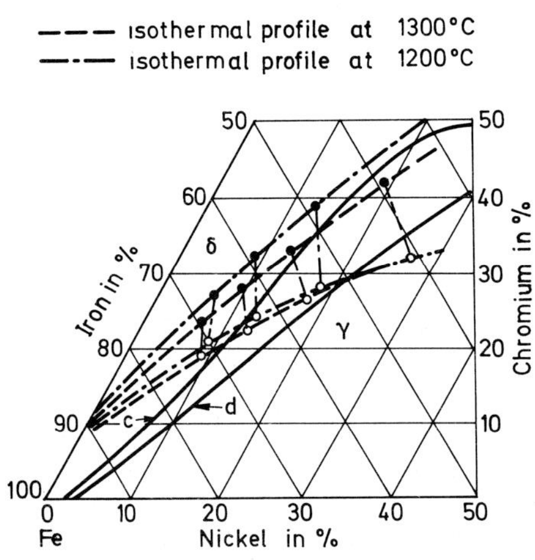

Fig. 71. Configuration of δ and γ sector in the ternary iron-chromium-nickel system. The lines *c* and *d* represent the lower boundaries of the three-phase sector $(L + \delta + \gamma)$ to Jenkins et al. [11]. The dashed and dash-dotted lines are isothermal profiles at 1300° C and 1200° C according to Mundt and Hoffmeister [149]

delta ferrite is completely transformed into austenite at temperatures below 1400° C. Therefore, at 1300° C the ($\delta + \gamma$) sector begins in the binary iron-chromium system (Fig. 1, page 2) at the phase boundaries of the constricted γ area. The shifting of the γ area with decreasing temperatures towards lower nickel contents means that primary solidified ferrite which is precipitated from the melt left of the line *c* is transformed (in part or completely) during cooling to austenite. A 20/10 chromium-nickel alloy e.g. AISI 308, for example, will solidify to the left of line *c* mainly to primary delta ferrite crystals which are transformed, however, in the equilibrium during cooling to 1200° C completely into austenite.

Changes in concentration will also occur during δ-γ transformation and it is important, therefore, to know the direction of segregation. In Fig. 71, some conodes are plotted according to Mundt and Hoffmeister [149] which indicate the direction of segregation during δ-γ transformation. According to this diagram, austenite crystals are produced during the transformation of primary delta ferrite crystals which are enriched with nickel and substantially depleted of chromium. The differences are more pronounced with higher nickel and chromium contents than with lower ones. In order to obtain a rough idea of the possible segregation processes that occur during δ-γ transformation, concentration profiles with constant iron content (see Fig. 6, page 8) can also be used. In electric arc welding, δ-γ transformation is normally passed through rather quickly and an impeded equalization of alloy concentration is to be expected. Impediment is greater on the side of precipitated austenite crystals than on the ferrite side because at the temperatures of δ-γ transformation, diffusion coefficients of chromium and nickel in austenite are lower than those in ferrite by a factor of 10^2 (see Table 1, page 50). The δ-γ transformation is also supercooled with increasing cooling rates. In the case of a weld metal alloy, e.g. with 20% chromium and 10% nickel (AWS E 308), the as welded structure which should be fully austenitic under equilibrium conditions consists after cooling to ambient temperature of 90—95% austenite and 5—10% residual delta ferrite which has been supercooled from the primary precipitated δ ferrite crystals. This residual delta ferrite is dispersed between the newly formed secondary austenite crystals which also determine its shape.

In electric arc welding, the process of δ-γ transformation occurs over a very short period of time, typically over one second or less. The new formation of secondary austenite crystals starts at the grain boundaries of the primary delta ferrite crystals, which are most frequently nickel enriched and chromium depleted due to segregations during solidification (see Fig. 68 a, page 75). The grain boundaries form ideal nuclei for the secondary austenite crystals which are precipitated during δ-γ transformation. Residual delta ferrite is greatly enriched by chromium and depleted by nickel. The chromium and nickel contents in residual delta ferrite which where determined by way of micro analysis are normally in the range of 23—26% Cr and 5—8% Ni for a weld metal alloy containing 20% Cr and 10% Ni. It is quite difficult here, however, to determine which content is produced by the primary segregations during solidification and which is produced by the secondary segregations during δ-γ transformation. According to Perteneder, Blöch and Wedl [148], it is quite certain that these processes overlap each other, with δ-γ transformation accounting for the majority of the segregation of residual ferrite that is present in the structure. The formation of residual delta ferrite from delta ferrite crystals primarily

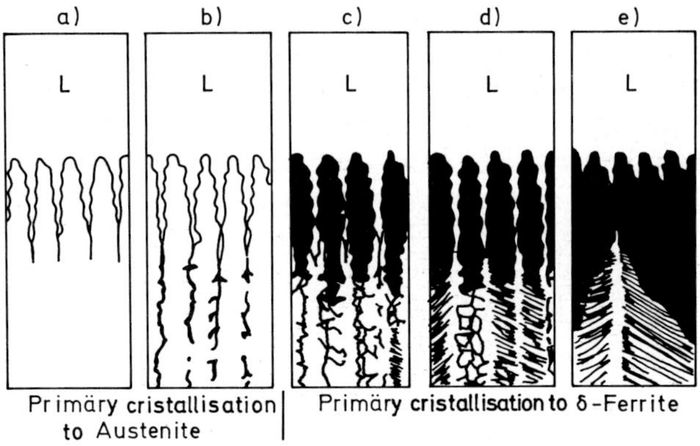

Fig. 72. Schematic representation of the crystallization of stainless steel weld metal during solidification and δ-γ transformation according to Kujanpää, Suutala, Takalo and Moisio [150], *a* and *b* primary solidification to austenite, *c, d* and *e* primary solidification to delta ferrite. Light area: austenite, dark area: delta ferrite

precipitated from the melt can be seen from the structures shown in Figs. 58–60. The fine ferrite is only barely visible as a bright skeleton-like residue in the area of the originally blue coloured primary delta ferrite. As shown in Figs. 6, 9 and 71, in the case of stainless steel weld metal there is a direct correlation between the various areas of different solidifications, δ-γ transformation and the chromium-nickel ratio. Kujanpää, Suutala, Takalo and Moisio [150] provide a schematic representation of the processes which occur during the solidification and δ-γ transformation of stainless weld metal, using instead of chromium and nickel contents the respective chromium and nickel equivalents which are also used in the Schaeffler diagram. Fig. 72 shows a schematic representation taken during the solidification and δ-γ transformation of a stainless steel weld metal. Figs. 72 a and b show the processes that occur during primary solidification to austenite crystals which grow, according to Kujanpää et al. [150] as shown in the schematic representation into the liquid weld metal above. The Cr_{equ} : Ni_{equ} ratio is less than 1.48. In Fig. 72 a, there is no further transformation, in Fig. 72 b, the alloy passes through the three-phase sector $(L + \delta + \gamma)$. During this process, small amounts of secondary ferrite are precipitated from the residual melt. The actual microstructures obtained after cooling to ambient temperature are shown in Figs. 61–65 (pages 68 and 69).

Figs. 72 c and d illustrate the processes during primary solidification to δ ferrite with the chromium and nickel equivalent ratios of 1.48 to 1.95 respectively. When the alloy passes the three-phase sector (Fig. 72 c), primary austenite is formed from the residual melt between the primary precipitated δ ferrite dendrites which represent the nuclei for the δ-γ transformation which starts almost simultaneously (see also Fig. 64, page 69). The formation of secondary austenite quickly progresses from the boundaries to the core of the delta ferrite dendrites with a marginal amount of residual delta ferrite remaining between the newly formed secondary austenite crystals. If the chromium-nickel ratio is slightly higher (Fig. 72 d), lath residual delta ferrite will also form between the austenite crystals. The actual micro structures for alloys with low ferrite contents of about 8–10% are shown in Figs.

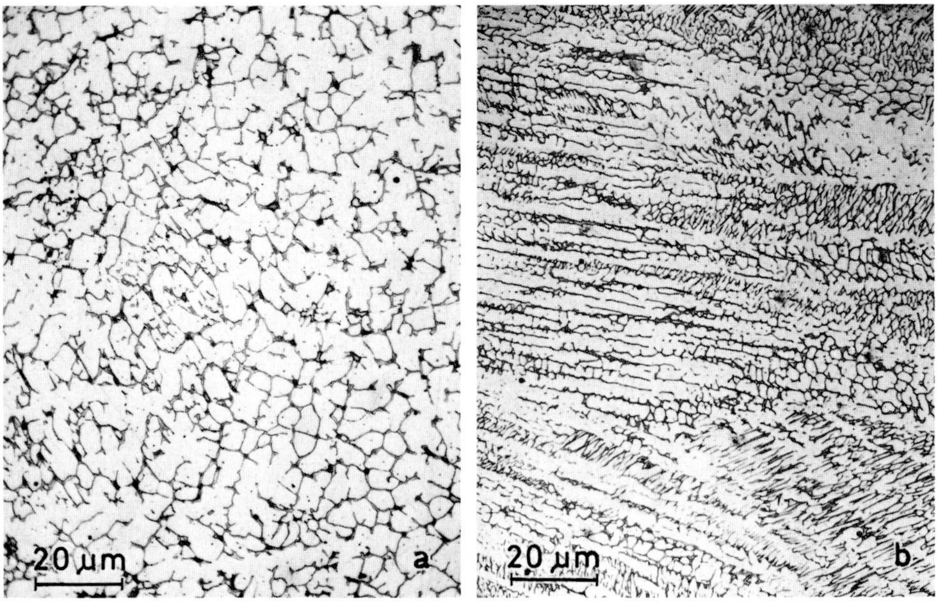

Fig. 73. Configuration of residual delta ferrite in stainless steel weld metal. *a* with 12% delta ferrite (approx. 15 FN), *b* with 17% delta ferrite (approx. 22 FN). Method—Potentiostatic ferrite etching with NaOH, ferrite is shown as dark colouration. Chemical composition of weld metal in %: *a* C = 0.034, Si = 0.90, Mn = 1.00, Cr = 22.53, Mo = –, Ni = 12.20, N = 0.07; *b* C = 0.032, Si = 0.63, Mn = 0.58, Cr = 20.83, Mo = 2.71, Ni = 9.15, N = 0.15

58—60 (page 65) and those for ferrite contents of approx. 12 and 17% in Fig. 73, with micrographs being produced by a ferrite etchant and not by colour etching. In this process, ferrite assumes a dark colour. At 12% residual delta ferrite (Fig. 73 a), ferrite is arranged in a reticular pattern between relatively fine austenite crystals and at 17% residual vermicular delta ferrite is formed (Fig. 73 b) between the longish austenite crystals. Fig. 72 e shows a schematic representation of the processes that take place with a Cr_{equ} : Ni_{equ} ratio of greater than 1.95. Here, liquid weld metal solidifies completely to δ ferrite, i.e. it no longer passes the three-phase sector ($L + \delta + \gamma$) during cooling. The crystallization process during the precipitation of primary δ ferrite is no longer interrupted, i.e. δ ferrite dendrites can grow until they reach the solidus line and their growth is no longer impeded by primary austenite precipitation in the three-phase sector ($L + \delta + \gamma$) and the immediately following δ-γ transformation.

After solidification has been completed, it is followed immediately by the new formation of delta ferrite grains caused by recrystallization. Depending upon how late δ-γ transformation commences during cooling, the coarseness of the newly formed ferrite crystals will increase accordingly.

Fig. 74 a shows the micro structure of a weld metal which has barely touched the last corner of the three-phase sector ($L + \delta + \gamma$) during cooling, the residual delta ferrite remains between secondary precipitated somewhat coarser austenite crystal and shows a vermicular configuration. Fig. 74 b shows a coarser lath like structure of the precipitated austenite with a higher residual delta ferrite content and complete primary solidification to delta ferrite (note the different magnification

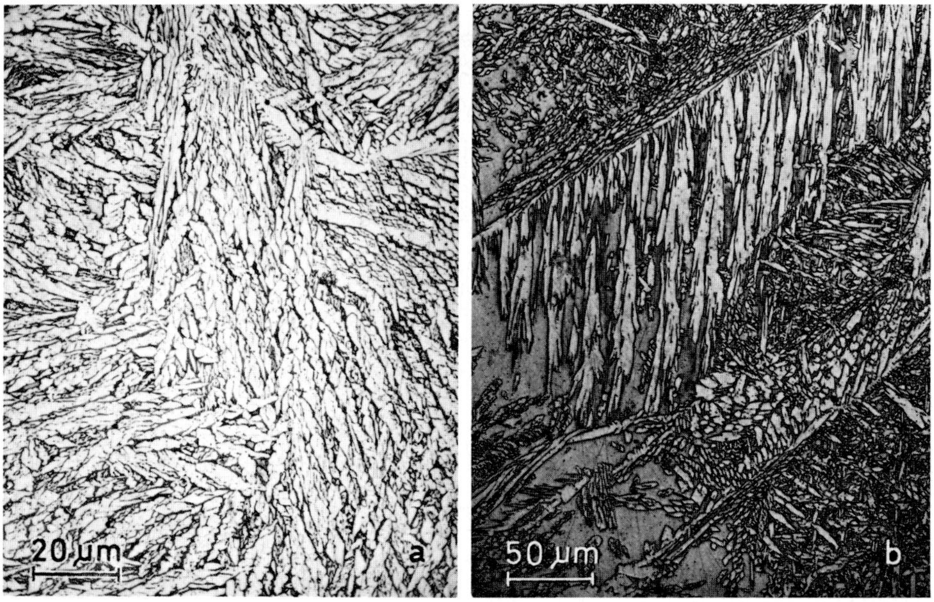

Fig. 74. Configuration of residual delta ferrite in stainless steel weld metal. *a* with approx. 22% delta ferrite, *b* with approx. 45% delta ferrite. Method—Potentiostatic ferrite etching with NaOH, ferrite is shown as dark colouration. Chemical composition of weld metal in %: *a* C = 0.03, Si = 1.0, Mn = 0.8, Cr = 22.5, Mo = 2.7, Ni = 12.0, N = 0.07; *b* C = 0.03, Si = 0.7, Mn = 0.7, Cr = 22.8, Mo = 2.8, Ni = 8.2, N = 0.05

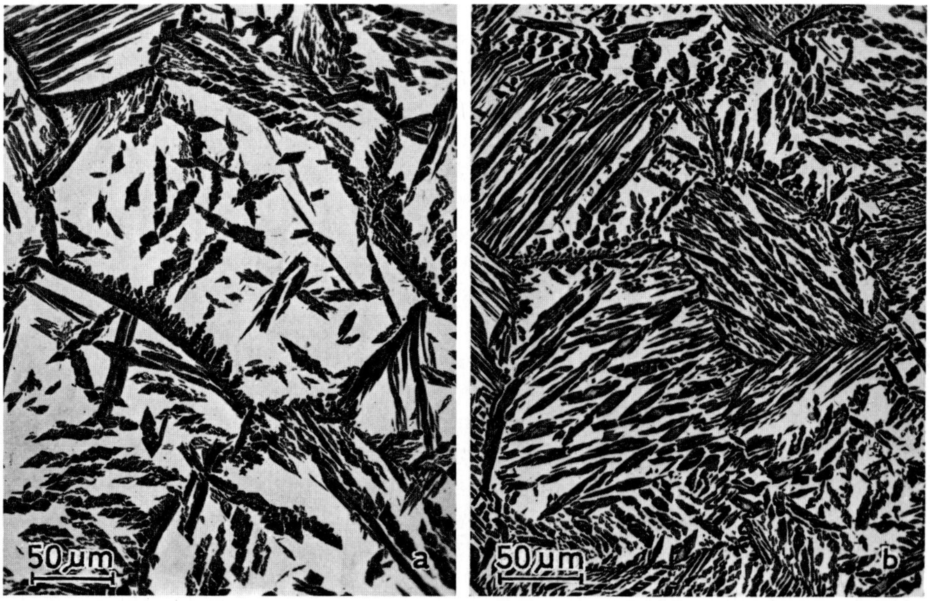

Fig. 75. Micro structure of a GTA weld bead. Etching same as Fig. 76. *a* Lower section of the bead with an average of approx. 70% delta ferrite, *b* middle section of the bead with an average of approx. 50% delta ferrite. Chemical composition of weld metal in %: C = 0.037, Si = 0.49, Mn = 1.67, Cr = 22.52, Mo = 3.10, Ni = 8.31, N = 0.14

factors of Figs. 74 a and 74 b). With increasing chromium-nickel ratios, δ-γ transformation starts at progressively lower temperatures. The primary ferrite grains have increasingly more time to grow and the structure becomes coarser with increasing residual delta ferrite contents. As an example, Fig. 75 shows the micro structure of a GTA weld bead with over 50% delta ferrite. At the bottom of the bead which cools most rapidly, more ferrite (light structural constituent) is present than in the remaining cross section. The dark coloured austenite has precipitated in an acicular configuration in the core and at the ferrite grain boundaries. Differences in delta ferrite content produced by the different cooling rates within a weld bead can be reduced by the addition of nitrogen to the weld metal alloy. Fig. 76 shows the macro structure of a nitrogen alloyed duplex stainless steel weld metal. When compared with the structure shown in Fig. 70 (page 78) which is of a weld metal of similar analysis but with less nitrogen, the positive influence of nitrogen on even ferrite distribution within a weld bead is quite apparent.

The micro structures of stainless steel weld metals with ferrite contents above 20% are rather varied. They depend not only upon the chemical composition of the weld metal and the respective ferrite content in the structure but are also additionally influenced by the varying cooling rates which occur during welding. For this reason, it seems difficult to classify the various types of structures produced or give them specific designations.

In recent years, numerous publications have appeared on the δ-γ transformation of stainless steel weld metal and the morphology of delta ferrite. The test results contained in these publications essentially coincide with the contents of this chapter. The pertinent references can be taken from the list of supplementary literature given at the end of this book using the following entries, without of course any claim of completeness being made [13, 15, 19, 31, 34, 116, 131, 135, 137, 138, 143, 144, 145, 147–161, 431, 435, 443, 471, 485].

Fig. 76. Configuration of delta ferrite in a multi-layer weld deposit showing the different structural constituents within the individual layers. Colour etching to Lichtenegger [120] represented in black and white. Ferrite is shown as light coloured area, austenite as dark area. Chemical composition of weld metal in %: C = 0.036, Si = 0.86, Mn = 0.74, Cr = 22.21, Mo = 2.82, Ni = 8.51, N = 0.13

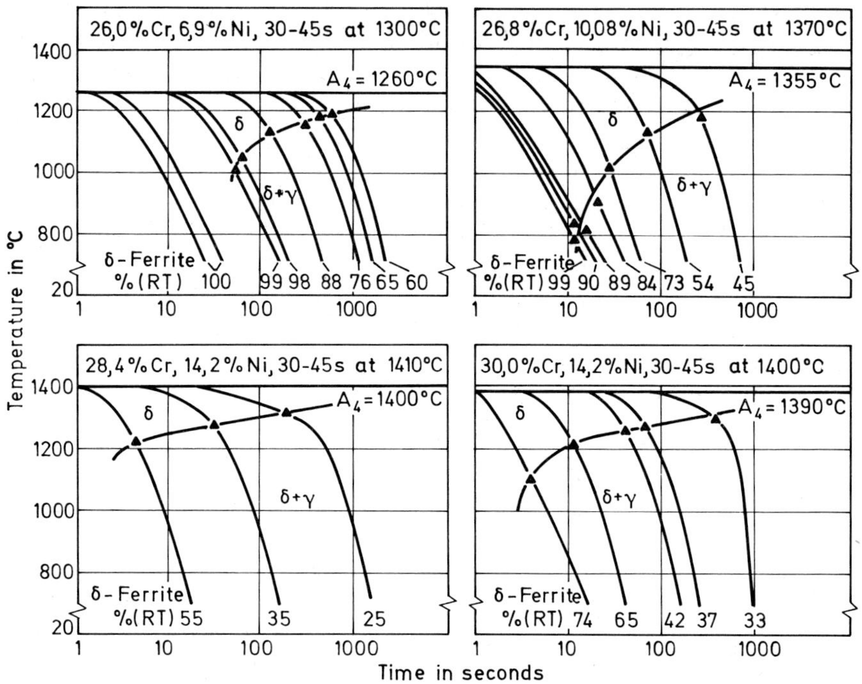

Fig. 77. TTT curves for the start of δ-γ transformation of austenitic-ferritic alloys with increasing chromium and nickel contents according to Mundt and Hoffmeister [161]. The chromium and nickel content is stated for each diagram. The carbon, silicon and manganese contents are 0.02%, 0.01% and 0.01—0.04% respectively (A_4 = Ferritizing temperature)

The mechanism of the solidification process which can be established from the information presented so far, in particular that of weld metal with primary delta ferrite crystallization, shows that also the δ-γ transformation may be strongly influenced by the cooling rate. Lippold and Savage [157] draw attention to this fact and note for example that in the case of rapid cooling vermicular residual delta ferrite is replaced by acicular delta ferrite. Suutala, Takalo and Moisio [155] also arrived at similar results. In GTA weld deposits of alloys which solidify completely to primary δ ferrite, Hoffmeister [137] noted a marked increase in the residual delta ferrite content as the welding speed increased. Mundt and Hoffmeister [161] determined that the phase boundaries of the γ area which represent the beginning of δ-γ transformation for iron-chromium-nickel alloys with relatively high residual delta ferrite contents are dependent upon the cooling rate and have shown them in the form of TTT curves. Fig. 77 shows the TTT curves of four austenitic-ferritic alloys with 26—30% chromium and rising nickel contents of about 7—14% which were taken from this publication [161]. These curves indicate the start of δ-γ transformation, with the percentage of residual delta ferrite being stated at the end of the cooling curves. It is apparent that with increasing cooling rates, the temperature of the start of transformation is shifted to progressively lower temperatures and δ-γ transformation becomes increasingly supercooled, which results in increases in residual delta ferrite contents. The difference in ferrite content can be quite high. Specimens were cooled from temperatures above A_4 (A_4 is equivalent to that temperature at which the structure is completely transformed into delta ferrite) with different cooling rates. Fig. 78 shows according to Hoffmeister and Mundt [19] the strong correlation between residual delta ferrite content and cooling rate and the time it takes to pass through the temperature range between 1200 and 800° C, designated as $t_{12/8}$, for an alloy with about 23% chromium and 7.6% nickel. Depending upon the cooling rate, the content of residual delta ferrite may vary between 40 and 100%.

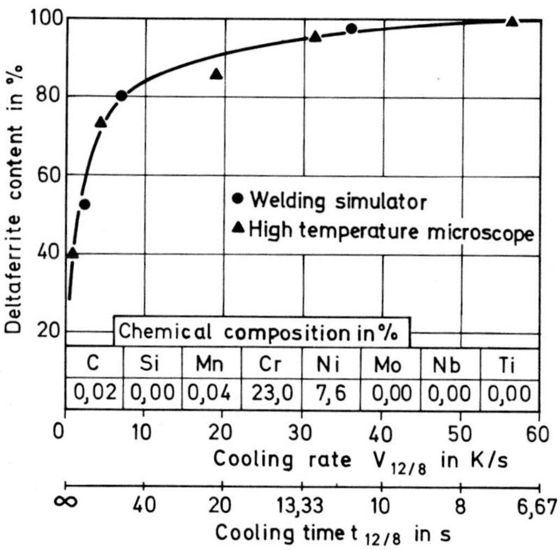

Fig. 78. Delta ferrite content as dependent upon the cooling rate $v_{12/8}$ and cooling time $t_{12/8}$ between 1200 and 800° C for an austenitic-ferritic alloy to Hoffmeister and Mundt [19]

During δ-γ transformation, austenite crystals precipitate from delta ferrite along the borderline between the $(\delta + \gamma)$ field and the γ area (see also Fig. 6, page 8). These austenite crystals are enriched by nickel and depleted by chromium. In this process of segregation, diffusion of the two elements plays a decisive role and according to Eckstein [197] this becomes more and more time dependent with decreasing temperature of the δ-γ transformation. In the case of welding, the inferior temperature range of δ-γ transformation below 1000° C is normally passed through so quickly that such segregations are of minor significance. They are important, however, if heat treatment is to be carried out in this temperature range. Carbide precipitations within the delta ferrite structure, e.g. $M_{23}C_6$, which according to Jolly and Hochmann [198], are quite possible at below 1000° C speed up δ-γ transformation, according to Eckstein [197], probably because of their nucleating effect on the formation of austenite crystals.

The influence on the range of δ-γ transformation by other alloying elements than chromium and nickel are dealt with in Chapter 1 by way of constitution diagrams. The shifts which can occur may be quite considerable (see Fig. 19, 20, 27 and 28). The changing of ferrite content by way of alloying additions for a 12% chromium and a 17/4% chromium-nickel steel is shown, according to Irvine, Crowe, Llewellyn and Pickering [188, 189], by the average values listed in Table 2. They give a rough idea of the technological effect that alloying elements have on the range of δ-γ transformation.

Table 2. *Changes in delta ferrite content in relative numbers (%) by the addition of alloying elements in amount of 0.1% and 1% to various basic compositions to Irvine, Crowe and Pickering* [188] *and Irvine, Llewellyn and Pickering* [189]

Element	Base: 0.1% C, 12% Cr		Base: 0.1% C, 17% Cr, 4% Ni	
	for 0.1%	for 1%	for 0.1%	for 1%
Nitrogen	−22		−20	
Carbon	−21		−18	
Nickel		−20		−10
Cobalt		−7		−6
Copper		−7		−3
Manganese		−6		−1
Silicon		+6		+8
Molybdenum		+5		+11
Chromium		+14		+15
Vanadium		+18		+19
Aluminium		+54		+38
Tungsten				+8

3.3 Schaeffler Diagram, Ferrite Measurement, DeLong Diagram

As indicated by Fig. 69 (page 76), stainless steel chromium-nickel weld metal with primary solidification to austenite shows due to constitutional reasons a stronger tendency towards the formation of low melting phases and as shown by practical experience also a tendency towards the formation of hot cracks (solidification and liquation cracks). Stainless steel chromium-nickel weld metal with primary solidification to δ ferrite performs much better in this respect. Therefore, an important measure to take to improve the hot cracking resistance of austenitic stainless steel weld metal is to control the solidification with the aim of obtaining the precipitation of primary δ ferrite crystals from the melt. The major characteristic of a primary δ ferrite solidification is the presence of residual delta ferrite which can be found in the structure after the weld metal has been cooled down to ambient temperature (see also Figs. 58–60, page 65). As indicated by Fig. 69 (page 76), it is mainly the chromium-nickel ratio which determines primary δ ferrite solidification. Other alloying elements that are present in standard weld metals, as well as the dilution of the weld metal with the base material, will also influence the processes during primary solidification. Therefore, these influences must also be considered in order to arrive at data which is useful in practical applications for controlling the solidification to primary δ ferrite with the help of the chemical composition of the weld metals.

Based on the works of Strauß and Maurer [1] and Scherer, Riedrich and Hoch [162] and practical welding tests, Schaeffler [163–165] first developed the chromium and nickel equivalents which permit the establishment of a relationship between the effects of other weld metal alloys on the base elements of chromium and nickel. With the help of the newly developed Schaeffler diagram, it is possible to roughly predict the delta ferrite content of a weld metal from its chemical composition and determine the structure under which a given weld metal which has been deposited under specific conditions can be classified, i.e. martensitic, austenitic or ferritic structure or mixtures of these types of structure.

The Schaeffler diagram was originally developed for the joining of dissimilar steel grades using stainless austenitic electrodes, in order to make predictions as to which alloying group the originally austenitic weld metal would be transformed as a result of the dilution with the differently alloyed base materials and to estimate the amount of residual delta ferrite to be expected in the austenitic weld metal. Schaeffler carried out his tests for determining the boundary lines of the Schaeffler diagram by using coated austenitic electrodes with a uniform core wire diameter of 3/16 inch (4.76 mm) and depositing them as single pass non-weave beads on various base materials, followed by subsequent metallographic and chemical testing of the deposited weld metal diluted with the base material. Thus, the Schaeffler diagram represents a technological chart which is based on practical tests but it is not an equilibrium diagram. In addition, only a single electrode diameter was used in the tests carried out by Schaeffler which means, for example, that the influence of different cooling rates during welding was not considered. The Schaeffler diagram is a valuable practical aid, but a possible scattering of the results must be considered in its application. Schaeffler too alludes to this fact [165]. This publication describes the proper application of the Schaeffler diagram stating a value of ±4 percent ferrite for the scatter of ferrite contents. According to the authors experience, this value applies only up to the range of about 15 percent volume. With higher ferrite contents, there is an increasing influence of the cooling rate on the δ-γ transformation with the possible deviations becoming larger and larger according to Mundt and Hoffmeister [168], i.e. the stated scatter values given above may be exceeded (see also section 3.2). With lower ferrite contents of between 1–5%, Suutala [127] found that the ferrite content was also influenced by the welding speed, however, this was found to be relatively marginal.

Fig. 79 shows the Schaeffler diagram as it is represented in the AWS Welding Handbook [166]. It largely coincides with the original by Schaeffler [165]. The diagram also contains rectangular zones which indicate the positions of various weld metal grades. These zones indicate that a scattering of the chemical analysis which is possible according to the applicable standards, may partly produce quite varying ferrite contents within one and the same weld metal type. With regard to the nickel equivalent, higher nitrogen contents have not been considered (except for type 22/5/3). According to Schaeffler [165], the diagram applies only for nitrogen contents in the range of 0.05–0.10%. In addition, a minimum content of 0.03% carbon and 0.3% silicon is also assumed. However, with manganese contents above 2.5%, the effect of manganese stated by the nickel equivalent no longer applies.

The Schaeffler diagram has proven its worth as a valuable practical aid for the rough determination of delta ferrite content and structural classification of weld metals, e.g. as in the welding of dissimilar steel grades and alloys, and for the rough evaluation of the influence of low delta ferrite contents on the hot cracking resistance of austenitic weld metal. Apart from the publications by Schaeffler [163–165], Ornig [167] also gives valuable practical information on the correct application of the Schaeffler diagram (see also Chapter 12 and Fig. 10, page 11).

Hull [169] tested about 1400 small cast specimens in order to determine the influence that the 13 alloying elements used (apart from chromium and nickel) in

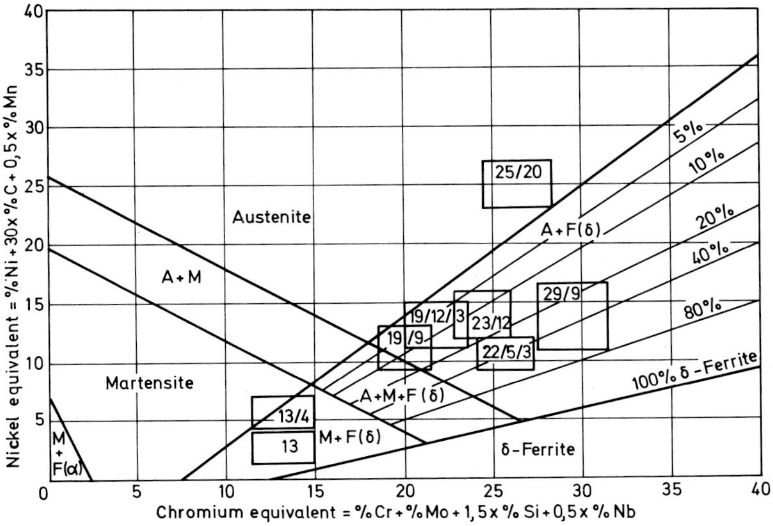

Fig. 79. Schaeffler diagram for stainless weld metal according to the AWS Welding Handbook [166]. The diagram shows the zones of various weld metals as deposited by coated electrode types. The alloy contents in % are stated in the order of Cr, Ni, Mo (short designation see appendix, page 241).

the chromium and nickel equivalent formulae have on the evaluation of the delta ferrite content. The cooling rates chosen for these cast specimens were selected so that they would also be representative of those expected for weld metal.

Hull arrived partly at deviating formulae for the calculation of chromium and nickel equivalents, probably because of the fact that some elements may influence each other with regard to their austenitizing or ferritizing effect and that their effect may vary with increases in content. The elements being considered were Al, C, Co, Cu, Mn, Mo, N, Nb, Si, Ti, Ta, V and W. Hull [169] also supplies slightly modified lines in the Schaeffler diagram for delta ferrite contents of 0–5%. Espy [170] reported on an extended Schaeffler diagram which also considered the influence of the alloying elements copper, nitrogen, vanadium and aluminium. Like Hull, he arrives at different chromium and nickel equivalents and a diagram which slightly differs from that of Schaeffler.

Apart from the Schaeffler diagram which permits a rough evaluation of the delta ferrite content from the chemical composition of the weld metal, there exists a number of other empirical methods for determining the ferrite content by measurement in austenitic or austenitic-ferritic weld metals. Krainer [171] supplies a thorough analysis of the measuring procedures on the basis of magnetic, metallographic and radiographic structural analysis, dealing in particular with the specific problems of each of these procedures. A very detailed report on the magnetic measuring processes for the determination of delta ferrite content and the parameters which must be considered is given by Bungart, Dietrich and Arntz [172]. Rettenbacher and Fuchs [173] deal with the significance of ferrite measurement and the possible deviations when measuring ferrite content by different magnetic measuring devices. They also supply calibration curves which make it possible to compare the values obtained with measuring devices which operate according to different principles. Apart from magnetic measurement, the method of metallographic determination of the ferrite content is also widely used. Guidelines for the execution of this method have been laid down by the International Welding Institute [174]. If ferrite has been partly transformed into sigma phase or molybdenum containing phases, special etching methods must be used to be able to distinguish ferrite under the microscope from its disintegration products. Such processes are described by Lüning and Tauchert [175] and by Blöch and Huszar [176].

A problem encountered with the empirical determination of delta ferrite is the considerable scattering of delta ferrite values measured by different methods and in different laboratories. Ratz and Gunia [177] reported on joint tests carried out by eleven experienced companies and institutes. They arrived at the following conclusion. Neither empirical formulae nor common measuring methods are able to determine delta ferrite content more accurately than ±3 ferrite percent for contents below 10% and ±6 ferrite percent for contents between 10–24%. Similar results were obtained by a joint study of 22 laboratories carried out by the International Welding Institute [178]. Here, the following scattering of delta ferrite measurements could be established. At an arbitrary value of 3.5%, measured values were between 2.0 and 6.5%, at 7.5% between 5.5 and 12.5% and at 15.5% between 11 and 29%. A joint study of five different laboratories carried out by the Welding Committee of Eisenhütte Österreich, reported by Ornig, Kohl, Rabensteiner, Rettenbacher and Weberberger [179] arrived at similar results. The graphic representation of the scatter values in [179] shows clearly the increase in the scattering of the measuring data with rising ferrite content.

Based on these experiences, DeLong [180] deals very thoroughly with both the importance of low delta ferrite contents in austenitic weld metals for the improvement of hot cracking resistance and the problems resulting from delta ferrite measurement. DeLong suggests a new procedure which is based on the following consideration. Since there is no method in the present state of the art which permits the measurement of the ferrite content in austenitic stainless steel weld metal in absolute terms with only little scattering, a measuring procedure which is based on a ferrite standard calibrating specimen is agreed upon, supplying comparable values which may, however, not always correspond to the actual ferrite content. To indicate the difference compared to other existing measurement methods, the measured ferrite value is not designated as the ferrite percentage but as the ferrite number, in short FN.

DeLongs suggestion has been adopted by the International Welding Institute as a standard procedure. The recommendations for the execution of this method have been published in [181]. The measuring procedure used for determining the ferrite number of austenitic Cr-Ni stainless steel weld metal is as follows: It is based on the measurement of the force required to tear off a defined permanent magnet from the surface of an austenitic weld metal specimen. With an increasing content of delta ferrite, which is ferro-magnetic, the "tearing force" will increase, whereas other structural constituents such as austenite, carbide, sigma phase and inclusions which are not ferro-magnetic will have no influence at all. Thus, the tearing force is an indirect measurement of the delta ferrite content. It has been agreed upon that the calibration of the tearing force is carried out with the aid of "primary calibration standards", consisting of a defined specimen made from unalloyed steel which is provided with copper layers of varying thicknesses. Since copper is non ferromagnetic, the tearing force decreases as the thickness of the copper layers increase.

A certain thickness of copper layer refers now to a certain ferrite number. A thickness of 1.778 mm, for example, is equivalent to ferrite number 3.0 (FN 3), 1.194 mm is equivalent to FN 5 and 0.610 mm is equivalent to FN 10. In the US, such calibration standards are prepared by the National Bureau of Standards. This calibration method is only applicable for instruments which measure the tearing force of a permanent magnet, a type of instrument widely used in the US.

In other countries and in Europe in particular, other ferrite measuring devices are used. These are based on various magnetic measuring methods. Instead of the static permanent magnets, the latter normally operate with alternating magnetic fields of higher frequency. To also be able to calibrate them in ferrite numbers, "secondary calibration standards" have been developed. Their preparation and calibration is reported on by Robinson, Rabensteiner and Neff [182]. In order to arrive in the case of secondary calibration standards at the same ferrite numbers with different measuring methods, all standards must be made from austenitic weld metal of specific ferrite contents. In addition, they must be highly homogenous to keep scattering to a minimum within the measuring area used for calibration. A very suitable process for the production of such calibration standards is submerged arc strip cladding. The International Welding Institute has introduced the production of such secondary calibration standards. They are produced today in eight graduations with ferrite numbers of about 3, 5, 8, 11, 14, 17, 23 and 27 and can be obtained through national welding associations or directly from the Welding Institute, Abbington Hall, Cambridge, England. Their development and application is reported on by Ornig, Kohl, Rabensteiner, Rettenbacher and Weberberger [183], and by Merinov, Rettenbacher, Rabensteiner and Neff [183 a]. When measuring austenitic chromium-nickel weld metal with different devices which have been calibrated prior to measurement with the secondary calibration standards described above, a possible deviation in ferrite readings of ± 10 percent could be reached, i.e. a reading of FN 10 will indicate a value of between FN 9 and FN 11. These deviations are substantially smaller than those obtained by the method of measuring ferrite percentage without the calibration of measuring equipment using secondary calibration standards.

DeLong [180] also suggested together with ferrite numbers the use of an improved Schaeffler diagram which also permits readings of approximate ferrite numbers from the chemical composition of the weld metal. This so-called DeLong diagram is widely used today. In the formula for the nickel equivalent, the alloying constituent nitrogen has been incorporated which similar to carbon possesses a strong austenitizing effect. Fig. 80 shows the DeLong diagram as it was published by DeLong [180]. Up to ferrite number 6, the numerical values of the ferrite numbers coincide well with the ferrite percent, above FN 6, a given weld metal will show higher ferrite numbers than ferrite percent. The publication of Long and DeLong [184] deals with the influence of nitrogen pick-up from the surrounding atmosphere when welding with a long arc and also the influence of carbon burn-off on the ferrite number.

Another criterion which is of special importance for the practical measurement of ferrite content using ferrite numbers is the measuring accuracy of these devices when they are used on production welds of common weld metal grades such as 308, 308 L, 316, 316 L. A task group formed to investigate this problem came to conclusions which were reported on by Hebble et al. [445] as follows: "The mean values of ferrite numbers which were measured on production welds represent a reasonable measure for the evaluation of ferrite content for values below 14 FN (upper limit of measuring data). This applies also for different welding processes, base materials, filler metals and material thicknesses. The values measured in production welds coincide relatively well with data obtained from qualification test welds. Even though a wide scatter of data exists, the percentage of production welds with ferrite values less than 3 FN was very low (0.14%) for qualification test welds with 5 FN or more."

The influence of manganese and nitrogen on the type of solidification has been investigated by Suutala [185] and correlated to a number of different formulae for the calculation of chromium and nickel equivalents. For the determination of ferrite number, the formulae according to DeLong and to Schaeffler (Fig. 80 and Fig. 79 respectively) are quite suitable for common stainless steel weld metals. To determine the boundary between primary ferritic and primary austenitic solidification, Suutala re-

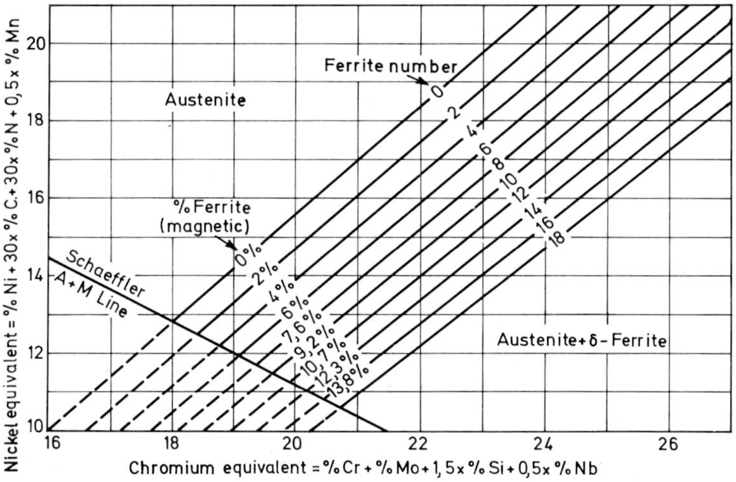

Fig. 80. DeLong diagram revised January 1973 for the determination of ferrite numbers in austenitic stainless steel weld metal according to DeLong [180]. If the nitrogen content for determining the nickel equivalent is not known, a value of N = 0.06% for coated electrodes and GTA weld metal and N = 0.08% for GMA weld metal should be assumed. If the chemical composition is accurate, the diagram predicts the WRC Ferrite number within plus or minus FN 3 in approximately 90% of the tests carried out for 308, 309, 316 and 317 types of alloy.

Comparison with the Schaeffler diagram.

1. A nickel equivalent of 30 × N has been added.
2. Ferrite Numbers for 308, 308 L and 347 coated electrodes are similar. The higher alloyed 309, 316 and 317 types have Ferrite Numbers on this diagram which are higher by about 2 to 4.
3. Generally, this diagram correlates better with GTA and GMA weld metals because it allows for nitrogen pick-up.
4. The Schaeffler austenite-martensite boundary has been included here for reference.

commends the following formula, developed by Hammar and Svennson [444]. According to this kind of calculation, the boundary is found at a Cr_{equ} : Ni_{equ} ratio of 1.52.

$$Cr_{equ} = \%\,Cr + 1.37 \times \%\,Mo + 1.5 \times \%\,Si + 2 \times \%\,Nb + 3 \times \%\,Ti$$
$$Ni_{equ} = \%\,Ni + 0.31 \times \%\,Mn + 22 \times \%\,C + 14.2 \times \%\,N + \% \times Cu$$

With higher manganese contents, however, the formula given by Hull [169] is more suitable. With contents of up to 2.5%, manganese excerts an austenitizing effect which may change, however, into a ferritizing effect with higher contents. Similar results were obtained by Szumachowski and Kotecki [440] and Ritter, Henry and Savage [443]. A comprehensive survey of the present state of ferrite measurement and its influence on the evaluation of austenitic weldments which includes 169 references is given by Olson [446]. He arrives at the conclusion that commonly used measuring techniques permit a fair evaluation of weldments which have been executed with conventional standardized base materials and filler metals. With regard to the evaluation of new stainless materials, however, their application is limited. With the advent of new stainless steels, a wider range of alloys must now be considered when predicting methods for ferrite measurement. In his paper, Olson reviews the evolution of many of todays diagrams for predicting the weld metal structure and suggests some new analytical forms for future predictive techniques.

3.4 γ-α Transformation of Stainless Steel Alloys

In chromium alloyed steel grades, the resistance towards many types of aggressive media rises markedly, according to Rocha and Lennartz [186], at approx. 11–12% chromium. This is the reason why stainless steels and weld metals are normally characterized by a chromium content above 11%. Depending upon the contents of chromium, nickel and carbon present in the alloy, the structural constituents delta ferrite, austenite and martensite may be produced after cooling down from the welding heat to ambient temperature.

The γ-α transformation is dependent upon the formation of transformable γ crystals. Therefore, the nickel content should not be too high in order to avoid the formation of stable austenite at ambient temperature (see also Fig. 14, page 14). In the ternary iron-chromium-nickel system, γ-α transformation is limited to the gamma sector in the iron corner. The alloying element chromium constricts the gamma sector in the binary iron-chromium system at about 11.5% chromium, at a content where stainless steel alloys are just beginning. To be able, nevertheless, to obtain γ-α transformation in these stainless steel grades, it is necessary to add other alloying elements which expand the gamma sector such as carbon, nickel and nitrogen. The alloying contents of elements which, like chromium, constrict the gamma sector, e.g. molybdenum, niobium and silicon, must be limited to prevent the sector of γ-α transformation from being constricted more and more and that primary delta ferrite is maintained down to ambient temperature. The effect of different elements on the gamma sector in the iron-chromium-nickel constitution diagram is shown in Figs. 10, 11, 16, 17, 25, 26 and 35 (expanding gamma sector) and Figs. 30, 31, 38 and 41 (constricting gamma sector).

The Schaeffler diagram (Fig. 79, page 90) which illustrates the technological effect of the major alloying elements on the micro structure of stainless weld metals with the common cooling rates encountered during practical welding operations, conveys a good idea of the alloying range in which γ-α transformation takes place. The sector which is designated in the Schaeffler diagram as martensite (M) characterizes this range. The chromium content reaches up to a chromium equivalent of approx. 15, with the nickel equivalent being limited, however, to 5—10 for stainless steel weld metal with more than 12% chromium (right corner of martensite range in Fig. 79). If this range is either exceeded or not reached at all, we also find besides martensite (M), austenite (A) and ferrite (F) or all three of these constituents (designated A + M + F) in the weld metal.

The kinetics of γ-α transformation and the influence of the different alloying elements can best be illustrated by way of time-temperature-transformation (TTT) and time-temperature-precipitation (TTP) diagrams. The processes occurring during γ-α transformation of stainless steels are investigated by Kulmburg, Sölkner, Korntheuer and Schmid [187], Brezina [93], Irvine, Crowe and Pickering [188], Castro and Tricot [190], Hull [169], Tösch, Perteneder and Rabensteiner [94, 191], Koepke, Skuin and Herfert [192], Woltron [193], Kulmburg, Korntheuer, Koren, Gründler and Hutterer [194], Leymonie, Ottmann, Lecocq, Risacher and Thauvin [195, 196] and for steels with more than 20% chromium by Eckstein [197] and by Jolly and Hochmann [198]. Below, we shall deal more closely now with the phenomena which are of particular importance for stainless steel welding operations.

Fig. 81 shows the TTT diagram of a steel grade with approx. 12% chromium and 0.13% carbon. The curves 1, 2 and 3 characterize the ranges in which stainless weld

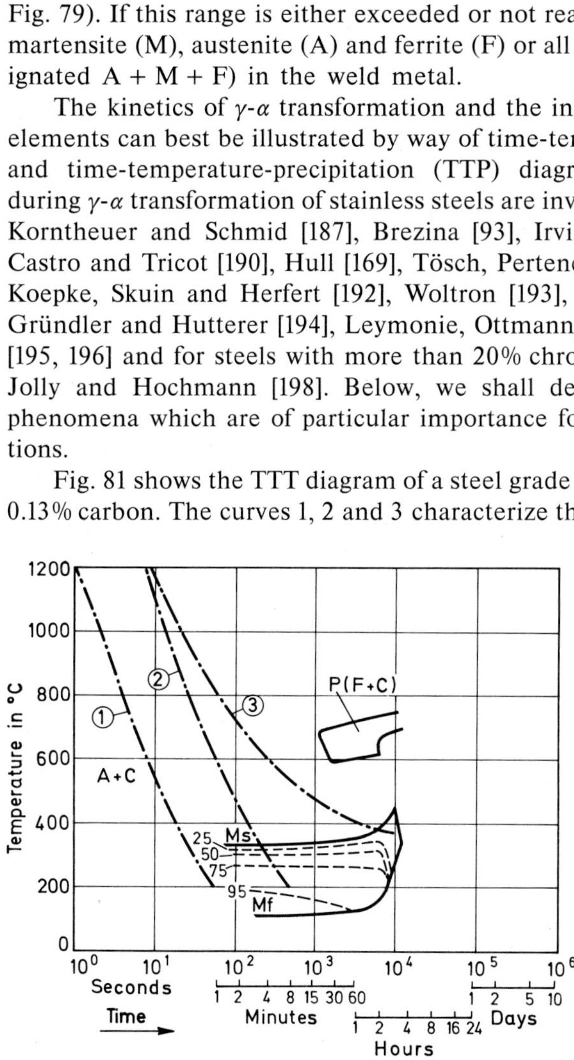

Fig. 81. TTT diagram of γ-α transformation of a 12% chromium steel. Chemical composition in %: C = 0.13, Si = 0.28, Mn = 0.38, Cr = 11.96, Mo = 0.10, Ni = 0.40. The curves *1*–*3* characterize the cooling rates of electric arc welding operations: *1* — cooling during the execution of a bead on a plate of 20 mm thickness with a heat input of 5 kJ/cm, *2* — same as *1* but with 10 mm plate thickness and 30 kJ/cm, *3* — same as *2* but with plate preheated to 350° C

metal normally cools through during electric arc welding. Because of the high chromium content of this grade, the beginning of pearlite transformation, where α ferrite and carbide are precipitated from γ crystals is found only after rather long periods of time, which means that the weld metal and the heat affected zone are practically always transformed during welding in the martensite range, except in the case of preheating above the martensite transformation temperature (M_s temperature, in this case approx. 360—420° C) (see curve 3 in Fig. 81). Martensite transformation takes place within a relatively wide temperature range, i.e. between approx. 330 and 110° C. Within the range of cooling curves 1 and 2, however, transformation to martensite is not complete, and small amounts of residual austenite between 2 and 6% are still present. Since these processes have a decisive influence on the mechanical properties, crack resistance and corrosion resistance of stainless steels and their weld metals, they shall be discussed in detail below.

3.4.1 Transformation in the Pearlite Range

The effect of carbon which in the case of low alloyed steels greatly influences the beginning of pearlite transformation is largely offset in the case of stainless materials containing 12% Cr by the high chromium content. Therefore, with the carbon contents normally found in weldable stainless steels and weld metals which are mostly between 0.02 and 0.15%, the influence of carbon on pearlite transformation is marginal in comparison to that of chromium. Constituents such as nickel and molybdenum will shift the pearlite transformation in 12% chromium steels to even longer times. A content of about 2% nickel or 1.5% molybdenum will shift the beginning of perlite transformation to times beyond 10^5 seconds = approx. 28 hours. Even with slow cooling, austenite is now no longer transformed in the pearlite range. It is transformed into martensite.

With such materials, e.g. 13/4 chromium-nickel alloys, it is hardly possible to obtain a forced isothermal transformation in the pearlite or bainite stage by welding with excessive preheating above the M_s temperature or by prolonged holding of the component at temperature, as is done, for instance, with low alloyed steels. Between the pearlite and the martensite range, there is a range of very stable austenite between 400 and 600° C (see Fig. 81). This means that with preheating temperatures of between 350—400° C the austenite formed during cooling of the weld metal due to the δ-γ transformation is more or less maintained and transformed into martensite only during subsequent cooling from the preheating temperature.

With higher chromium contents, delta ferrite is only partially transformed into austenite during δ-γ transformation. With 16—18% chromium, more than 50% delta ferrite is normally present in the structure (see Figs. 9, 10 and 16). Only about 50% of the austenite takes part in the γ-α transformation. Here, the transformation in the pearlite stage is not—as would be expected—shifted to longer times but to shorter times. It may start already after approx. 10—20 seconds in the range between 630—730° C. According to Castro and Tricot [190], the following processes take place: First, in the temperature range of approx. 850° C, carbide is precipitated from delta ferrite at the grain boundaries with austenite. This precipitation starts rapidly within a few seconds. Then, the carbide forms a pearlite-like phase with the austenite, called D-phase, which starts the actual γ-α transformation. Now, austenite is transformed into α ferrite, with the formation of a mixture of α ferrite and carbide $M_{23}C_6$ which can no longer be distinguished by optical means. This is called the G-phase. In the case of higher nitrogen contents, the G-phase may be followed by the precipitation of a pearlite-like mixture of ferrite and nitrides which is sometimes called pearlite (see also constitution diagrams in Figs. 24 and 25). Such structures can be found in steel castings and in weld metal where they are hardly identifiable because of the fine grain structure of the latter.

In steels with 16—18% chromium, the addition of nickel will shift the processes described above with the exception of carbide precipitation to considerably longer times, i.e. with 1.6% nickel to about 10^4 seconds and with 5% nickel beyond 10^5 seconds. Here, the structure at ambient temperature consists of delta ferrite and martensite with carbide precipitations found mainly at the delta ferrite grain boundaries.

In steels with more than 20% and in particular more than 25% chromium, the transformation processes become more complicated, since transformation from delta ferrite to austenite (δ-γ transformation) is shifted to lower temperatures, below 1000° C (see Fig. 77, page 86) in the ranges where carbide precipitation also starts in delta ferrite. Here, δ-γ transformation is often intermingled within the γ-α transformation. According to Eckstein [197], the following transformations take place in chronological order in the temperature range between 1000 and 700° C:

— start of carbide precipitation from delta ferrite
— start of austenite precipitation from delta ferrite
— start of the eutectic precipitation of austenite and carbide
— further precipitation of austenite from delta ferrite
— below 800° C start of transformation of austenite into bainite.

These processes may be superimposed by the precipitation of additional phases, e.g. sigma phase. According to Eckstein [197], the kinetics of transformation and precipitation processes are highly dependent upon the heat treatment, e.g. ferritizing temperature and time, and therefore the representation by way of TTT diagrams is rather problematic.

3.4.2 Transformation in the Martensite Range

As indicated in Fig. 81, a 12% chromium steel will be transformed in the course of martensitic transformation under the normal cooling rates encountered during electric arc welding (area between curve 1 and 2) to a level of only 95—98% martensite with approx. 2—5% residual austenite remaining at ambient temperature. Alloying elements which are normally used in stainless steels have an influence on both the beginning (M_s temperature) and end (M_f temperature) of martensitic transformation. Fig. 14 (page 14) shows the effect of the two main alloying elements of stainless steels, i.e. chromium and nickel, on the M_s temperature according to Eichelmann and Hull [28]. This temperature is lowered by both elements to roughly the same extent. With approx. 18% chromium and 8% nickel (AISI 304), the M_s temperature is lowered to about minus 50° C, which means that austenite formed at the solution annealing temperature of 1050° C will be maintained down to ambient temperature. This process is supported by quenching from the solution annealing tem-

Table 3. *Changes in the M_s temperature and the transformation point A_{C1} by alloying additions in steels with 12% chromium according to Irvine, Crowe and Pickering* [188]

Element	Drop in the M_s temperature by a 1% element addition in ° C	Change of the A_{C1} point by a 1% element addition in ° C
Carbon	− 474	—
Silicon	− 11	+ 20 − 30
Manganese	− 33	− 25
Chromium	− 17	—
Nickel	− 17	− 30
Molybdenum	− 21	+ 25
Aluminium	—	+ 30
Cobalt	—	− 5
Vanadium	—	+ 50
Tungsten	− 11	—

perature. Quite to the contrary, the austenite in 12% chromium and low carbon martensitic 13/4 chromium-nickel steels (AISI 410 and 410 S), is largely transformed to martensite.

Kulmburg, Sölkner, Korntheuer and Schmid [187] supply the following formula for the determination of the M_s temperature in ° C which takes into consideration the influence of the alloying elements carbon, manganese, chromium and nickel:

$$M_s = 492 - 125 \times \%\,C - 65.5 \times \%\,Mn - 10 \times \%\,Cr - 29 \times \%\,Ni.$$

The scatter of calculated values as compared with measured values is reported to be ± 20 K on average. Irvine, Crowe and Pickering [188] report on some average values, as shown in the left hand column of Table 3, regarding the influence of the major alloying elements of stainless steels on the beginning of martensite transformation in 12% chromium steels. In addition, the authors point out that the M_s temperature should always be above 200° C to obtain maximum transformation in the martensite stage because the temperature range between the beginning and the end of martensite transformation is about 150° C.

The influence of 15 alloying elements on the M_s temperature together with the simultaneous influence of cold straining phenomena has been investigated by Hull [169]. His aim was to determine the effect of these elements on the boundary between the martensite and austenite range in the Schaeffler diagram. He used mainly alloys which are austenitic at ambient temperature and change into martensite only through cold straining processes or cooling to subzero temperatures. For this reason, the average values established by him deviate substantially in part from those values stated in Table 3 which apply only to the range of 12% chromium steels.

Fig. 82 shows the beginning of martensite transformation as observed during the cooling of a weld deposited by electric arc welding according to Tösch, Pertenender and Rabensteiner [94, 191]. Here, the approximate ferrite content of welds containing 12–13% chromium and 1–6% nickel were measured with the aid of a magnetic probe. The beads were deposited with 4 mm diameter electrodes and with heat inputs of 8–10 kJ/cm on multilayer weld deposits of 20 mm thickness. All of the specimens were of similar composition and the readings observed from the magnetic probe were recorded during cooling. During the above procedure, the temperatures were measured using a thermocouple. The hatched area indicates the content of supercooled delta ferrite which was not transformed further during δ-γ transformation. Martensitic transformation which is indicated by a quick rise in the magnetic reading starts with 13/1 Cr-Ni weld metal at approx. 290° C, and with 13/4 and 13/6 Cr-Ni weld metals at 240 and 220° C respectively. This indicates quite clearly the influence of nickel on the M_s temperature. After the welds were cooled to ambient temperature, the contents of residual austenite measured in the as welded condition were found to be below 2% for 13/1 Cr-Ni weld metal, 5% for 13/4 Cr-Ni weld metal and 9% for 13/6 Cr-Ni weld metal. Pre-eutectic carbide precipitations may have an influence on martensite transformation according to Kulmburg et al. [187] because carbon which precipitates as carbide is lacking in the matrix, a fact which changes the martensite point. These processes are however only of minor importance for weld metal alloys and weldable stainless steels, where the carbon content is normally below 0.1%.

In steels with higher delta ferrite contents in the range of 50%, e.g. 17% chromium steels, carbide precipitation starts at the γ-δ phase boundaries, according to Irvine, Llewellyn and Pickering [189] after relatively short periods of time, in the range of 1–10 seconds, at temperatures of 800–900° C. With such alloys, the temperature range of martensite transformation is therefore more strongly influenced by the austenitizing temperature. Irvine, Llewellyn and Pickering [189] measured, for example, in a steel with 17% chromium and 4% nickel at an austenitizing tem-

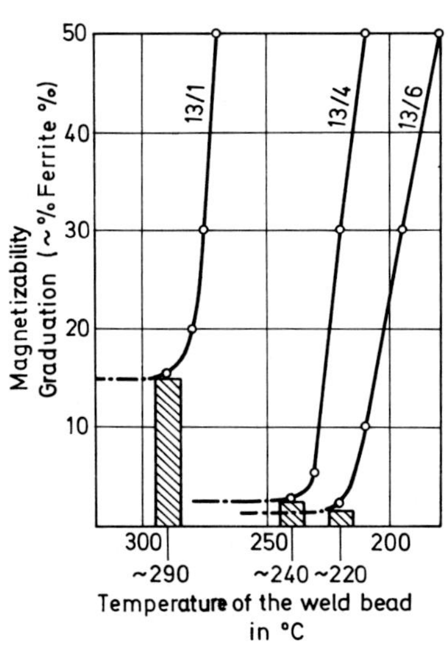

Fig. 82. γ-α transformation from austenite to martensite during the cooling of a weld bead with 12% chromium and nickel contents of about 1, 4 and 6% to Tösch, Perteneder and Rabensteiner [94, 191]. Measurement of magnetizability was carried out with a magnetic probe of low penetration which guarantees that only the weld metal cooling from the welding heat would be measured. The hatched area indicates residual delta ferrite which has not been changed into austenite during the previous δ-γ transformation but supercooled

perature of 1050° C, a temperature range of martensite transformation of + 10° to − 140° C. With austenitizing at 900° C, this range is shifted to approx. + 120° to 0° C, i.e. considerably higher temperatures. With austenitizing at 950° C and additional annealing at 700° C, it rises by another 20° C. These phenomena must be considered in the heat treatment of such steels and weld metals or weldments because the mechanical properties of weldments can be markedly influenced by differing contents of residual austenite and martensite. If chromium contents are increased beyond 20%, martensite formation is shifted further towards lower temperatures. According to Eckstein [197], alloys with more than 22% chromium show no martensite transformation at all even with cooling down to − 196° C. In this case the basic structure consists of delta ferrite in which increasing amounts of austenite are interspersed with increasing contents of austenite forming elements (nickel, nitrogen, carbon, manganese). Because of their two-phase structure, such alloys are also called duplex steels.

3.4.3 Formation of Stable Austenite During Tempering

The second column in Table 3 (page 98) shows the changes in the A_{C1} temperature (transformation temperature of α iron into γ iron during heating) for 12% chromium steels by the addition of alloying elements according to Irvine, Crowe and Pickering [188]. With additions of nickel, manganese and cobalt, the A_{C1} is shifted to lower temperatures. This effect is also apparent in the constitution diagrams of nickel and manganese with iron which are shown in Figs. 2 and 35 (page 3 and 29). The basic element chromium will also reduce the A_{C1} temperature, as indicated by Fig. 1 and in particular Fig. 10 (page 11). The combined effect of chromium and nickel on the A_{C1} temperature can be seen from the concentration profile in the ternary iron-chromium-nickel system at a chromium-nickel ratio of 3 : 1, as shown in Fig. 9 a (page 10). The $(\alpha + \gamma)$ sector extends at 12—15% chromium and 4—5% nickel down to about 600° C. Low carbon martensitic stainless Cr-Ni steels are found in this range. According to Niederau [199], they owe their excellent toughness properties to the fact that new finely dispersed austenite which cannot be detected by the optical microscope is formed from martensite during tempering in the temperature range of around 600° C. This finely dispersed austenite is made stable by segregation processes to a point where it will no longer disintegrate even when cooled down to − 196° C. This process is described by Kulmburg, Korntheuer, Koren, Gründler and Hutterer [194] as follows: From dilatometer tests, it is apparent that the formation of stable austenite starts at 570° C. Stable austenite reaches a maximum content at 615° C when the volume amounts to about 30% for the alloy in question. From this temperature upwards, a second type of austenite is formed during further heating which no longer remains stable but changes to martensite during cooling at an M_s temperature of 70° C. The austenite which forms during further heating will disintegrate to martensite at increasingly higher M_s temperatures during cooling until a maximum value for the M_s temperature of 230° C is reached at a tempering temperature of 750° C. This tempering temperature is also roughly equivalent to the end of α-γ transformation. With the temperature increase beyond 615° C, the content of stable austenite decreases again, i.e. it is practically dissolved again by the unstable austenite.

Due to the stable austenite which is finely dispersed in the martensitic matrix, the toughness is improved but the 0.2% yield strength is reduced. Fig. 83 shows an example of the effect of the tempering temperature on tensile properties, impact energy and the structural configuration according to Kulmburg et al. [194]. There are three different types of austenite here, i.e. about 7% stable residual austenite Au_1, left over from the γ-α transformation to martensite, finely dispersed stable austenite Au_2, precipitated during tempering, which reaches a maximum of 28% at approx. 615° C, and unstable austenite Au_3 which precipitates above this temperature and which is transformed to martensite during cooling after tempering. The bottom diagram in Fig. 83 shows the content of austenite still found in the structure after tempering at different temperatures. The positive effect of the finely dispersed stable tempering austenite on the impact energy (IE) is quite apparent. Leymonie, Lecocq and Ottmann [196] arrived at similar results for 16/5 chromium-nickel steels with additions of copper and molybdenum.

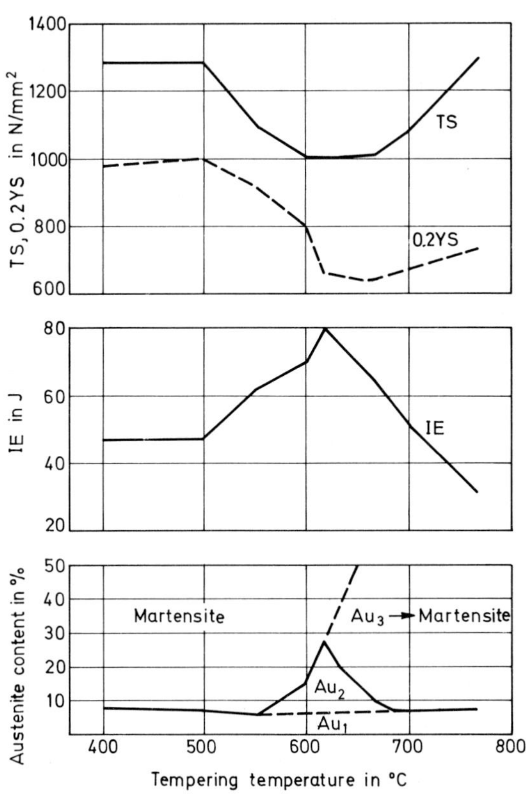

Fig. 83. Influence of tempering temperature on the 0.2% yield strength (0.2 YS), tensile strength (TS), impact energy (IE) and austenite content of steel grade 12 Cr/6 Ni/1.5 Mo according to Kulmberg et al. [194]. Chemical composition in %: C = 0.039, Si = 0.35, Mn = 0.69, Cr = 11.82, Mo = 1.49, Ni = 5.23. Au_1 stable residual austenite supercooled during γ-α transformation, Au_2 stable, finely dispersed tempering austenite, Au_3 unstable austenite which is changed into martensite during cooling from the tempering temperature

Brezina [93] supplies in a very comprehensive work the results obtained from a total of 12 different types of low carbon martensitic steels and shows the effects that heat treatment processes have on material properties. The determination of different types of structures and their evolution are also described. With regard to carbides, the presence of carbide $M_{23}C_6$ and carbonitride M(CN) in 13/4 chromium-nickel steels could be established, with the latter being precipitated in the center of the grains and $M_{23}C_6$ both inside the grain and along the grain boundaries. In the case of the latter, intergranular corrosion may occur. Rabensteiner, Perteneder and Tösch [191] also report on the properties of 13/4 chromium-nickel weld metal. They show, that with precise setting of chemical composition together with a suitable heat treatment an optimum compromise, between maximum toughness and minimum drop in 0.2% yield strength is produced, making maximum use of the formation of finely dispersed stable tempering austenite (see also chapter 7).

4 Precipitation Phenomena in Stainless Steels and Weld Metals

As shown by the constitution diagrams which characterize metallurgical equilibrium, stainless steels possess at ambient temperature a very low solubility for carbon (see Chapter 1). Due to the high affinity of the major alloying constituent chromium towards carbon, there is a strong tendency towards carbide formation. The ranges of intermetallic phases, e.g. sigma phase, will often reach under equilibrium conditions far into the alloying ranges of stainless steels (see, for example, the iron-chromium system in Fig. 1 (page 2). Therefore we must expect the precipitation of these phases associated with the normal cooling rates experienced when working stainless steels, e.g. after rolling or forging. In most cases, however, they are undesirable because of their negative influence on both the corrosion resistance and the mechanical properties of the alloy.

For this reason, austenitic stainless steels are generally subjected prior to being put into service to a solution annealing treatment carried out between 1050 and 1150° C which brings precipitated carbides as well as most other intermetallic phases back into solution. This condition is maintained by rapid cooling (quenching) from the solution annealing temperature to ambient temperature, which forces the elements responsible for the formation of carbides and intermetallic phases to stay in solution by supercooling. It is accepted here that the material remains in a condition of metallurgical non-equilibrium. At ambient temperature, the diffusion potential of most elements is very low which means that the condition of non-equilibrium may be maintained for a practically infinite period of time. If such materials are subjected, however, to a heat treatment or are operating at elevated temperatures, the diffusion potential of the alloying elements is greatly increased (see Table 1, page 50). In this way, carbides or intermetallic phases may precipitate again.

During welding also, heating of the parent material takes place in the heat affected zone. This heating is, however, rather brief. Contrary to heat treatment carried out at defined temperatures, the material in the heat affected zone is heated during welding to a variety of temperatures ranging from ambient temperature to the melting range of the alloy in question. The duration of the heating cycle and the extent of the HAZ are mainly dependent upon the welding parameters and material thickness, with the cooling rate being largely determined by heat input and by material thickness (see also section 2.6).

The weld metal itself cools under similar conditions to those of the heat affected zone of the base metal. A major difference, however, derives from the fact that the initial metallurgical state of the weld deposit differs from that of the parent steel. Weld metal cools directly from the liquid weld pool and is—contrary to the base metal—in most cases not subjected to a solution annealing treatment. It is rarely possible to heat treat finished weldments to temperatures of around 1050—1150° C, and if so, this heat treatment is limited to relatively small components only. Thus, the weld metal will generally behave differently from the parent steel. In multilayer welds, there are many heat affected zones at the transition between individual layers which, similar to the heat affected zone of the base metal may influence the corrosion resistance and the mechanical properties of the weldment.

Whether or not harmful precipitations occur during welding is largely determined by the kinetics of the precipitation behaviour of the alloy. Therefore, the latter is of special importance for the welding and heat treatment procedures applicable to stainless steels.

4.1 Carbide Precipitations in Stainless Steels with Special Consideration to Intergranular Corrosion

As mentioned earlier, stainless steels and their weld metals show at ambient temperature a very low solubility for carbon, as indicated by Fig. 18 (page 17) according to Tama, Vycklicky and Löbl [29]. At ambient temperature the carbon solubility is about 0.006% in austenitic steels and even lower in ferritic steels according to Wiester and Pier [200]. With higher carbon contents, excess carbon is precipitated in the form of chromium-iron carbides, mainly as $M_{23}C_6$, and in rare instances also as M_7C_3 or M_6C. According to Bungardt, Kunze and Horn [21], the chromium contents of the carbides $M_{23}C_6$ and M_7C_3 are in the range of approx. 42—65%. Compared to the normal composition of stainless steels, these mixed carbides are highly enriched with chromium.

If excessive quantities of carbon are dissolved in the stainless steel alloy by a solution annealing treatment with subsequent quenching, the carbon tries to precipitate again as carbide at that moment when the diffusion rate of carbon in iron permits it. Carbide precipitation is a time-temperature dependent process which is largely determined by carbon diffusion in the lower temperature range and by the rising carbon solubility in the base material at higher temperatures (see constitution diagrams in Figs. 17, 18, page 17).

Fig. 84 shows the time-temperature-precipitation diagram of the mixed carbide $M_{23}C_6$ according to Herbsleb, Schüller and Schwaab [201] for the austenitic steel grade 18 Cr/9 Ni (AISI 304) with 0.042% carbon in which the latter has been dissolved in excess by solution annealing at 1050° C and subsequent quenching. The $M_{23}C_6$ precipitation starts at 500° C after a duration of about 40 hours (curve 1). At 800° C, the precipitation rate reaches a maximum. Above this temperature, precipitation slows down again due to the fact that at the same time, the solubility of carbon in the austenitic matrix increases again (curve 4). This amounts to 0.05% carbon at 950° C which is roughly equivalent to temperature T_L in Fig. 84. According to Wiester, Schüller and Schwaab [202], curve 1 characterizes the time of incubation for the precipitation of the chromium carbide $M_{23}C_6$. At elevated temperatures, the continuation of curve 1 merges as curve 4 with the horizontal line 3 which corre-

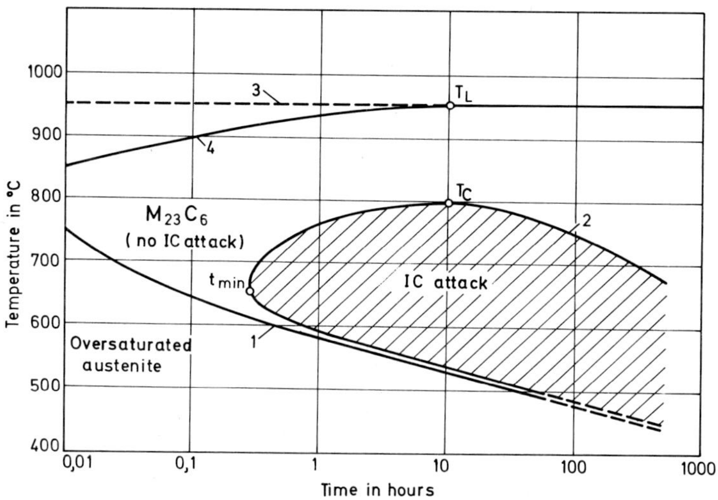

Fig. 84. TTP diagram of the precipitation of carbide $M_{23}C_6$ and the area of intergranular corrosion attack of stainless steel grade AISI 304 according to Herbsleb, Schüller and Schwaab [201]. Chemical composition in %: C = 0.042, Si = 0.59, Mn = 1.23, Cr = 17.46, Ni = 10.58, N = 0.046

sponds with the solubility temperature (T_L) of austenite for the respective carbon content.

Since the carbide $M_{23}C_6$ contains two to four times the amount of chromium than the base material, the immediate surroundings of precipitated carbides are depleted of chromium. These zones can be enriched again by chromium diffusion at elevated temperatures from the rest of the matrix back to the original level of chromium content. As indicated, however, in Table 1 (page 50), the diffusion coefficient for the diffusion of chromium in iron is considerably lower than that of carbon and equalization of the chromium contents in the zones depleted of chromium will only begin at higher temperatures which are above those of the $M_{23}C_6$ precipitation (curve 2 in Fig. 84). The boundary of this area can be viewed as a curve of equal chromium depletion. The carbide $M_{23}C_6$ is predominantly precipitated at the grain boundaries of crystals, with the zones in the immediate vicinity of the grain boundaries being depleted of chromium. Thus, they become susceptible to corrosion because their chromium content may markedly drop below the resistance limit of approx. 11.5% chromium. At the same time, according to Rocha and Lennartz [186], the activation potential steeply increases, marking the beginning of the attack by intergranular corrosion, progressing along the chromium depleted grain boundaries, which results in the destruction of the boundaries by the breaking away of crystal grains. This phenomena is also sometimes called "weld decay". Whereas the $M_{23}C_6$ precipitation is limited towards higher temperatures by the increasing carbon solubility of the material, the area of intergranular corrosion attack is determined by the diffusion of chromium into the depleted grain boundary areas. Fig. 84 shows clearly that the area of intergranular corrosion is smaller than the area of $M_{23}C_6$ precipitation, i.e. that sufficient chromium re-diffusion (curve 2) takes place at lower temperatures than the increase in carbon solubility of the material (curve 4). In other words, between the highest intergranular corrosion attack temperature T_C and the solubility temperature T_L and also to the left of the area of intergranular corrosion attack at t_{min} (respective temperature = 650° C), we find areas of carbide $M_{23}C_6$ precipitation which are free from intergranular corrosion attack. At lower temperatures, the curve indicating the beginning of intergranular corrosion attack follows closely that of the beginning of the $M_{23}C_6$ precipitation (curve 1) with a theoretically small delay in time. Because of limited measuring accuracy, both curves coincide according to Herbsleb, Schüller and Schwaab [201].

In stainless steels, the precipitation of the mixed carbide $M_{23}C_6$ has a detrimental effect, especially because of the increased susceptibility to intergranular corrosion attack. This explains why the literature contains a great number of intergranular corrosion attack diagrams (TTC diagrams[5]) and only rather few TTP diagrams[5] of $M_{23}C_6$ precipitation. The test basis used for intergranular corrosion attack diagrams is normally in accordance with the Strauß test, standardized for example by ASTM 262, practice E or DIN 50 914. In this context, we would like to draw your attention to the book "Corrosion Engineering" [233].

A study carried out by the International Welding Institute investigated a number of national standards for their application of the Strauss test, which varied only slightly, and found some differences in the determination of the area of intergranular corrosion attack (see IIW documents II-C-501-76, II-C-594-79 and II-C-602-79). From the very comprehensive body of literature now available on the intergranular corrosion of stainless steels, we shall only mention some of the basic works. The establishment and evaluation of intergranular corrosion (TTC) diagrams are discussed by Zitter [128], Wiester, Schüller and Schwaab [202] and Rocha [203]. A critical interpretation of the causes of intercrystalline corrosion through chromium depletion at the grain boundaries is given by Bäumel, Bühler, Schüller, Schwab, Schwenk, Ternes and Zitter [204], Stawström and Hillert [205], Fontana and Green [233] and by Hall and Briant [447]. All authors agree that the chromium depletion theory is sufficient to easily explain all known phenomena. For more details on this subject, please refer to the pertinent literature and the lists of supplementary literature included in the above publications.

4.1.1 Influence of the Type of Structure on the Tendency Towards Carbide Precipitation in Stainless Steels

Carbon diffuses much easier in a ferritic structure than it does in austenitic one (see Table 1, page 50). In addition, ferrite has a lower solubility for carbon than does austenite (see Figs. 17 and 18, page 17). Both effects facilitate the precipitation of carbide $M_{23}C_6$ in a ferritic matrix, with the possibility that the precipitation times are shortened to such a degree that, according to Bäumel [206, 207], ferritic chromium steels may already become susceptible to corrosive attack during electric arc welding and after air cooling from temperatures above 900° C. Moreover, because of the low carbon solubility of ferrite, the precipitation range is shifted towards higher temperatures. Fig. 85a shows the TTP and TTC diagrams of a 17% ferritic chromium steel (AISI 430) and an 18/8 chromium-nickel steel (AISI 304), both with

[5] TTP = time-temperature-precipitation diagram, TTC = time-temperature-corrosion (intergranular corrosion attack) diagram.

0.05% carbon according to Bäumel [206]. The shifting of the $M_{23}C_6$ precipitation and intergranular corrosion attack to shorter times in the ferritic steel as compared to the austenitic 18/8 Cr-Ni steel is clearly visible. The upper boundary of the $M_{23}C_6$ precipitation runs in accordance with the solubility limits in ferrite and in austenite (see Figs. 17 b and 18, page 17). In addition, the common cooling range of a one-layer electric arc weld has been plotted and this is located between curves 1 and 2. With a ferritic matrix the cooling range extends into the area of intergranular corrosion attack which illustrates the susceptibility of such grades to corrosion. The cooling curves of the single layer welds give apply to base metals which have not been preheated. Since ferritic chromium stainless steels are often preheated to 300° C prior to welding we may expect a cooling curve which coincides with curve 1′ in Fig. 85a. It follows then that with the ferritic steel presented here, any type of welding may lead to corrosion in aggressive media.

4.1.2 Influence of Alloying Elements on Carbide Precipitation and Intergranular Corrosion (IC) in Austenitic Stainless Chromium-Nickel Steels

4.1.2.1 Influence of Carbon and Nickel

With increasing carbon contents, the area of intergranular corrosion attack is shifted to shorter times along the lower boundary of the area of $M_{23}C_6$ precipitation. Fig. 85b shows the shift of an AISI 304 steel grade according to Rocha [203] in the representation according to Bäumel [206].

Rising nickel contents will increase carbon activity and decrease carbon solubility (see Fig. 18, page 17). The influence of nickel is normally felt, however, when the nickel content is above 20%. Diebold and Weingerl [208] recommend, therefore, that for the highly corrosion resistant 25/20 chromium-nickel steels with additions of molybdenum and nitrogen the carbon content should be limited to 0.020% to guarantee sufficient intergranular corrosion resistance. The negative influence of nickel on intergranular corrosion resistance is particularly felt in nickelbase alloys with high molybdenum contents.

4.1.2.2 Influence of Silicon

Like nickel, silicon also increases the carbon activity, but its influence is much stronger. Fig. 85c shows the area of intergranular corrosion attack for the low carbon steel 18 Cr/15 Ni with increasing silicon contents according to Bäumel [206]. According to Horn and Kügler [63], with 5.4% Si the area of intergranular corrosion attack may be shifted to even much shorter times of about 0.01 hour (= 36 seconds) due to segregations. Here, apart from the carbide $M_{23}C_6$, also a carbo-nitride (π phase) is precipitated which means that with increasing silicon contents the $M_{23}C_6$ precipitation is slowed down more and more and is substituted by the π phase. Its effect on intergranular corrosion is similar to that of the carbide $M_{23}C_6$. According to Masumoto and Imai [64], the π phase represents a carbo-nitride with the formula $M_{11}(CN)_2$. The cause for its precipitation seems to be the simultanous effect of silicon on the carbon and nitrogen activity. To reduce the strong tendency towards

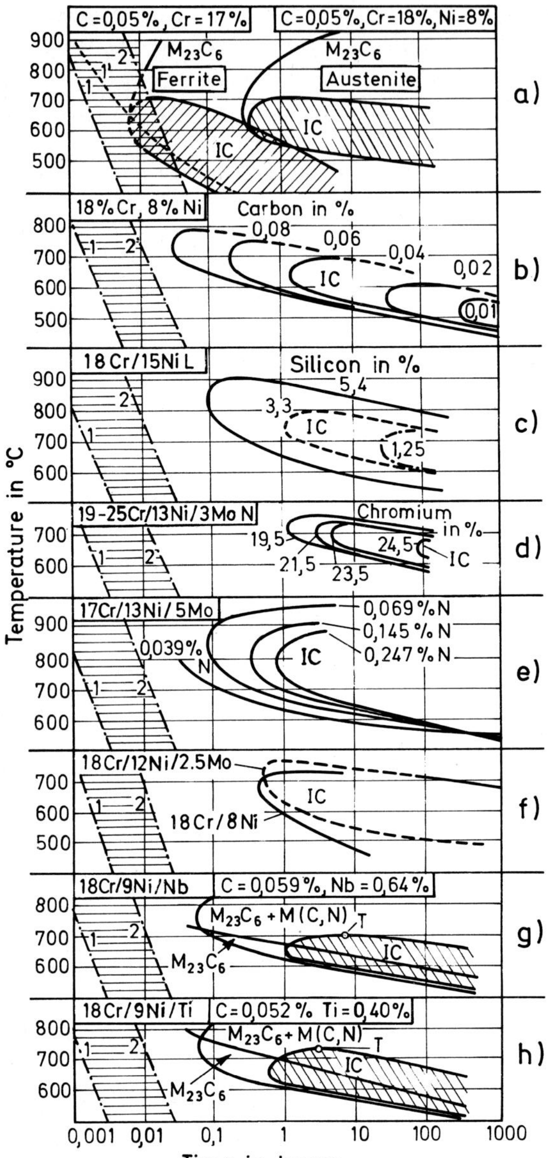

Fig. 85. Influence of structure and alloying additions on $M_{23}C_6$ precipitation and intergranular corrosion attack (IC), determined by the Strauss test. *a* influence of ferritic and austenitic structure according to Bäumel [206], *b* influence of carbon content according to Rocha [203], *c* influence of silicon according to Bäumel [206], *d* influence of chromium according to Lennartz [226], *e* influence of nitrogen according to Thier [38], *f* influence of molybdenum according to Edström and Ljundberg [217], *g* influence of niobium according to Herbsleb, Schüller and Schwaab [201], *h* influence of titanium to Herbsleb, Schüller and Schwaab [201]; horizontally hatched area: common cooling range of electric arc welding without preheating (line 1 and 2); *1* heat input 5 kJ/cm on 20 mm stainless steel plate, *2* heat input 30 kJ/cm on 10 mm stainless steel plate; *1'* same as *1* but plate preheated to 300° C

intergranular corrosion of austenitic steels with more than 4% silicon, Bäumel [206] recommended limiting the carbon content to 0.020% max. After 1980 this value has further been reduced to 0.012% C max.

4.1.2.3 Influence of Chromium

Increasing chromium contents have only a noticeable influence on the area of $M_{23}C_6$ precipitation in austenitic stainless steels. They excert a strong influence, however, on intergranular corrosion (IC). With higher chromium contents, the time to reach the resistance limit of chromium depletion at the grain boundaries is shifted to longer periods of time. This is more noticeable with lower carbon contents of around 0.03% than with higher carbon contents. In addition, higher chromium contents facilitate the diffusion of chromium into the depleted grain boundary areas. This produces a shift of the upper boundary of the area of intergranular corrosion attack towards lower temperatures. Fig. 85d shows the influence of chromium on the area of intergranular corrosion attack for steels with about 0.03% carbon, 3% molybdenum, 19.5–24.5% chromium, about 16% nickel, 0.2% niobium, and 0.3–0.4% nitrogen according to Lennartz [226]. The beginning of intergranular corrosion is shifted to much longer times by the increase in the chromium content from 19.5% to 24.5% and also the area of intergranular corrosion attack is narrowed. Wedl and Kohl [211] determined that for austenitic chromium-nickel steels with about 0.01% carbon and 0.25–0.42% nitrogen an increase in the chromium content from 18% to 23% will shift the beginning of grain decay by nitrogen induced intergranular corrosion by as much as a factor of 10^5 towards longer times. The susceptibility to this special type of intergranular corrosion can thus in practical terms be eliminated.

4.1.2.4 Influence of Nitrogen

The effect of nitrogen is of particular importance for the precipitation behaviour of stainless steels. Therefore, this alloying constituent has—apart from its use for improving the yield strength of austenitic steels—greatly gained in importance. According to Grützner [209, 210], nitrogen may produce intergranular corrosion at the grain boundaries by the precipitation of chromium-nitrides, but only in the presence of extremely low carbon contents in the range of approx. 0.01% and comparatively high nitrogen contents of more than 0.15%. Conventional stainless steels and weld metals have normally higher carbon contents, with carbides such as $M_{23}C_6$ being precipitated prior to nitrides. Therefore, carbides are mainly responsible for the occurrence of intergranular corrosion. According to Thier [38, 41], the $M_{23}C_6$ precipitation in chromium-nickel-molybdenum steels like 17 Cr/13 Ni/5 Mo is shifted with increasing nitrogen contents to longer times, as indicated by Fig. 85e. According to investigations by Blazejak, Herbsleb and Westerfeld [213], however, $M_{23}C_6$ precipitation in molybdenum-free austenitic stainless steel containing 0.019–0.029% carbon, about 18.5% chromium and about 11% nickel are only marginally influenced by nitrogen, with chromium nitride Cr_2N also being precipitated apart from carbide $M_{23}C_6$. When testing these steels for intergranular corrosion in accordance with the Strauß test, Herbsleb and Westerfeld [214] found a slight shift in the area of intergranular corrosion attack towards higher temperatures and longer times, apparently because of impeded $M_{23}C_6$ precipitation due to the nitrogen content. With molybdenum containing austenitic stainless steels, the above authors [214] found in agreement with other publications [38, 39, 41] that a content of 0.13% nitrogen will shift the lower boundary of the area of intergranular corrosion attack to higher temperatures and longer tempering times. An influence on the area of intergranular corrosion attack by the precipitation of chromium nitride could not be established by the Strauss test, neither for molybdenum free nor for molybdenum containing steels. However the precipitation of chromium nitride was severely impeded by the presence of molybdenum.

In molybdenum free steels with an increased chromium content of 23%, a nickel content of 15% and carbon contents of 0.02—0.06%, Wedl and Kohl [211] have found a marked reduction in $M_{23}C_6$ precipitation through the addition of nitrogen, with 0.35% nitrogen, for example, producing a shift in the beginning of intergranular corrosion attack towards longer times. According to these findings, the effect of nitrogen is much stronger at 23% chromium than it is in the range of around 18% Cr. With extremely low carbon contents of around 0.01%, intergranular corrosion attack can only be detected with the modified Strauss test (severe corrosion conditions, see lit. 233) after 300—1000 hours at 550° C. The authors arrive at the conclusion that steels with increased chromium contents of about 23% and with nitrogen contents up to approx. 0.4% are not susceptible to nitrogen induced intergranular corrosion attack. In steels with 0.013—0.077% carbon, 16—18% chromium, 8.5—11.6% nickel, with and without 2% molybdenum, Briant, Mulford and Hall [212] found that with increasing nitrogen contents a slowdown took place in the beginning of intergranular corrosion attack when testing with the modified Strauss test. With the addition of 2% manganese, this effect is supported at 600° C but not at 650 and 700° C. Wessling and Bock [215] noted that nitrogen exhibits a strong delaying effect upon the process of intergranular corrosion especially in the case of long term tempering of up to 10^5 hours. They draw the conclusion that in cases where resistance to intergranular corrosion is required, the service temperature of nitrogen alloyed austenitic steels with carbon contents of 0.03% max. can be raised from 350° C with molybdenum free materials to 400° C and with molybdenum and nitrogen alloyed steels to 450° C. Lorenz and Medavar [216] arrived at similar results for austenitic chromium-nickel and chromium-nickel-molybdenum steels with 0.20—0.25% nitrogen. They also found that nitrogen produces a favourable influence when testing such alloys in boiling 65% nitric acid (Huey test).

4.1.2.5 *Influence of Molybdenum*

According to Edström and Ljundberg [217], molybdenum influences the process of carbide precipitation in stainless chromium-nickel austenitic steels, since this alloying element reduces the solubility of carbon in austenite. According to Wiegand and Doruk [33], the carbide precipitation is accelerated at higher temperatures, where the reduced carbon solubility of austenite is felt. At lower temperatures below about 700° C, carbide precipitation is slowed down according to Edström and Ljundberg [217], as shown in Fig. 85f for the steel grade AISI 316. According to the

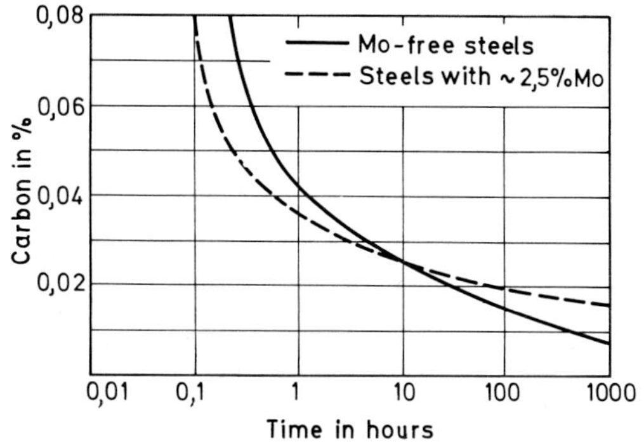

Fig. 86. Correlation between carbon content and the beginning of intergranular corrosion attack at the respective critical temperature in molybdenum free and molybdenum containing austenitic chromium-nickel steels according to Edström and Ljundberg [215]

same authors, however, this slowing down effect produced by molybdenum is only noticeable at low carbon contents of below 0.03% as shown in Fig. 86.

Hall and Briant [477] investigated chromium depletion through carbide precipitation relating to nitrogen alloyed chromium-nickel-molybdenum austenitic steels of the type AISI 316 L containing approx. 17.5% Cr, 9.8% Ni, 2.5% Mo and 0.08–0.16% N after an annealing treatment in the range of 650–700° C with durations of 50–300 hours. The most severe chromium depletion occurs right next to the grain boundaries of the precipitated carbides of the type $M_{23}C_6$. The chromium depletion is higher at 650° C than at 700° C. With increasing annealing temperatures and longer annealing times, the area of depletion expands. Apart from chromium depletion, there is also a molybdenum depletion due to the incorporation of this element into the carbide $M_{23}C_6$. Methods of calculation have been developed based on thermodynamic considerations, with the calculated profiles for chromium and molybdenum coinciding well with the empirical data. The results obtained support previous experience with regard to the resistance of stainless steels towards intergranular corrosion attack. The latter may always occur if chromium and molybdenum contents in the depleted zones drop below certain minimum values and if the number and size of precipitated carbides is large enough to produce a respective depletion in Cr and Mo. These assumptions are based, of course, on the presence of an aggressive corrosive medium.

In molybdenum containing chromium-nickel austenitic steels, according to Kügler [218], the chromium-iron mixed carbide $M_{23}C_6$ is precipitated first at temperatures of 750–850° C. With prolonged annealing times, this carbide picks up molybdenum which, being a carbide forming element, becomes increasingly dissolved in $M_{23}C_6$, with a simultaneous drop, however, in carbon content. First, iron-chromium-molybdenum mixed carbides are formed which are then converted to the Chi phase ($Fe_{36}Cr_{12}Mo_{10}$)[6] with about 1% carbon. For unstabilized

[6] This formula is stated according Bechtold and Vacher [43]. In more recent literature, the formula for the composition of Chi phase is often stated as $Fe_{36}Cr_{12}Mo_{12}$.

chromium-nickel-molybdenum austenitic stainless steels, Kügler [218] describes the precipitation scheme of carbon containing phases as follows:

$$
\begin{array}{c}
(Fe,\ Cr)_{23}C_6 \\
| \\
+\ Mo \\
\downarrow \\
(Fe,\ Cr)_{21}Mo_2C_6 \\
| \\
+\ Mo \\
\downarrow \\
(Fe,\ Cr,\ Mo)_6C \\
| \\
+\ Mo \\
\downarrow \\
(Fe_{36}Cr_{12}Mo_{10}) \ = \ \text{Chi-Phase.}
\end{array}
$$

According to Wessling and Bock [215], nitrogen shifts the area of intergranular corrosion attack of low carbon austenitic chromium-nickel steels containing 2.2% molybdenum to longer times and to approx. 50° C higher temperatures. According to Lennartz and Oppenheim [219], low carbon steels of the type 316 L when tested in accordance with the Strauss test are resistant to intergranular corrosion attack even after long time annealing at temperatures of 500° C and for durations of more than 23 000 hours.

With increasing molybdenum contents, $M_{23}C_6$ precipitation and intergranular corrosion attack becomes increasingly influenced by the precipitation of intermetallic phases. Therefore, the combined effect of all precipitations and their influence on each other must also be taken into consideration. These processes shall be discussed in more detail in section 4.2.

4.1.2.6 Influence of Titanium and Niobium

The alloying elements titanium and niobium are very strong carbide forming elements. They are used, therefore, for the retention of carbon in the form of "stable" titanium and niobium carbides, and by doing so reduce the precipitation of $M_{23}C_6$ at the grain boundaries and thus limit the alloys susceptibility to intergranular corrosion. According to Fig. 18, a steel with 18% chromium and 10–20% nickel will dissolve about 0.1% carbon at 1050° C, i.e. during solution annealing at this temperature, the carbon which is linked to iron and chromium will be dissolved up to a content of 0.1%. After quenching and tempering in the range of 500–700° C, it precipitates again at the grain boundaries as $M_{23}C_6$ and may induce intergranular corrosion attack. If carbon is retained, however, by titanium or niobium, dissolution of carbon by the austenitic matrix is shifted towards substantially higher temperatures, as indicated in Fig. 39 by curves 1–4 for niobium. After one hour at 1050° C only approx. 25%, at 1200° C approx. 50% and at

1300° C approx. 70–80% of the carbon linked to niobium is dissolved again in the matrix, with the remainder being retained by niobium as niobium carbide.

With an annealing temperature of 1050° C, however, carbon is largely retained by the titanium or niobium content and only a small amount of carbon remains in solution in the austenitic matrix after the final quenching treatment. In other words, the stable retention of carbon by titanium or niobium has a similar effect as to the effect that the reduction in carbon has on unstabilized steels. It is important, however, that the titanium and niobium contents are high enough to retain carbon as titanium carbide (TiC) or niobium carbide (NbC) at the temperatures normally used for solution annealing treatments. In accordance with the stoichiometric ratio, the titanium must be at least four times the carbon content and the niobium content at least eight times the carbon content (see also section 1.3.5 and 1.3.6). The ratio of the stabilizing element content to the carbon content is called the "stabilization ratio". For practical purposes, however, it is necessary to consider that about one fifth of the titanium and niobium additions are retained by nitrogen which is normally present in stainless steels in contents of approx. 0.05%. Therefore, the minimum content of titanium is set at 5 × % C and of niobium at 10 × % C to compensate for these losses.

With increasing solution annealing temperatures the effect of titanium and niobium is progressively diminished due to the increased dissolution of TiC and NbC (see Fig. 39, page 32) until it is almost completely lost at 1300° C. The same mechanism applies to the areas of the heat affected zone of a weld bead that are subjected to high temperatures during welding. In such cases, it is necessary to overstabilize, i.e. the stabilization ratio must be increased. This situation is discussed in more detail in section 4.1.4 which deals with the stabilization of weld metal.

The corrosion and precipitation behaviour of stabilized austenitic stainless steels has been the subject of numerous investigations by Wiester and Pier [200], Wiester, Schüller and Schwaab [202], Rocha [203], Ruttmann, Gerlach and Kautz [52], Bungardt and Lennartz [221, 222, 223] and Bungardt, Lennartz and Wetzlar [224]. Herbsleb, Schüller and Schwaab [201] investigated a possible correlation between carbide and nitride precipitation and their effect on intergranular corrosion attack of titanium and niobium stabilized steels. Fig. 85g (page 108) shows the effect of niobium on the $M_{23}C_6$ precipitation and on intergranular corrosion attack in the steel grade AISI 347 containing 0.059% carbon and 0.64% niobium after a solution annealing treatment at 1050° C with subsequent rapid cooling according to Herbsleb et al. [201]. The precipitation of $M_{23}C_6$ is overlapped by the precipitation of a niobium containing carbo-nitride of type M(CN). With shorter annealing times, however, neither types of precipitation produce sufficient chromium depletion to permit intergranular corrosion to occur, because IC only begins after a period of about one hour of annealing and after further precipitation of $M_{23}C_6$. A comparison with the unstabilized steel with 0.06% carbon in Fig. 85b (page 108) and 0.042% carbon in Fig. 84 (page 105) shows the effect of niobium. The area of intergranular corrosion attack is shifted towards longer periods of time with the upper temperature limit T_c being lowered. According to Reisenhofer and Weingerl [225] a greater excess of niobium results so that this element no longer precipitates as carbo-nitride but appears in separate phases in the form of NbC and NbN.

Fig. 85h shows the precipitation behaviour of a titanium stabilized stainless steel with 0.052% carbon and 0.4% titanium according to Herbsleb, Schüller and

Schwaab [201]. This grade shows a similar pattern to the niobium stabilized grade, with the only difference being that a titanium containing carbo-nitride is precipitated.

4.1.2.7 Influence of Manganese

According to Bäumel [206], the addition of manganese to stainless steels reduces the carbon activity and increases the carbon solubility. In this way, the carbide precipitation rate is slowed down. The effect of manganese is of special importance in fully austenitic weld metals, to which this alloying constituent is often added in contents of 4—6% to reduce the hot cracking sensitivity. In stainless steels, manganese is sometimes used to increase the nitrogen solubility of high tensile austenitic alloys (see Fig. 37). In general, manganese has no marked influence on the corrosion resistance of stainless steels. Thus, the reduction of the $M_{23}C_6$ precipitation rate is accepted according to Bäumel [206] as a favourable side effect in both weld metals and steels.

4.1.3 Carbide Precipitation in Ferritic, Low Carbon Martensitic and Austenitic-Ferritic (Duplex) Steels

4.1.3.1 Ferritic Stainless Chromium Steels

In ferritic chromium steels with about 0.05—0.1% carbon corrosive attack occurs under somewhat different conditions compared to that of austenitic chromium-nickel steels. By rapid cooling from temperatures above 900° C, the former become susceptible to corrosion but become resistant again after tempering in the range of 650—800° C. According to Baerlecken, Fischer and Lorenz [40], the intergranular corrosion attack behaviour is mainly influenced by carbon. Susceptibility to corrosive attack occurs in both non-transformable and transformable chromium steels after quenching from elevated temperatures if the carbon content is higher than 0.01%. The carbon content above this threshold value will be precipitated as carbide, even in the case of rapid cooling. As indicated by Fig. 85a, carbide precipitation takes place much faster in ferrite than in austenite. In addition, it is also shifted towards lower temperatures. A ferritic steel with 17% chromium becomes susceptible to intergranular corrosion attack after only approx. 6 minutes at 400° C compared to about 100 000 hours at the same temperature for an austenitic stainless steel containing 18% chromium. If the chromium content of the ferritic steel is near the resistance limit of about 12%, corrosive attack occurs over the total surface of the crystal grains when tested in accordance with the Strauss test. Stabilization with titanium or niobium is only employed with higher chromium contents from about 17% onwards.

The corrosion behaviour of 13% chromium steels (AISI 410) has been tested by Bäumel [227] in boiling 5% and 20% acetic acid, since the Strauss test is too aggressive for these steel grades. The weight loss is determined by chromium depletion due to the precipitation of chromium carbide. Intergranular corrosion does not occur since corrosion affects the total surface. The weight losses are greatest if hardening is followed by tempering at 400—600° C which is in the range of carbide precipitation (see Fig. 85a, page 108). The latter takes place with such rapidity that even the rapid cooling that occurs during electric arc welding is not sufficient to avoid corrosion susceptibility. Low weight losses can only be obtained by a heat treatment which consists of annealing at temperatures around 950° C followed in most cases by cooling in air. The result is a largely martensitic structure which must be tempered. The most suitable tempering temperature is determined by the chromium and molybdenum contents. This is normally located in the range around 700° C. If the annealing temperature is lowered, the annealing time must be extended. According to Bäumel [228], during tempering, martensite is decomposed, with carbide precipitation taking place in the order of M_3C—M_7C_3—$M_{23}C_6$. The subsequent chromium depletion and possible corrosion occurs over the total area of the original martensite grains. It is equalized by re-diffusion of chromium during tempering at around 700° C. Molybdenum when present in contents of about 1% improves the corrosion resistance of the alloy in both the as hardened and tempered condition.

17% chromium steels with about 0.1% carbon (AISI 430) are susceptible to corrosive attack after rapid cooling from temperatures above 900° C and during electric arc welding. Their structure consists partially of primary delta ferrite and martensite. According to Bäumel [228], the preferential corrosion will occur at the martensite grains. This is as a result of the segregation of carbon and chromium between austenite and ferrite grains during partial δ-γ transformation. The segregations so formed produce partial corrosive attack on the carbon enriched and chromium depleted martensitic grains transformed from austenite. Only after rapid cooling from temperatures above 1100° C partial corrosion may be superimposed by regular intergranular corrosion. By tempering between 700 and 850° C, this susceptibility towards corrosive attack is again eliminated. According to Herbsleb and Schwenk [229] for a 17% chromium steel, the ideal temperature for such a "stabilization tempering" is 750° C, with a holding time of 30 minutes being sufficient to completely eliminate susceptibility to corrosive attack. This temperature is also used in most cases for tempering treatment of welded components. 17% chromium steels show a relatively strong tendency towards grain coarsening by recristallization at temperatures above 900° C. The addition of molybdenum and titanium helps, according to Woltron [193], to greatly reduce the rate of crystal growth and also improves corrosion resistance. 17% chromium steels are therefore often alloyed with 1—2% molybdenum.

17% chromium steels can also be stabilized with either titanium or niobium. Therefore, the area of corrosive attack is shifted by the stabilization to longer periods of time but the shift is much smaller than in austenitic steels due to the ferritic structure of the matrix. According to Bäumel [207], stabilized 17% chromium steels are resistant to intergranular corrosion attack when electric arc welded with coated electrodes. If however they are GTA welded, they may become susceptible to intergranular corrosion because of possible slower cooling rates associated with the GTAW process. The precipitation and corrosion behaviour of stabilized ferritic chromium steels with 17% Cr and 0.03—0.07% carbon have been investigated by Bond and Litzlovs [230] and Bäumel [231, 232]. Testing in accordance with the Strauss test (test method see ASTM 262, practice E) shows sufficient resistance to corrosion with both titanium or niobium stabilized alloys. When tested, however, with boiling strong nitric acid in the Huey test titanium stabilized 17% chromium steels show evidence of corrosive attack, since the titanium carbide formed is not corrosion resistant. Niobium carbide, however is resistant to nitric acid and niobium stabilized steels are thus resistant to corrosion when subjected to the Huey test. Therefore, niobium stabilized 17% chromium steels are more advantageous for applications in oxydizing media than titanium stabilized ones.

According to Bäumel [232], the precipitation mechanism of the unstabilized steel grade 17 Cr/1.5 Mo with 0.08% C which has been quenched in water from 1200° C and subsequently tempered at 600° C, leads to the precipitation of chromium carbides after tempering times of up to 50 hours. In the case of longer periods of time, increasing quantities of sigma phase will be precipitated, which consists roughly of equal parts of iron, chromium and molybdenum. Due to the depletion of Cr and Mo in the vicinity of these precipitations we find that, when the material is exposed to a strong solution of boiling nitric acid (Huey test), severe corrosion takes place over the whole grain surface. In the Strauss test, however, sigma phase precipitations produce no such corrosive attack. In the case of an overstabilization of such steels we find, according to Bäumel [231], that with an increasing stabilization ratio, apart from the carbide $M_{23}C_6$ an increasing amount of niobium containing carbides of the type $(Fe, Cr)_3Nb_3C$ appear. If the steel is also nitrogen alloyed, niobium carbo-nitride Nb (C, N) and z-phase $Cr_{3.5}Fe_{3.5}Nb$, i.e. a chromium-iron niobide, are precipitated.

From the investigations made by Baerlecken, Fischer and Lorenz [40] it appears that the alloying constituent chromium neither improves nor worsens the toughness properties of stainless chromium steels. Toughness is first of all determined by the elements carbon and nitrogen, whose solubility in ferrite drops more and more with increasing chromium contents. This means that these two elements will increasingly precipitate as carbides, nitrides or carbo-nitrides and the toughness of the ferritic matrix will decrease more and more. Therefore, the application range of the common ferritic chromium steels becomes increasingly limited with rising chromium contents. As early as 1960 attempts were made to improve the toughness and corrosion resistance of stainless ferritic chromium steels by reducing the carbon contents to below 0.01% and the nitrogen contents to below 0.005%, with the total of carbon and nitrogen together remaining below 0.01%. Lennartz and Kiesheyer [234] report on the corrosion behaviour of such steels with chromium contents of around 28–35%, carbon contents in the range of 0.002–0.006% and nitrogen contents around 0.002%. They are fully resistant to intergranular corrosion and stress corrosion cracking. In recent years, numerous works have been published on these new steels which are also called "extra low interstitial" or ELI steels. Please refer in this context to the volume of the conference proceedings "Stainless Steel 77" [235].

4.1.3.2 Low Carbon Cr-Ni Martensitic Stainless Steels (Soft Martensitic Steels)

Low carbon martensitic steels were developed in the late fifties in order to improve the toughness and weldability of the stainless ferritic chromium steels. By lowering the carbon content to 0.04% max., a relatively soft and tough martensitic matrix is obtained. By the addition of nickel, normally in contents of between 4–6%, the

γ area is expanded and the development of delta ferrite largely eliminated (see Fig. 11, page 12). The combined effect of chromium and nickel on the transformation processes in the iron-chromium-nickel constitution diagram can be seen from Fig. 9a, page 10.

Low carbon soft martensitic steels are always quenched and tempered. During cooling from the hardening temperature or during arc welding the structure initially consists mainly of austenite (which possesses a higher carbon solubility at high temperatures than does ferrite) until martensite transformation begins at the M_s temperature. Because of the rapid cooling, carbon stays in forced solution. During tempering the carbon is precipitated as carbide. According to Irvine, Crowe and Pickering [188], precipitations like M_3C, $M_2(C, N)$, M_7C_3 and $M_{23}C_6$ may appear. According to Brezina, Erdoes, Geiger, Habel, Lorenz and Wintsch [236], tempering will produce relatively coarse grain boundary carbides $M_{23}C_6$ and finer carbides of the same type in the grain cores next to the very fine irregularly dispersed carbo-nitrides of the type $M_2(C, N)$, with chromium depletion occurring both at the grain boundaries and within the grain areas. This also explains why actual intergranular corrosion rarely occurs in low carbon martensitic chromium steels. According to Niederau [199] and Folkhard [237], low carbon martensitic 13/4 and 13/6 chromium-nickel steels are superior to conventional martensitic 13% chromium steels in terms of corrosion resistance mainly because of their lower carbon content and the subsequent reduction of carbide precipitation. In a 20% boiling acetic acid solution, an increase in the rate of corrosive attack can be seen in the tempering range of between 450 and 580° C, the increase in material corrosion rate is smaller, however, than in the case of pure chromium steels. In quenched and tempered low carbon martensitic steels, Heimann and Hoock [238] and Gümpel, Hook and Strom [461] noted during long term aging at 425° C a strong rise in the quantity of precipitations, with fine M_6C carbides being precipitated in great numbers in the crystal cores next to the previously precipitated $M_{23}C_6$ carbides.

With the standard Strauss test being too aggressive for the 13% ferritic chromium steels and low carbon martensitic 13/4 chromium-nickel steels, Süry and Brezina [239] developed a modified corrosion test which allows the relative influence of heat treatments to appear more clearly. The influence on the corrosion resistance due to a molybdenum content of about 1.5% can also be established in this test. According to Brezina [93], the attack in this corrosion test is not only intergranular but also affects crystal cores as well, being a result of chromium depletion caused by chromium rich carbides, which have precipitated during the transformation to martensite. A special type of intergranular corrosion can also occur in the form of a local attack on the phase boundaries between delta ferrite and tempered martensite. Its appearance is rare, however, and can be avoided by suitable heat treatment procedures. After tempering in the range of around 600° C, 13/4 chromium-nickel steels show not only excellent toughness properties but also good corrosion resistance.

Apart from this grade, there are a number of other low carbon martensitic chromium-nickel steels with nickel contents of up to 6% and chromium contents of up to 17%, some with additions of molybdenum, copper and niobium. With regard to carbide precipitation, they behave in a similar manner to the 13/4 chromium-nickel steels. For further details on these alloys please refer to the pertinent publications [189, 190, 196, 240–243].

4.1.3.3 Austenitic-Ferritic (Duplex) Steels with Low Carbon Contents

Duplex stainless steels[7] with higher carbon contents have been known for a long time. In comparison to conventional austenitic grades, they possess good resistance to stress corrosion cracking in chloride containing media as well as in a number of other corrosive substances. They are often susceptible, however, to intergranular corrosion attack. Due to their strong tendency towards coarse grain formation, they may become during welding rather brittle in the HAZ and thus difficult to weld. This fact has led to the development of austenitic-ferritic steels with low carbon contents which show good resistance to stress corrosion cracking, high resistance to intergranular corrosion attack as well as excellent mechanical properties and good weldability. At the same time, these stainless low carbon duplex steels show good resistance to pitting in chloride containing media and sea water which makes them suitable for a relatively wide field of applications in oil pumping installations, waste water plant and marine engineering. They possess chromium contents of 18.5–27%, nickel contents of around 4–8% and are normally alloyed with additions of 2–4% molybdenum, often with nitrogen and sometimes also with copper. Their carbon content is normally below 0.03%. The structure is two-phase, consisting of austenite with a relatively high proportion of delta ferrite. In the Schaeffler diagram (Fig. 79, page 90), they are located in the middle of the austenite + ferrite field with chromium equivalents of 20–30 and approx. 50% delta ferrite (see e.g. alloy 22/5/3 in Fig. 79). Table 4 represents a survey giving the average analysis values for typical common duplex steels.

Table 4. *Typical chemical composition of stainless austenitic-ferritic duplex steels (see also Table 16, page 187)*

Similar to UNS Designation*	Chemical composition in %						
	C (max.)	Cr	Mo	Ni	Cu	N	Others
S 31500	0.03	18.5	2.7	5.0	–	–	Si = 1.7
–	0.025	21.0	2.5	6.5	1.5	–	–
S 31803	0.03	22.0	3.0	5.5	–	0.15	–
–	0.03	24.0	1.5	5.0	1.0	0.10	–
–	0.03	24.0	2.8	7.0	0.4	–	–
S 31200	0.03	25.0	1.5	6.0	–	0.17	–
S 32550	0.04	25.5	3.4	5.5	2.0	0.17	–
S 31250	0.03	25.0	3.0	6.5	0.5	0.20	W = 0.3
–	0.03	25.0	4.5	7.0	–	0.18	cast steel
–	0.05	25.0	3.0	8.0	1.0	0.18	cast steel
–	0.03	25.5	2.3	3.7	–	0.33	Mn = 5.8
–	0.03	26.5	1.5	6.0	–	0.16	–

* New designation established in accordance with ASTM E 527 and SAE J 1086, Practice for Numbering Metals and Alloys (UNS).

[7] By duplex stainless steels, we mean two-phase austenitic-ferritic chromium-nickel steels with a relatively high delta ferrite content in the range of around 30–60%.

The precipitation behaviour of nitrogen alloyed chromium-nickel-molybdenum duplex steels has been investigated by Herbsleb and Schwaab [245]. The TTP diagram of such steels is shown in Fig. 87, steel 1, is compared with the behaviour of a steel grade that has a similar chemical composition but contains less nitrogen (steel 2), according to investigations by Jolly and Hochmann [198]. After a solution annealing treatment at 1050° C for 1/2 hour followed by quenching in water, the nitrogen alloyed steel 1 shows a precipitation free structure with about 60% delta ferrite and 40% austenite. Precipitations start after only 2 minutes at about 800° C with the formation of chromium nitride Cr_2N at the ferrite-austenite grain boundaries together with the somewhat coarser globular particles of Chi phase. In addition, very small amounts of chromium-iron carbide $M_{23}C_6$ are precipitated. After about 20 minutes, sigma phase appears for the first time in the temperature range of 800—900° C.

In the Strauss test, the solution annealed steel 1 is fully resistant to corrosion at all temperatures between 300—1000° C with annealing times of up to 30 hours. An increased susceptibility towards corrosive attack does not occur since carbide precipitations are rather low and thus negligible and the chromium depletion due to the precipitation of chromium nitride is not sufficient to produce intergranular corrosion because of the increased chromium level of this grade. Testing with the Huey test in a strong solution of boiling nitric acid (for test procedure, please refer to [233]) shows, however, strong corrosive attack with the appearance of the Chi phase.

If we compare here the precipitation behaviour of steel 2 (Fig. 87) which is lower in nitrogen, we note that the precipitation ranges are located at much shorter times and that carbide precipitation is much more intensive. Such steels are thus more susceptible to intergranular corrosion attack. The effect of nitrogen in duplex steels with low carbon contents is not limited, however, to the substitution of a major share of carbide precipitation by nitride precipitation. It also has a marked influence

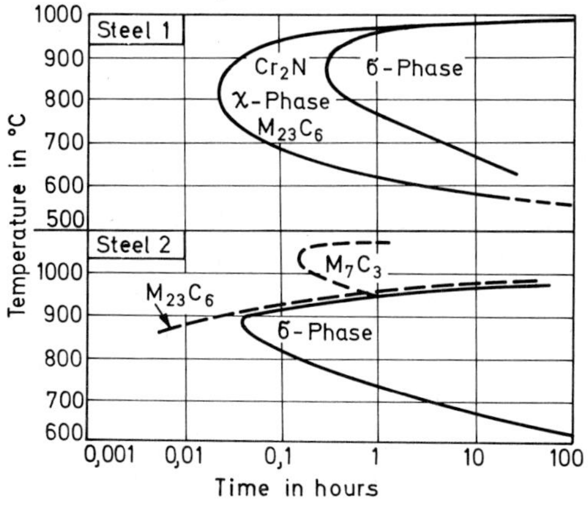

Fig. 87. TTP diagrams of two austenitic-ferritic duplex steels. Steel 1 according to Herbsleb and Schwaab [245], steel 2 according to Jolly and Hochmann [198]. Solution annealing: steel 1 – 1050° C/30 min., steel 2 – 1150° C/60 min. Chemical composition in %: 1 – C = 0.028, Si = 0.45, Mn = 1.63, Cr = 21.80, Mo = 3.12, Ni = 5.00, N = 0.113, 2 – C = 0.028, Si = 0.50, Mn = 0.70, Cr = 20.90, Mo = 2.30, Ni = 7.40, N = 0.073, Cu = 1.4

on the improvement of weldability. The tendency of duplex stainless steels towards coarse grain formation in the heat affected zone is greatly reduced by an increased nitrogen content. In addition, the heat affected zone is no longer susceptible to intergranular corrosion attack as may be the case with duplex steels without the addition of nitrogen. Also precipitations are dispersed by the fine grained structure over a much larger area which makes them less harmful. This effect improves not only the resistance to intergranular corrosion attack but also the toughness of the heat affected zone. Wehner and Speckhardt [246] draw attention to a special feature of the duplex steels. The elements carbon and nitrogen are mainly dissolved in austenite and will precipitate to a great degree in this crystal phase. Chromium and molybdenum containing phases, on the other hand, precipitate first in delta ferrite since the diffusion potential of chromium and molybdenum is much greater in ferrite than in austenite (see Table 1, page 71).

Because of the low carbon content and the increased chromium content, duplex steels of UNS type S 31803 (22 Cr/5 Ni/3 Mo) are fully resistant to intergranular corrosion in the as welded state. Wessling, Bock and Fuchs [247] point out the good weldability of such steels. During welding operations, the heat affected zone does not become predominantly ferritic, as is the case of duplex steels without the addition of nitrogen, instead the more favourable austenite-ferrite ratio of the base material is largely maintained in the heat affected zone which means that the toughness of the latter can be retained.

Kohl, Hochörtler, Kriszt and Koren [248] report on chromium-nickel-molybdenum-manganese alloyed duplex steels with high nitrogen contents of 0.35% and improved yield strengths and tensile strengths. Welded joints show full resistance to intergranular corrosion over the whole cross section of the weld, when subjected to the Strauss test.

4.1.4 Carbide Precipitations in Stainless Steel Weld Metal and in the Heat Affected Zone (HAZ)

If welds are heat treated in the same way as the parent stainless steel, i.e. solution annealed and sometimes tempered, they perform in most cases with regard to carbide precipitation and susceptibility to intergranular corrosion in the same manner as a steel grade of similar analysis. In such instances the phenomena discussed in sections 4.1.1—4.1.3 would also apply to the weld metal and the HAZ. The only exceptions to this being welds with strong segregations and very coarse grained heat affected zones, as may occur when welding ferritic stainless steels or, more generally, in welding processes with very high heat inputs, e.g. electroslag welding. Segregations which are produced under such circumstances can only be eliminated by diffusion annealing and not by solution annealing.

In most cases, however, it would not be possible to subject stainless steel weldments to such annealing procedures. In such cases, the weldment would be put into service in the as welded state. Therefore the weld deposits may under service conditions perform differently to the equivalent parent steel grade with regard to carbide precipitations. This may also apply to weld metals which for other reasons (e.g. hot cracking resistance) show somewhat different chemical compositions and structural configurations to those of the equivalent steel grades. In this context, please note that the manufacturers of filler metals often keep the carbon content of their products as low as possible in order to obtain minimum carbide precipitation. At the same time, the chromium content is often increased to about 20% to delay intergranular corrosion attack. To obtain sufficient hot cracking resistance, the chemical composition of austenitic steel weld metal is often set in such a way that the austenitic matrix contains low delta ferrite contents in the range of 5–10 FN (see also section 2.2.5). Because of solution annealing and subsequent quenching, the base metal shows a purely austenitic structure.

The effect of alloying elements on carbide precipitation and intergranular corrosion in weld metal is basically the same as shown in Fig. 85. Due to the effect of heat during welding operations, however, structural changes may occur which diminish the corrosion resistance in the vicinity of the weld. Bäumel deals thoroughly with this phenomenon [206, 207, 249, 250]. Stalder [251] summarizes the possible effects of welding operations by establishing the following demands:

— The corrosion resistance of the base metal to be joined by welding should not be changed by the welding operation.
— The material fused during welding (base metal and filler metal) must possess at least the same corrosion resistance as the base material which is not affected by the welding heat.

The metallurgical differences between base metal, heat affected zone and weld metal together with the effect of heat input during welding may have an influence on carbide precipitation and the resulting corrosion sensitivity. This may lead to the occurrence of special types of intergranular corrosion attack. The latter shall be discussed in detail in the following pages.

4.1.4.1 Knife Line Corrosion in Weldments Made from Stabilized Austenitic Stainless Steels after Heat Treatment in the Temperature Range of 500–700° C

This special type of corrosion may occur during the welding of stabilized austenitic stainless steels since carbon which is normally stably retained by titanium or niobium may become re-dissolved in the HAZ. Corrosive attack will normally only occur after annealing in the range of 500–650° C. According to Schabereiter, Ablasser, Folkhard, Ornig and Neff [252], corrosive attack is limited to a narrow area in the heat affected zone which is heated during welding to temperatures above 1300° C. Here because of this high temperature, niobium and titanium carbides are brought back into solution in the immediate vicinity of the weld deposit. Fig. 39 (page 32) shows the portion of niobium which is re-dissolved within the base material at temperatures between 900 and 1450° C at holding times of 30 to 60 minutes and 5 seconds respectively (roughly the equivalent to the holding time in the heat affect zone during arc welding).

During the welding operation, an important portion of the niobium carbide may be brought back into solution in the HAZ. During the rapid cooling of the latter, a part of the dissolved carbon no longer precipitates as niobium carbide but stays in the oversaturated solution in the austenite. During annealing, carbon is predominantly precipitated in the temperature range of 500–650° C as chromium rich carbide $M_{23}C_6$ along the grain boundaries, producing intergranular corrosion attack due to chromium depletion in the surrounding area. Corrosion is therefore limited to zones where niobium carbide has been re-dissolved, i.e. zones in the immediate vicinity of the weld. These zones are rather narrow and this type of intergranular corrosion attack is thus called knife line corrosion. Its appears only in stabilized austenitic materials in both the HAZ of the base metal and the HAZ of the weld metal in multi-layer welds or where there are two or more parallel weld deposits.

Fig. 88 shows the appearence of some typical welded joints where intergranular corrosion has taken place according to Schabereiter [253]. After welding, these specimens were annealed in the temperature range of 550–600° C and then subjected to the Strauss test. In the unstabilized, corrosion susceptible materials (upper micrographs), intergranular corrosion attack occurs over the whole area of the weld metal (Fig. 88a) and the base metal (Fig. 88b). The specimens were taken from the top run of a single V but weld of 20 mm thickness. Intergranular corrosion attack is only visible in the deformed part of the bent specimen, where the area attacked by intergranular corrosion has opened up. In niobium stabilized weld metal (ratio of stabilization— % Nb : % C = approx. 10 : 1), intergranular knife line corrosion

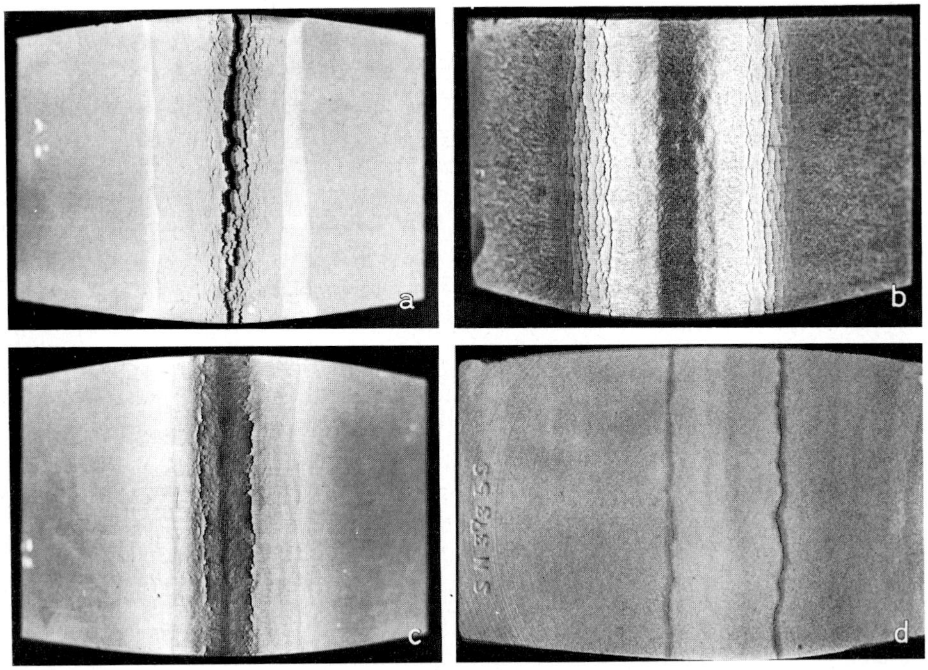

Fig. 88. Different types of intergranular corrosion in weldments after prolonged annealing in the temperature range of around 550° C obtained in the Strauss test according to Schabereiter [253]. Upper micrographs—Unstabilized chromium-nickel steels of the type AISI 308; *a* intergranular corrosion in the IC prone weld metal, *b* intergranular corrosion in the heat affected zone of the IC prone base metal. Lower micrographs—Niobium stabilized chromium-nickel steel of type AISI 347. *c* knife line corrosion in the heat affected zone between the three beads of the top layer of intergranular corrosion prone weld metal, *d* knife line corrosion in the HAZ of the intergranular corrosion prone base metal

appears in the heat affected zone between the three beads of the top layer of the weld (Fig. 88c). The bead in the middle which had been deposited last, produced knife line corrosion in the heat affected zone of the previously deposited outer beads of the top layer. If the weld metal is sufficiently resistant to intergranular corrosion attack, but the base metal is not, corrosive attack will be limited to the heat affected zone of the base metal on both sides of the weld (Fig. 88d).

The occurrence of knife line corrosion in the HAZ after annealing in the temperature range of 500–700° C can be avoided by lowering the carbon content to below 0.04% with a simultaneous overstabilization to at least one and a half times the normal stabilization ratio (with titanium $> 8 \times$ % C, with niobium $> 15 \times$ % C). Heat treatments of 10–15 hours in the temperature range of 500–600° C are most critical. If annealing takes place at higher temperatures, a smaller degree of overstabilization will normally be suffice since re-diffusion of chromium in the depleted zones is already apparent in such cases.

Fig. 89 shows the areas of grain decay caused by intergranular corrosion in austenitic weld metal with different chromium contents and varying stabilization ratios which clearly shows the influence of alloying elements. The behaviour of stabilized austenitic weld metal is mainly governed by the combined effect of the three elements carbon, niobium and chromium which are interdependent with regard to their effect on knife line corrosion. An additional factor to be considered in this context is the fact that niobium contents above 1% will increase the hot cracking sensitivity and chromium contents above 20% will increase the tendency towards sigma phase precipitation.

Therefore, an increase in these two elements is limited and for a change in content we must always consider the overall property requirement of the weld metal.

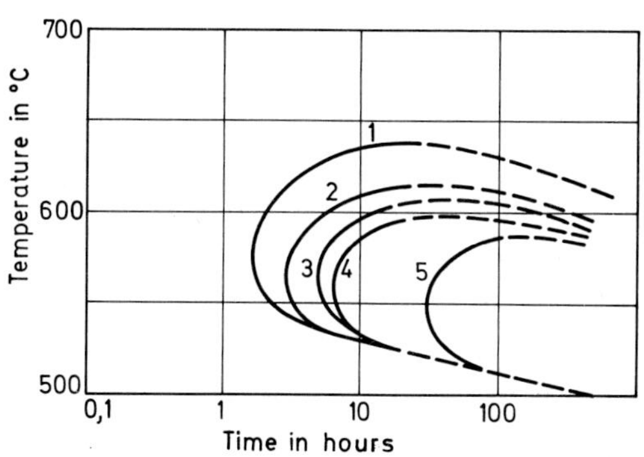

Fig. 89. TTC diagram of the beginning of intergranular corrosion attack with consideration to knife line corrosion in austenitic weld metals with different contents of chromium and niobium. Chemical composition in %: No. 1: C = 0.027, Si = 0.63, Mn = 1.55, Cr = 17.97, Ni = 9.00, Nb = –; No. 2: C = 0.027, Si = 0.74, Mn = 1.71, Cr = 20.07, Ni = 10.02, Nb = –; No. 3: C = 0.033, Si = 0.70, Mn = 1.75, Cr = 21.20, Ni = 11.08, Nb = 0.30; No. 4: C = 0.035, Si = 0.85, Mn = 1.58, Cr = 17.98, Ni = 8.97, Nb = 0.55; No. 5: C = 0.031, Si = 0.62, Mn = 1.78, Cr = 19.96, Ni = 10.14, Nb = 0.65

4.1.4.2 Increased Intergranular Corrosion in Weld Metal Due to Segregations and Phase Separations

If the alloying elements that promote carbide precipitation also show a stronger tendency towards segregation, there will be an additional increase in the danger of intergranular corrosion attack. A particular element in this respect is silicon. Fig. 85c shows that this element will shift the area of intergranular corrosion attack to much shorter times, even in case of very low carbon contents. This effect can be further increased by segregations in places with increased silicon contents which means that the area of intergranular corrosion attack may well be extended into the range of the cooling rates experienced during welding. According to Horn and Kügler [254], such materials must be welded with low heat and low interpass temperature input to avoid intergranular corrosion attack.

Austenitic weld metal often contains small proportions of delta ferrite to increase the hot cracking resistance. As shown by Fig. 85a, carbide precipitation takes place much faster in ferrite than in austenite. One ought to assume, therefore, that ferrite containing weld metal must show an increased tendency towards intergranular corrosion attack in the ferritic zones. In practical applications, however, this effect is not noticable in austenitic weld metal with low ferrite contents. It seems that the chromium enrichment of delta ferrite during solidification and δ-γ transformation of up to about 23–26% in 20/10 chromium-nickel weld metal (AWS E 308) compensates the unfavourable effect of the ferritic structure (see also Table 6, page 133).

4.1.4.3 Ferrite Track Corrosion in Austenitic Weld Metal

This rather rare type of corrosion is related by Bäumel [255] to weld metals of the type 19 Cr/8 Ni and 19 Cr/10 Ni/3 Mo and which were produced by submerged arc strip cladding operations. This type of corrosions occurs only during the annealing of unstabilized weld metal with the simultaneous appearance of a coherent netlike ferritic and relatively coarse structure. This kind of structure is normally only formed in austenitic weld metals with ferrite contents above 12–15%. According to Bäumel [255], this special type of intergranular corrosion attack is caused by chromium carbide precipitations at the ferrite-austenite grain boundaries, along which the corrosive attack may progress into the interior of the material. In niobium stabilized weld metal, ferrite track corrosion does not occur. Since weld metal alloys with about 15–20% delta ferrite of the low carbon type 23 Cr/12 Ni/3 Mo (AWS E 309 MoL) which also have a distinct ferritic network structure, show good resistance to intergranular corrosion, it appears that the ferrite track corrosion detected by Bäumel represents a very rare phenomenon which appears under very specific circumstances only. According to Thier [397], ferrite track corrosion occurs only if weld metals with delta ferrite contents above 12% (approx. 14 FN) are allowed to cool rather slowly because of a too high preheating temperature or too much heat input during welding, where the delta ferrite then partially transforms again into sigma phase and austenite. Since molybdenum containing weld metal is more susceptible in this respect, Thier recommends rapid cooling after welding.

4.2 The Precipitation of Intermetallic Phases in Stainless Steels and Weld Metals

Apart from carbides and nitrides, intermetallic phases may also precipitate in stainless steels. In general, the latter represent metallic compounds with different crystal structures to those of the base metal, the precipitation of which may have a detrimental effect on toughness as well as corrosion resistance. In the case of stainless steels which are exposed to elevated temperatures, the precipitation of such phases may diminish service life, in particular creep rupture strength, by longtime embrittlement if the operating temperature reaches into the precipitation range of these intermetallic phases. In such cases, it is important to determine the precipitation behaviour at the operating temperature over an extended period of time of up to 100 000 hours. If welded components are subjected to a heat treatment, the precipitation kinetics at the selected annealing temperature will largely determine the properties of the weldment. Precipitations may also appear, however, during cooling from the solution annealing temperature and cooling from the welding temperature in the HAZ of the base metal and the weld metal. Here, the short-time behaviour of the material as it passes through the critical temperature ranges is of decisive importance.

4.2.1 The Precipitation of Sigma Phase[2] in Austenitic Stainless Steels and Weld Metals

A general survey of the appearance of sigma phase in metallic alloys is given by Hall and Algie [256]. In stainless steels, sigma phase appears as an iron-chromium compound FeCr with a tetragonal crystal structure. In the pure iron-chromium system, it contains approx. 47% chromium. Its precipitation produces a drop in toughness and an increase in hardness. In some cases it may also increase the susceptibility of the material to corrosive attack. As indicated by the iron-chromium constitution diagram (Fig. 1, page 2), sigma phase starts to precipitate from temperatures below 820° C, with the precipitation range being greatly expanded with falling temperatures. In the ternary iron-chromium-nickel system (see Figs. 12 and 13, page 13), the area of sigma phase precipitation starts at the iron-chromium line and extends to about 40% nickel, i.e. it is constricted by nickel.

The formation of sigma phase in stainless steels is decisively influenced by the possibility of local chromium enrichment of about 20% in the steel (average content), to about twice that amount in the phase. The chromium diffusion which is necessary for this enrichment to take place also seems to determine the lower temperature limit of sigma phase precipitation. This limit often overlaps the upper limit of the area of grain decay by intergranular corrosion attack which is also governed by chromium diffusion. In stainless steels, sigma phase is normally only precipitated with chromium contents of more than 16%. Because of the higher diffusion potential of chromium, precipitation is more rapid in ferrite than in austenite. This also explains why in austenitic weld metal, which for reasons of improved crack resis-

[2] See page 3

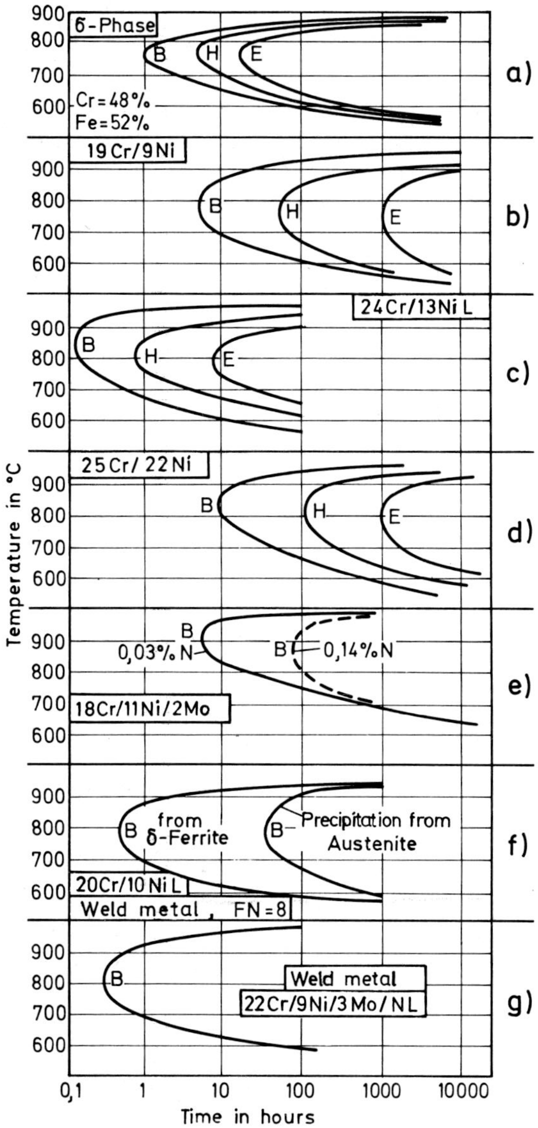

Fig. 90. TTP diagrams of sigma phase precipitation in solution annealed and quenched stainless steels and in weld metal (as welded state). $B=$ beginning of precipitation, $H=$ half way through precipitation, $E=$ end of precipitation, *a* pure sigma phase according to Baerlecken and Fabritius [257], *b* austenitic 19/8 Cr-Ni steel (AISI 304), *c* austenitic 24/13 Cr-Ni steeel (AISI 309 S) with approx. 15% delta ferrite, *d* austenitic 25/22 Cr-Ni steel (AISI 310), *e* austenitic 18/11/2 Cr-Ni-Mo steels (AISI 316) with varying nitrogen contents, *f* austenitic weld metal with low contents delta ferrite (FN = 8) of the type 20 Cr/10 Ni L (AWS E 308 L), *g* austenitic-ferritic duplex weld metal with 24—27% delta ferrite of the type 22 Cr/9 Ni/3 Mo LN.

tance often shows ferrite contents of 5–12%, sigma phase is predominantly precipitated in the dispersed ferrite crystals (see Fig. 90f).

After prolonged annealing at 650 and 700° C, Vitek and David [448] determined in the steel grades 308 and 308 CRE (Controlled Residual Elements) similar chromium contents (30–36%) in the precipitated sigma phase as in the ferrite from which it was precipitated. They draw the conclusion that sigma phase transformation does not require any sizable chromium diffusion in ferrite containing austenitic metals. Therefore, the latter is not significant. According to these authors, sigma phase transformation is mainly governed by the formation of crystallization nuclei. Quote: "The new results indicate that diffusion necessary for the chromium enrichment found in sigma phase does not play a major rate-limiting role during nucleation. Rather it is suggested that the sigma phase transformation is nucleation controlled and that structural requirements for establishing sigma phase nuclei are of primary importance."

There are other elements such as molybdenum, nickel, manganese, silicon, titanium and phosphorus which may also be incorporated into sigma phase, therefore the latter's actual chemical composition in steels may differ considerably from the theoretical alloying content according to the iron-chromium constitution diagram (Fig. 1).

Baerlecken and Fabritius [257] investigated the transformation kinetics of pure sigma phase in an alloy with 48% chromium and 52% iron after a solution annealing treatment at 1000° C. Fig. 90a shows the respective TTP diagram. The transformation maximum is located at 770° C with the start of transformation taking place after a period of about 60 minutes. The difference in time between the beginning and the end of transformation is considerable, being in the range of 10 to 100 hours.

In stainless steels, the area of sigma phase precipitation as well as the transformation kinetics are influenced by the alloying additions. Hull [55] investigated a great number of standard carbon and nitrogen free iron-chromium-nickel alloys with regard to the influence of alloying additions upon the precipitation of sigma and Chi phase at a temperature of 816° C and precipitation times of up to 1000 hours. From the data obtained, he calculated the chromium equivalent for the material embrittlement at 816° C, from which the influence of alloying elements can be determined as follows:

$$\begin{aligned} Cr_{\text{Äqu}} = {} & \%\,Cr + 0.31 \times \%\,Mn + 1.76 \times \%\,Mo + 1.70 \times \%\,Nb + 1.58 \times \%\,Si + \\ & + 2.44 \times \%\,Ti + 1.22 \times \%\,Ta + 2.02 \times \%\,V + 0.97 \times \%\,W - 0.266 \times \%\,Ni - \\ & - 0.177 \times \%\,Co. \end{aligned}$$

This formula indicates that elements with positive signs speed up embrittlement at 816° C by the precipitation of sigma phase. Only nickel and cobalt slow it down. The elements carbon and nitrogen are not included in the investigation of Hull [55]. According to Wiegand and Doruk [33], the presence of carbon will considerably slow down sigma phase precipitation, which can start only, if most of the carbon which is retained in the austenite as a result of the solution annealing treatment, has been precipitated as the carbide $M_{23}C_6$. A possible explanation for this phenomenon may be the fact that the carbon solubility of sigma phase is rather small. Therefore, sigma phase can only be formed from austenite crystals if the latter no longer contain any dissolved carbon. If carbon is precipitated as carbide, e.g. $M_{23}C_6$, the surrounding area of these carbides will also be depleted of chromium and the chromium content of the carbon-free zones so created may drop below the precipitation limit for sigma phase which is located at about 16% chromium. This means a further slowdown in the precipitation of sigma phase. Precipitation can only start if the chromium depletion has been equalized by chromium diffusion from the surrounding areas. This process may also serve to explain why the area of sigma phase precipitation borders on the upper boundary of the area of intergranular corrosion attack.

If carbon is retained by titanium or niobium as a stable carbide, the effect of carbon on sigma phase precipitation is largely lost. In Nb or Ti containing stabilized steels, therefore, carbon has only a weak slowing down effect on the precipitation of sigma phase and stabilized steels show a somewhat higher tendency towards embrittlement. L (ELC) steels with low carbon contents below 0.03% in which the effect of carbon is rather minor, also show a similar behaviour.

Fig. 90b–d shows a number of TTP diagrams of the sigma phase precipitation of solution annealed chromium-nickel steels. In comparison to steel type 19 Cr/9 Ni (AISI 304) (Fig. 90b) with a purely austenitic structure produced by solution annealing, steel type 24 Cr/13 Ni L (AISI 309 S) shows a much stronger tendency towards the precipitation of sigma phase (Fig. 90c). Three major reasons for this difference are: 1) a higher chromium content, 2) a lower carbon content and 3) the presence of about 15% delta ferrite in the structure. Because of increased carbon (C = 0.11%) and nickel contents and the fully austenitic structure of steel 25 Cr/ 22 Ni (AISI 310, Fig. 90d), sigma phase precipitation is slowed down to such an extent that it is shifted to considerably longer times.

The effect of nitrogen is similar to that of carbon (Fig. 90e). According to Thier, Bäumel and Schmidtmann [41], the precipitation of all phases not capable of dissolving nitrogen is shifted by nitrogen towards longer times. According to Brandis, Heimann and Schmidtmann [258], the upper phase boundary is also lowered. The matrix must first be "freed" of nitrogen for sigma phase precipitation to start. In the case of nitrogen, however, its precipitation is considerably more difficult than with carbon which precipitates quite easily as carbide $M_{23}C_6$. Nitrogen has a much higher solubility in austenite than carbon (see Fig. 23, page 21). Precipitation is often possible only in the form of complex nitrides of the type M_2N which according to Freißmuth [259] will be precipitated at 0.15% nitrogen only after rather long periods of time in the range of 70 hours at a temperature of 800° C. In the presence of niobium, nitrogen may also precipitate as z phase (nitrogen containing chromium-iron niobide) or as niobium carbonitride. Chromium nitride Cr_2N will only precipitate after prolonged aging times. If the precipitation of nitrogen is not possible, it would be necessary to displace it from the austenite crystals by diffusion and enrich it in other places. Because of the good solubility of nitrogen, the latter process is only possible with a high precipitation potential towards the formation of sigma phase. Such a potential normally only exists when there is a greater surplus of elements which promote sigma phase precipitation (see also the formula provided by Hull, page 127). This means that nitrogen shifts the beginning of sigma phase formation to higher chromium contents and to considerably longer times. The precipitation inhibiting effect of nitrogen is only slightly reduced by stabilizing elements such as niobium, because they are first retained by carbon and therefore can only add to the formation of nitrogen containing niobium phases in the event of overstabilization. Fig. 90e shows the effect that an addition of 0.14% nitrogen has on the beginning of sigma phase precipitation in a low carbon austenitic chromium-nickel-molybdenum steel (AISI 316).

Molybdenum exerts a relatively strong influence on the precipitation of sigma phase. The sigma phase range is not only shifted to lower chromium contents but also to higher temperatures. This means that molybdenum containing steels must be solution annealed at higher temperatures (normally 1100° C). As shown in Fig. 34 (page 28), the precipitation range of sigma phase is also shifted to considerably higher nickel contents which means that molybdenum containing steels require more nickel to obtain freedom from precipitations. With higher molybdenum contents, freedom from precipitations can also be achieved by the addition of nitrogen. Molybdenum speeds up the precipitation of sigma phase. This applies in particular to ferritic structures. Kügler [46] has found that sigma phase precipitation leads in certain corrosive media to an increased removal of ferrite which can also be detected in the Huey test. Therefore, molybdenum containing steels are sometimes alloyed with nitrogen because of the resulting twin effect, i.e. a fully austenitic structure and a slowdown of sigma phase precipitation. Molybdenum possesses good solubility in sigma phase and molybdenum is often found in the sigma phase of molybdenum alloyed steels. With molybdenum contents beyond 2.5% other molybdenum containing intermetallic phases may also appear which influence or block the precipitation of sigma phase. They are discussed in detail in section 4.2.4.

According to Schüller [62], in stainless austenitic steels alloying additions of silicon will shift the range of sigma phase precipitation to lower chromium contents and to higher temperatures. In the case of silicon containing austenitic steels, therefore, solution annealing temperatures must be raised (see Fig. 43). With increasing silicon content, the beginning of sigma phase precipitation is shifted according to Horn and Kügler [63] to shorter times and extends over a wider range of temperatures. Sigma phase is only precipitated, however, if the carbon dissolved in the austenite has dropped to very low values. But silicon will also significantly speed up carbide precipitation and this effect also facilitates sigma phase precipitation. With silicon contents above 2.5% and with 18% chromium, sigma phase precipitation already starts after relatively short periods of time. With the addition of nitrogen, this can be slowed down again.

According to Hull [55], alloying additions of titanium and niobium basically promote the precipitation of sigma phase. Because of the fact, however, that initially they are retained as stable carbides by carbon, this effect is only achieved in the case of a surplus over the normal stabilization ratio (with titanium $> 5 \times \%$ C, with niobium $> 10 \times \%$ C). Since an excess of niobium, for example, will also induce the formation of other phases such as Laves Fe_2Nb, the direct effect on the precipitation of sigma phase is rather small. Yet both elements indirectly speed up the precipitation of sigma phase because of their retention of carbon. In other words, with regard to sigma phase precipitation, stabilized austenitic steels behave in a similar way to steels with very low carbon contents.

4.2.2 The Precipitation of Sigma Phase in Ferritic and Austenitic-Ferritic Stainless Steels

According to Bungardt, Borchers and Kölsch [260], sigma phase will appear in ferritic chromium steels containing about 18% chromium only after approx. 10^3—10^4 hours in the temperature range around 550° C. With rising chromium contents, the start of precipitation is shifted to shorter times and higher temperatures. Additions of silicon and in particular of molybdenum will speed up sigma phase precipitation in ferritic chromium steels much more than in austenitic steels. According to Bäumel [232], sigma phase will precipitate in a ferritic chromium steel of the type 17 Cr/2 Mo (AISI 430 Mo) at 600° C after approx. 200 hours in combination with a great increase in the corrosion rate when tested in a strong solution of boiling nitric acid (Huey test). Similar to austenitic steels, sigma phase precipitation in ferritic steels will only start after the completion of a previous carbide precipitation in the ferritic matrix which would then be practically free of carbon.

Of special importance is the sigma phase precipitation in low carbon austenitic-ferritic duplex steels with 22—26% chromium which, when compared with purely ferritic steels, are distinguished by higher toughness, better weldability and lower susceptibility to coarse grain formation (for chemical composition see Table 4, page 118 and Table 16, page 187). Because of the relatively high chromium contents and additional alloying with molybdenum in amounts of between 1.5 and 3%, sigma phase precipitation in these steels starts after only short annealing times in the range of about 10 to 30 minutes. In duplex steels, the effect of sigma phase on toughness may be considerably stronger than in austenitic steels. According to Norström, Petterson and Nordin [261], the presence of 1% sigma phase in the structure is already enough to reduce the impact energy values by about 50%. A complete embrittlement will occur at about 10% sigma phase.

In Fig. 87 (page 119) steel 1 shows the TTP diagram according to Herbsleb and Schwaab [245] of the beginning of sigma phase precipitation in a steel of the type 22 Cr/5 Ni/3 Mo LN (UNS S 31803). According to Schwaab [262], fine precipitations of chromium nitride Cr_2N and somewhat coarser globular particles of carbide containing Chi phase are formed at the ferrite-austenite grain boundaries after an annealing time of only about 2 minutes. Because of the low carbon content, small amounts of $M_{23}C_6$ will also be precipitated simultaneously. All three phases are formed practically at the same time and it is not possible to distinguish separate precipitation ranges. Precipitation of sigma phase starts after the precipitation of carbon from the matrix after about 20 minutes with a maximum at between 800 and 900° C. When testing in a strong solution of boiling nitric acid (Huey test), an increase in the corrosion rate already appears with the precipitation of the chi phase. Fig. 87 (page 119), steel 2 shows according to Jolly and Hochmann [198] the transformation behaviour of a steel of similar analysis but with a lower nitrogen content. Here, the precipitation of sigma phase already starts after shorter annealing times.

Due to the rapid cooling rates experienced during electric arc welding, no sigma phase is normally precipitated in the HAZ of duplex steels and duplex weld metals (Fig. 90g). If the steel and the weld metal are also alloyed with nitrogen, which is normally the case, there is a lesser tendency towards sigma phase precipitation and also little susceptibility to coarse grain formation, therefore a greater drop in toughness in the heat affected zone can thus be avoided. Due to the rapid sigma phase precipitation and 475° C embrittlement, as well as the unfavourable influence on corrosion resistance properties, a heat treatment below the solution annealing temperature is not recommended and should be avoided as far as possible for the duplex steels (see also supplementary literature [246–248, 261, 262, 448, 469, 470, 471, 476]).

4.2.3 The Precipitation of Sigma Phase in Austenitic, Austenitic-Ferritic and Ferritic Stainless Steel Weld Metals

Fully austenitic weld metal without a delta ferrite content performs in a similar way with regard to the precipitation of sigma phase as a steel of similar chemical composition. This is not true, however, for austenitic weld metal in the as welded condition which—for reasons of improved hot cracking resistance—contains small amounts of residual delta ferrite (FN approx. 5–10). During the solidification and the δ-γ transformation, delta ferrite is enriched with chromium and depleted of nickel. In 20/10 chromium-nickel weld metals, the chromium contents of the delta ferrite are normally between 22 and 26% and the nickel contents are between 6 and 7% (see Table 6, page 133). This means that delta ferrite shows a higher chromium content than the average composition of the weld metal and the austenite. This fact combined with the rapid precipitation of $M_{23}C_6$ in delta ferrite (see Fig. 85a) and the easier diffusion of chromium in ferrite lead to a quicker formation of sigma phase in delta ferrite as compared to austenite. Fig. 90f (page 126) shows the TTP diagram of sigma phase precipitation in a ferrite containing austenitic weld metal of the type 20 Cr/10 Ni L with 8 FN (AWS E 308 L). The precipitation of sigma phase in delta ferrite takes place much more rapidly than in the austenitic matrix. According to Schabereiter, Folkhard, Ablasser, Ornig and Neff [252], the precipitation of sigma phase in ferrite containing austenitic 20/10 chromium-nickel weld metal in the temperature range of between 700–800° C can be described by the following schematic representation:

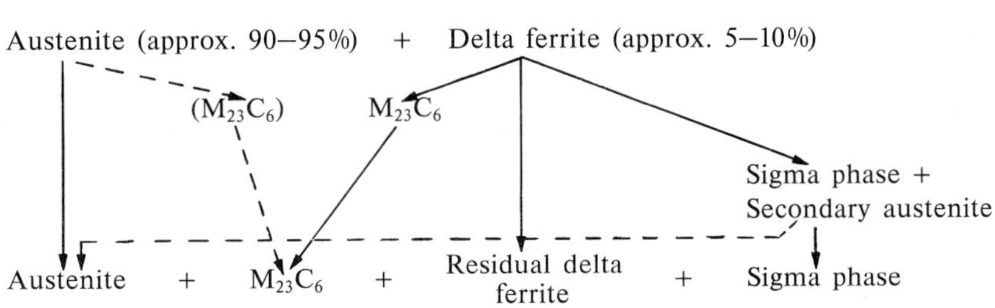

After precipitation of the carbide $M_{23}C_6$, sigma phase precipitation starts in the delta ferrite according to Fig. 90f. Because of the high chromium content and the low nickel content of the precipitated sigma phase, the chromium content of delta ferrite is reduced and the nickel content is increased. This means that part of the delta ferrite is thus transformed into secondary austenite which will normally be deposited at the boundaries of the austenitic matrix. With carbon contents above

0.05%, secondary austenite may already be formed in considerable amounts from the delta ferrite during $M_{23}C_6$ precipitation, because the latter carbide will deplete the surrounding area of chromium but does not absorb nickel. Depending upon the chemical composition, delta ferrite is transformed completely or only partially into sigma phase, with a small amount of residual delta ferrite remaining in most cases. With longer annealing times, sigma phase is also precipitated from austenite (Fig. 90f, page 126). Finally, the structure consists of four phases, i. e. austenite, carbide $M_{23}C_6$, sigma phase and residual delta ferrite. Sigma phase is not ferro-magnetic and its precipitation by way of delta ferrite decay can be measured with the aid of a magnetic ferrite measuring device. In weld metal grades with a low delta ferrite content between FN 5—FN 10, sigma phase will only precipitate during heat treatment between 600 and 900° C from delta ferrite, because the precipitation of sigma phase from austenite is shifted to longer times.

Both embrittlement and the reduction of the CVN-impact energy of the weld metal during heat treatment are dependent upon the chemical composition of the weld metal, on ferrite content, annealing temperature and annealing time. Fig. 91 shows some examples of the influence on impact energy values and ferrite contents for different weld metals after ten hours annealing in the temperature range between 500—1100° C with subsequent cooling in air. Fig. 91a shows the behaviour of unstabilized (No. 1 and 2) and niobium stabilized (No. 3 and 4) chromium-nickel

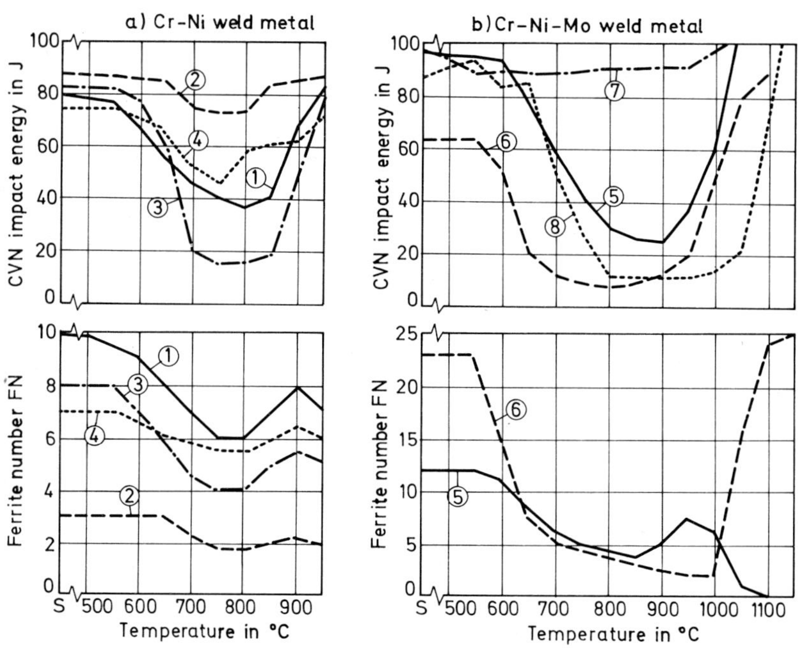

Fig. 91. Embrittlement and transformation of delta ferrite through precipitation of intermetallic phases in austenitic stainless steel weld metals after ten hours of annealing in the temperature range of 550—1000° C followed by cooling in air according to Schabereiter and Rabensteiner [263—266] with supplements. *W* = as welded condition. The chemical composition and the delta ferrite content of the above weld metal grades may be taken from Table 5 (page 133)

weld metal. The chemical composition and the ferrite contents as well as the ANSI/AWS classification according to ANSI/AWS A5.4–81 may be taken from Table 5. As a supplement, Table 6 shows the chromium and nickel contents of the structural constituents of delta ferrite and austenite for weld metals 1–4 as determined by electron probe microanalysis. A comparison with the average analysis of the weld metal shows clearly the chromium enrichment and nickel depletion of delta ferrite and the reverse tendency in austenite.

The diminishing effect on toughness produced by the precipitation of sigma phase from delta ferrite in the temperature range between 600 and 900° C after ten hours of annealing is clearly apparent in Fig. 91a. Niobium alloyed weld metal (No. 3 and 4) shows a somewhat less favourable behaviour. With a reduction in both the chromium content and the delta ferrite content (No. 2) the drop in toughness produced by the annealing treatment can be slowed down. With an annealing time of ten hours, precipitation of sigma phase takes place exclusively from delta ferrite which is clearly shown at a microscopic examination. The transformation of delta ferrite into sigmma phase (lower diagram in Fig. 91a) is roughly proportional to the reduction in toughness of the weld metal. With the progressive decay of delta ferrite and its transformation into sigma phase, there is a similar progression in the de-

Table 5. *Chemical composition and delta ferrite content of weld metal grades 1–8, of which the CVN-impact energy values and delta ferrite content after annealing are shown in Fig. 91*

Weld metal No.	AWS classi-fication	Chemical composition in %									Delta ferrite number FN
		C	Si	Mn	Cr	Mo	Ni	N	Nb	Others	
1	E 308 L	0.027	0.74	1.71	20.07	–	10.02	0.047	–	–	10
2	E 308 L	0.028	0.65	1.73	19.92	–	12.67	0.056	–	–	3
3	E 347 L	0.033	0.62	1.68	19.90	–	10.10	0.045	0.55	–	8
4	E 347 L	0.035	0.85	1.58	18.03	–	9.02	0.052	0.55	–	7
5	E 316 L	0.032	0.76	0.75	19.22	2.49	12.32	0.043	–	–	12
6	E 309 MoL	0.018	0.90	0.75	22.50	2.50	11.83	0.059	–	–	23
7	–	0.038	0.25	4.50	18.45	2.20	15.52	0.047	–	–	0
8	–	0.031	0.38	5.02	19.37	6.44	24.66	0.150	–	Cu = 1.48	0

Table 6. *Chromium and nickel contents of weld metals and of respective delta ferrite and austenite contents (for overall analysis of weld metals see Table 5)*

Weld metal No.	FN	Chemical composition in %					
		Weld metal average		Delta ferrite		Austenite	
		Cr	Ni	Cr	Ni	Cr	Ni
1	10	20.07	10.02	23.6	6.4	19.9	10.6
2	3	19.92	12.67	23.6	8.8	19.8	13.1
3	8	19.90	10.10	23.8	6.3	19.8	11.2
4	7	18.03	9.02	22.1	6.1	17.8	10.0

crease in impact energy values. A sigma phase content of 3–4% is already enough to produce a considerable embrittlement of the austenitic weld metal. At the annealing temperature of 900° C, only little delta ferrite will be transformed into sigma phase, because the rate of sigma phase precipitation has already slowed down considerably at this temperature. The re-dissolution of both the sigma phase and the delta ferrite in austenite starts at about 950° C.

Fig. 91b shows the behaviour of a ferrite containing (No. 5 and 6) and a fully austenitic (No. 7 and 8) chromium-nickel-molybdenum weld metal. With molybdenum additions, the sigma phase range is expanded to higher temperatures. In molybdenum-free weld metal, sigma phase precipitation is practically halted at 950° C, whereas in weld metals with 2.5% molybdenum it continues up to about 1000–1050° C. Therefore, molybdenum containing weld metals and steels must be solution annealed at higher temperatures than molybdenum-free ones. The weld metal No. 6 with the highest chromium and delta ferrite content shows the strongest tendency towards embrittlement, with embrittlement already beginning at 600° C. However at temperatures of 1100° C the delta ferrite content rises steeply again, because due to the high chromium content the alloy reaches into the (δ–γ) range which means a new formation of delta ferrite. The fully austenitic weld metal No. 7 with about 18.5% chromium and 2.2% molybdenum does not show any embrittlement at all after 10 hours annealing because it is free from delta ferrite. According to Schabereiter [265], the precipitation of sigma phase from the fully austenitic structure starts at 750° C only after about 250 hours and finishes after 2000 hours, with impact energy values dropping by about 50%. If the molybdenum content is increased to 6.5% (weld metal No. 8), we encounter, inspite of the fully austenitic structure in the as welded state, a severe embrittlement after anealing in the temperature range of 700–1100° C, caused by the precipitation of molybdenum containing intermetallic phases. According to Rabensteiner and Schabereiter [266], their re-dissolution is only possible at temperatures above 1100° C. The effect of pronlonged post weld heat treatment of up to 5000 hours on the transformation of delta ferrite into sigma and chi phase in austenitic ferrite containing ER 316L stainless steel weld metal in the temperature range of 600–800° C is discussed by Gill et al (485).

Low carbon austenitic-ferritic duplex weld metal with 22–25% chromium shows a relatively strong tendency towards the precipitation of sigma phase because of the increased chromium content. For reasons of improved toughness, the chemical composition of weld metals will often deviate slightly according to Perteneder, Tösch, Schabereiter and Rabensteiner [267] from those of the parent steel grades. The nickel content is increased to about 7–10%, the nitrogen contents are normally in the range of around 0.15%. The weld metal delta ferrite content of approx. 25–40% is lower than that of the duplex steels which is generally around 50%. Fig. 90g (page 126) shows the beginning of phase precipitation in a weld metal of the type 22 Cr/9 Ni/3 Mo NL. The maximum of this precipitation is located at 800° C and in the range of 15–20 minutes (for more details see Fig. 87, page 119 and chapter 8).

According to Beckitt [268], the precipitation of sigma phase from delta ferrite takes place in duplex steels with 25% chromium and 8% nickel in such a way that it is always combined with the formation of secondary austenite. The latter is already formed during the precipitation of carbide $M_{23}C_6$ from delta ferrite due to the chromium depletion and nickel enrichment of neighbouring zones at the austenite-ferrite grain boundaries. During the subsequent sigma phase precipitation, this process is repeated because sigma phase—due to its high chromium content—depletes the surrounding area of relatively high amounts of chromium and rather low nickel contents, because of the low nickel solubility of sigma phase which means that the surrounding area is enriched by nickel. Both the precipitation of $M_{23}C_6$ and the precipitation of sigma phase are combined with an increase in austenite content.

The sigma phase precipitation of purely ferritic weld metal is similar to that of a steel with roughly the same chemical compositon. In low carbon martensitic 13/4 chromium-nickel weld metal, there is practically no sigma phase at all because of the low chromium content. In 17% chromium weld metal with additions of 1—2% molybdenum, intermetallic phases only precipitate after prolonged times of normally more than 100 hours at temperatures of about 700° C, with carbon and molybdenum contents being the major parameter for the precipitation kinetics.

4.2.4 The Precipitation of Intermetallic Phases in Chromium-Nickel-Molybdenum Stainless Steels and Weld Metals

In the iron-molybdenum constitution diagram (Fig. 29), three phases appear between 45 and 55% molybdenum. The Laves phase Fe_2Mo with approx. 45% molybdenum, also known as η (eta) phase, reaches down to relatively low molybdenum contents. In the iron-chromium-nickel-molybdenum system we also find the χ (chi) phase ($Fe_{36}Cr_{12}Mo_{10}$) and the sigma phase. The chi, sigma and Laves phases (Fe_2Mo) are also found in stainless chromium-nickel-molybdenum steels. The precipitation of these phases is additionally influenced by the constituents carbon and nitrogen and precipitation kinetics can thus be represented by way of TTP diagrams of the pertinent steel grades. Regarding the influence of carbon and nitrogen, it can be noted that the precipitation kinetics of intermetallic phases is largely determined by the latters dissolution potential for these constituents. At elevated temperatures, austenitic stainless steels and weld metals show a relatively high solubility for carbon and nitrogen (Figs. 18 and 23). With a solution annealing treatment at 1050—1150° C it is normally possible to bring both elements completely into solution. By rapid cooling to ambient temperature, this state can be maintained in steels, i. e. austenitic stainless steels will generally contain constituents such as carbon and nitrogen in the state of supercooled dissolution. If an intermetallic phase shows now a good solubility for one or both of these constituents, it may absorb them in the course of precipitation without difficulty. This means that the precipitation is not slowed down. If, however, the solubility of an intermetallic phase for carbon or nitrogen is low or practically zero, these constituents which are dissolved in the matrix must first be precipitated and removed from the matrix as carbides, nitrides or carbonitrides, in order for the intermetallic phase to precipitate. In other words, precipitation will be slowed down or even prevented until the matrix is more or less free of dissolved carbon and/or nitrogen.

According to Thier, Bäumel and Schmidtmann [41], the precipitation of all phases which are unable to dissolve nitrogen, such as $M_{23}C_6$, sigma, chi and Laves phase, can be shifted in austenitic steels by the addition of nitrogen to longer periods of time. The precipitation of the carbide M_6C, however, which is able to dissolve nitrogen is promoted. According to Kügler [218], carbon slows down the precipitation of all intermetallic phases possessing only low or no carbon solubility at all, e.g. sigma and Laves phase Fe_2Mo, but will favour the precipitation of chi phase which shows good solubility for carbon. Thus, nitrogen appears to be a more

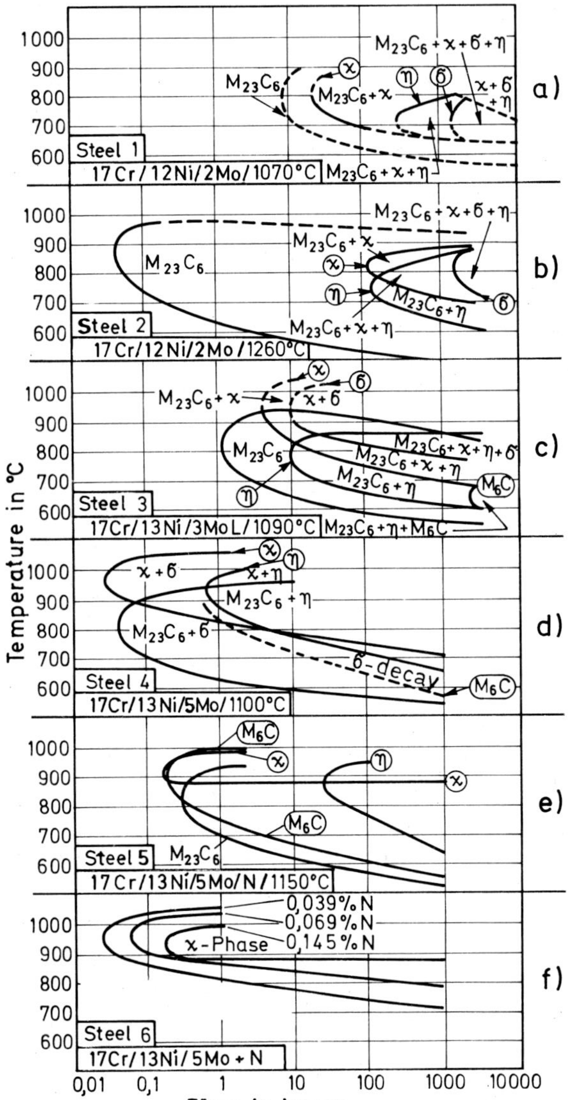

Fig. 92. TTP diagrams of the beginning of phase precipitations in molybdenum containing austenitic chromium-nickel steels. The pertinent chemical compositon may be taken from Table 7, page 137. σ = sigma phase, χ = chi phase, η = Laves phase Fe_2Mo. *a* to Wiegand and Doruk [33], *b* and *c* to Weiss and Stickler [45], *d–f* to Thier, Bäumel and Schmidtmann [41]. *f* shows the influence of increasing nitrogen contents of 0.039 to 0.247% on the beginning of chi phase precipitation in steel of type 17 Cr/13 Ni/5 Mo (at 0.247% N no chi phase is detectable)

Table 7. *Chemical composition of Steel 1–6, the TTP diagrams of which are shown in Fig. 92*

Steel No.	AISI classi-fication	Chemical composition in %						
		C	Si	Mn	Cr	Mo	Ni	N
1	316	0.05	0.48	1.21	17.12	2.07	11.56	–
2	316	0.066	0.21	1.57	17.4	2.05	12.3	–
3	316 L	0.023	0.73	1.74	17.3	2.66	13.1	–
4	–	0.048	0.64	0.80	16.80	4.89	13.55	0.039
5	–	0.050	0.49	0.67	17.05	4.73	13.45	0.145
6	–	0.048–0.050	0.47–0.64	0.67–0.80	16.80–17.25	4.69–4.89	13.35–13.55	0.039–0.247

Table 8. *Chemical composition of precipitated Mo-containing phases in steel 3 of Fig. 92 according to Weiss and Stickler* [45]

Phase	Chemical composition in %, determined by energy dispersion X-ray spectrography			
	Fe	Cr	Mo	Ni
$M_{23}C_6$	18	63	14	5
Sigma	55	29	11	5
Chi	52	21	22	5
Laves	38	11	45	6
Matrix Steel 3 average	64	17.3	2.66	13.1

potent "phase inhibitor" since it also inhibits the precipitation of chi phase. If both, carbon and nitrogen are present in sufficient amounts in stainless chromium-nickel-molybdenum steels, the precipitation of all intermetallic phases is slowed down and only the precipitation of carbide M_6C is promoted since the latter is able to dissolve both carbon and nitrogen. With the presence, however, of niobium or titanium, the slowing down effect of carbon on phase precipitation will diminish in proportion to that amount of carbon which is retained as stable niobium or titanium carbide.

Fig. 92 shows TTP diagrams of stainless austenitic chromium-nickel steel with molybdenum contents of about 2–5% which have been solution annealed and quenched in water. The chemical composition may be taken from Table 7, page 137. Average values of the chemical composition of precipitated Mo-containing phases for steel No. 3 shown in Fig. 92 are given in Table 8, page 137 according to Weiss and Stickler [45]. It must be noted, however, that the composition of the phases is influenced by the chemical composition of the base metal as well as by the time and temperature of the annealing treatment. For this reason, the values listed should only be seen as a rough survey of the alloying contents of the molybdenum containing phases.

Fig. 92a shows the precipitation behaviour of steel 1 of the type 17 Cr/12 Ni/2 Mo (AISI 316) according to Wiegand and Doruk [33]. The precipitation of the carbide $M_{23}C_6$ is quickly followed by the precipitation of chi phase. This is capable of dissolving carbon and therefore not dependent upon a previous complete precipitation of carbon from the matrix, i.e. it can precipitate simultaneously with the carbide. The precipitation of chi phase is followed by the precipitation of Laves phase Fe_2Mo and finally sigma phase.

Fig. 92b shows the precipitation behaviour of a steel (AISI 316) of similar analysis according to Weiss and Stickler [45] which has been brought into the $(\delta+\gamma)$ area, (i.e. which contains delta ferrite) by an increase in the solution annealing temperature to 1260° C. The $M_{23}C_6$ precipitation in ferrite takes place much more quickly than in the fully austenitic steel 1. The precipitation of the other phases, however, are similar to those of steel 1 without delta ferrite.

Fig. 92c shows the effect of a reduction in the carbon content to 0.023% and an increase in the molybdenum content to 2.66% (steel 3 [AISI 316 L] in Table 7, page 137) according to Weiss and Stickler [45]. Because of the higher molybdenum content and the reduced carbon content, intermetallic phase precipitations are shifted to shorter periods of time as compared to steel 1. The structure of this steel is fully austenitic which means that there is no acceleration of $M_{23}C_6$ precipitation by delta ferrite. After prolonged annealing times, carbide M_6C is also precipitated. Table 8, page 137 states average values of the chemical composition of precipitated phases.

According to Cieslak, Ritter and Savage (439) chi phase formation may even take place in the as welded condition in GTA weld metals of type ER 316 as a result of extremly strong segregations causing very high molybdenum contents in the eutectic delta ferrite in welds solidifying either as primary delta ferrite or as primary austenite (solidification mode through the three phase sector $(L+\delta+\gamma)$ in the ternary iron-chromium-nickel constitution diagram, see also Fig. 6 and 7, page 8). This type of chi phase forms as a eutectic residual melt constituent at the solidification interfaces between primary delta ferrite crystals and primary austenite crystals (see also sections 2.2 and 2.3).

Fig. 92d shows the effect of an increase in molybdenum content to about 5% (steel 4) according to Thier, Bäumel and Schmidtmann [41]. This increase has a marked influence on the precipitation behaviour of such a steel. After solution annealing and in spite of quenching in water, sigma phase is already present in the structure, the precipitation of which is accelerated extraordinarily by the higher molybdenum content (see also the isothermal profile at 816° C in Fig. 34, page 28). This steel is no longer fully austenitic. It primarily solidifies to delta ferrite, with the residual ferrite left over from δ–γ transformation being no longer completely dissolved during solution annealing. During quenching in water, the residual delta ferrite already precipitates sigma phase and is transformed in the course of this process, because of chromium depletion and nickel enrichment, into austenite. The sigma phase which has precipitated from the ferrite is no longer stable in the now fully austenitic matrix and will transform according to Thier [38] during prolonged annealing in the temperature range between 550 and 800° C into M_6C and austenite, and in the range between 800 and 1000° C into chi phase with the simultaneous formation of Laves phase Fe_2Mo (see dashed line in Fig. 92d "sigma decay"). The precipitation of $M_{23}C_6$ starts at 750° C after only about 3 minutes. Because of the increased molybdenum content, the precipitation of chi and Laves phase is also greatly accelerated.

As mentioned earlier, nitrogen has a strong inhibiting effect on the precipitation of all phases except carbide M_6C which is capable of dissolving nitrogen. Fig. 92e (steel 5) shows the influence of a nitrogen addition of 0.145% according to Thier, Bäumel and Schmidtmann [41] as compared to a similar steel 4 with almost the same chemical analysis but with the exception of the nitrogen content. The effect of nitrogen is here a multiple one. Because of its austenitizing effect, the steel becomes fully austenitic, and sigma phase precipitation from delta ferrite is no longer possible. With the exception of M_6C, the precipitation of other phases is also considerably slowed down by nitrogen.

The slowing down effect of increasing nitrogen contents on the precipitation of chi phase is shown in Fig. 92f for steel 6 according to Thier, Bäumel and Schmidtmann [41]. At 0.247% nitrogen, the precipitation of chi phase is completely suppressed. The inhibiting effect of nitrogen on the precipitation of $M_{23}C_6$ for the same type of steel can be taken from Fig. 85e. From Fig. 92d–f, it appears that an austenitic chromium-nickel steel with about 5% molybdenum can only be made precipitation free by alloying additions of nitrogen.

Stainless molybdenum alloyed chromium-nickel weld metal shows a similar behaviour as the steels discussed above. With alloying additions of 2–3% molybdenum it is still possible to use delta ferrite containing weld metal which shows very good hot cracking resistance. To obtain resistance to intergranular corrosion as well, requires either a reduction in the carbon content or the use of niobium stabilized weld metal. Molybdenum extends the critical temperature range to higher temperatures in the range of 550–1050° C, therefore heat treatments in this temperature range are not recommended. For reasons of corrosion resistance a ferrite free, fully austenitic weld metal structure is sometimes demanded. In such cases, however, attention must be paid not only to the absence of ferrite but for reasons of corrosion resistance, the precipitation of sigma and chi phase must also be avoided.

To obtain, for example, good pitting resistance in chloride containing media, austenitic weld metal with 5% molybdenum must be fully austenitic. Similar to the steel discussed above, alloying of filler metals with nitrogen is mandatory to obtain precipitation free weld metals during welding operations. Heat treatment in the critical temperature range of 550—1080° C should be avoided.

The behaviour of ferritic and austenitic-ferritic molybdenum alloyed steels and weld metals is mainly determined by $M_{23}C_6$ and sigma phase precipitation. In some cases, it may also be influenced by the precipitation of chi phase, in particular in low carbon duplex steels with molybdenum contents of between 1 and 3%. The precipitation behaviour of these materials has already been discussed together with $M_{23}C_6$ and sigma phase precipitation in sections 4.1.3, 4.1.4, 4.2.2 and in Fig. 87 (page 119). Furthermore chapter 8 deals with the practical behaviour of stainless duplex steels.

4.3 475° C Embrittlement in Stainless Steels and Stainless Steel Weld Metals

The 475° C embrittlement only occurs during the annealing of stainless steels and stainless steel weld metals in the temperature range of around 475° C if they contain ferrite. It is limited, in other words, to ferritic and austenitic-ferritic steels and to austenitic weld metals which contain delta ferrite. With a chromium content of 12%, it takes very long annealing times in the range of 10^5 hours to produce embrittlement. With increasing chromium contents, however, 475° C embrittlement is greatly accelerated. According to investigations by Baerlecken and Fabritius [269], De Nys and Gielen [270], Ettwig and Pepperhoff [271], Albritton and Lewis [272], and Solomon and Levinson [273], 475° C embrittlement is induced by segregation processes occurring in the ferrite, which segregate into a ferro-magnetic component which is rich in iron and a chromium rich para-magnetic component with approx. 80% chromium neither being detectable, by the optical microscope. In the binary iron-chromium system (Fig. 1), these two types of ferrite are indicated below 500° C as α and α'. Their composition is roughly in accordance with the outer boundaries of the $(\alpha + \alpha')$ area. The dotted line represents the upper temperature limit of the 475° C embrittlement. According to Heger [274], both time and temperature of maximum embrittlement are influenced by alloying elements. Additions of silicon and aluminium as well as increasing contents of chromium and molybdenum will increase and accelerate maximum embrittlement. Carbon reduces the tendency towards embrittlement by the formation of chromium carbide with resulting chromium depletion of the matrix. This effect disappears, however, in the presence of titanium or niobium to the extent that carbon is retained again in these elements as a stable carbide. Nitrogen seems to have no practical influence on 475° C embrittlement.

The segregation of ferrite into a chromium-rich and an iron-rich phase exerts a relatively strong influence on the mechanical properties of stainless materials with higher ferrite contents. Hardness, tensile strength and yield strength are increased, elongation, reduction of area and especially of impact energy values are reduced. The process of 475° C embrittlement can be tested relatively well with the aid of a hardness tests. Fig. 93a shows the influence of chromium content on the variation of hardness after 500 hours of annealing between 300 and 600° C according to Bandel and Tofaute [275]. Steel 1 with 16.3% chromium shows practically no hardness increase at all. With increasing chromium contents, however, the hardness also increases with an annealing temperature of 500° C. The influence of long annealing times on the hardness of steels with 26–30% chromium is shown in Fig. 93b according to Newell [276]. It appears that the process of hardness increase takes place rather slowly. Fig. 93c shows the behaviour of an austenitic-ferritic duplex steel UNS grade S 31803 according to Herbsleb and Schwaab [245], the chemical composition of which is listed in Table 9 (steel 4). Embrittlement takes place exclusively in the delta ferrite. The macro hardness of this duplex steel rises after annealing at 450° C from approx. 235 to 300 Vickers units, but micro hardness tests indicate that only ferrite is involved in this hardness increase.

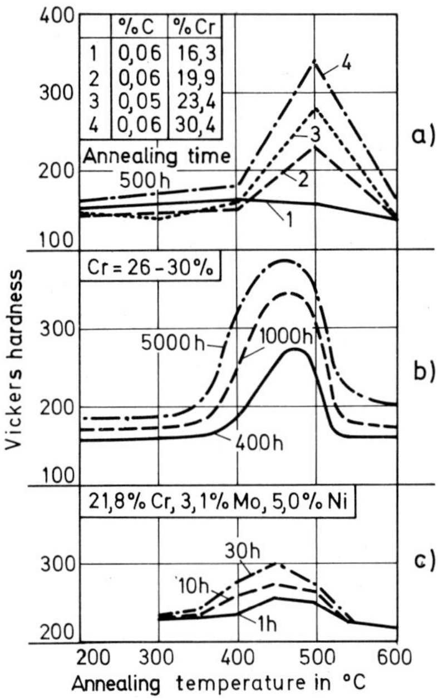

Fig. 93. Influence of annealing temperature and related 475° C embrittlement on the hardness of different steels. *a* iron-chromium alloys to Bandel and Tofaute [275], *b* chromium steels with 26–30% chromium to Newell [276], *c* austenitic-ferritic duplex steels to Herbsleb and Schwaab [245]

Table 9. *Chemical composition of steels 1–7, the behaviour of which in the temperature range of 475° C embrittlement is shown in Fig. 94*

	Chemical composition in %						
Steel No.	C	Si	Mn	Cr	Mo	Ni	Others
1	0.002	<0.01	<0.01	13.86	0.08	–	–
2	0.044	0.31	0.64	18.04	0.01	–	–
3	0.043	0.31	0.64	18.03	1.94	–	–
4	0.028	0.45	1.63	21.8	3.12	5.0	N = 0.113
5	0.028	1.37	0.50	21.6	0.25	8.0	–
6	0.041	0.28	0.37	16.21	0.43	5.76	–
7	0.038	0.41	0.72	12.83	0.31	4.03	–

The behaviour of ferritic chromium and chromium-molybdenum steels in the temperature range of 370—540° C has been investigated by Riedrich and Loib [277] and Grobner [278]. Figs. 94a and 94b show the TTT diagrams of the beginning and the end of 475° C embrittlement for 3 chromium-(molybdenum) steels. The chemical composition of the steels shown in Fig. 94 can be taken from Table 9. By increasing the chromium content to about 18% and with alloying additions of molybdenum, the embrittlement area is shifted to much shorter times.

Fig. 94c shows the beginning of 475° C embrittlement according to Herbsleb and Schwaab [245] for a duplex steel of type UNS S 31803 (steel 4). Embrittlement in the temperature range of around 475° C already starts after about 20 minutes. As mentioned already in section 4.2.4, the heat treatment of such steels should not take place below the solution annealing temperature, not only for reasons of sigma phase precipitation but also because of the rapid 475° C embrittlement. Trautwein and Gysel [279] have found that for thick-walled steel castings with 6—42% delta ferrite the 475° C embrittlement will progress about 10 times as fast as previously assumed. Below 400° C, the measured activating energy of ferrite segregation is practically independent of temperature. Therefore, we can establish a physical equivalent between embrittlement time and temperature, thus permitting predictions of aging and embrittlement behaviour for time and temperature situations outside of the measuring range. The time dependent course of embrittlement for the ferritic phase seems

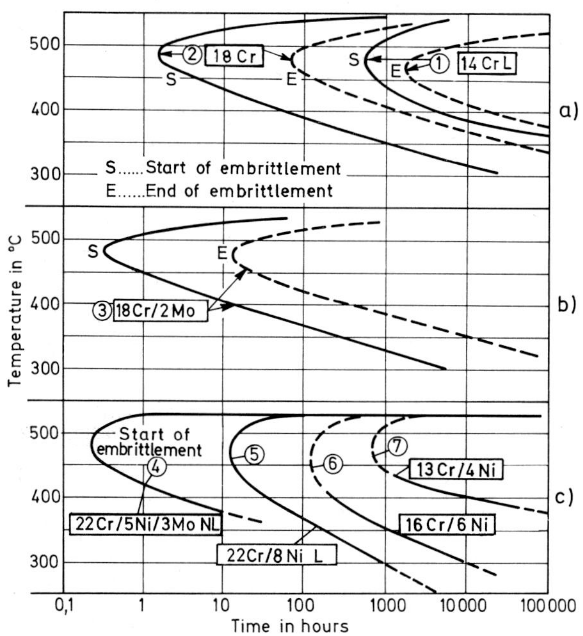

Fig. 94. TTT diagrams for the 475° C embrittlement of stainless steels. *a* and *b*: beginning and end of embrittlement in ferritic chromium and chromium-molybdenum steels according to Grobner [278], *c*: beginning of embrittlement in austenitic-ferritic duplex steels. Steel 4 to Herbsleb and Schwaab [245] and cast steel 5 according to Trautwein and Gysel [279]. 6 and 7 represent low carbon martensitic steels according to results compiled by Püber and Auer [280]. The chemical composition can be taken from Table 9.

to be largely independent from the ferrite content. The amount of embrittlement and the resulting drop in impact energy increases, however, with increasing ferrite content in the structure. Fig. 94c shows the TTT diagram for the beginning of 475° C embrittlement in a cast steel grade 22 Cr/8 Ni (steel 5) with a delta ferrite content of about 40% according to Trautwein and Gysel [279]. It appears that the segregation process of ferrite which leads to the 475° C embrittlement reaches down to much lower temperatures than previously assumed. This means that this phenomenon must be fully considered in the case of long time exposure to temperatures in the range of 250—300° C. The chemical composition of the cast steel can be set, however, in such a way that a 475° C embrittlement at 300° C is practically excluded even in the case of very long annealing times. This can be achieved, for example, by a lowering of the delta ferrite and silicon contents.

Stainless low carbon martensitic chromium-nickel steels also show a 475° C embrittlement, with the only difference being that the beginning of embrittlement is shifted to longer times because of the lower chromium content. Fig. 94c shows the TTT diagrams of the beginning of embrittlement according to Püber and Auer [280], for two low carbon martensitic cast steel grades containing 16% and 13% chromium (steel 6 and 7) defined by a drop in impact energy to 85% of the original values. Their chemical composition is listed in Table 9 (page 141). As expected, the embrittlement in cast steel No. 6 with 16% chromium starts earlier than in cast steel No. 7 with 13%. At 300° C, the embrittlement of cast steel No. 6 starts after approx. 1 year and after 4 years the impact energy values have dropped to about 70% of the original value. In the low carbon martensitic 13/4 chromium-nickel cast steel No. 7, on the other hand, there are no losses in toughness up to a temperature of 300° C, even after prolonged aging times. However the losses in toughness may be quite important at long time aging in the temperature range between 350—500° C according to Gümpel, Hook and Strom [461]. It seems that the reason for this decrease in impact energy is beside a possible segregation process the precipitation of fine dispersed carbides in the ferritic structure.

Weld metal behaves with regard to 475° C embrittlement in the same way as a steel of similar analysis. 475° C embrittlement occurs only in ferrite containing weld metals, with the degree of embrittlement being determined by the ferrite content. With increasing delta ferrite contents, for example, embrittlement will also increase accordingly. In the case of austenitic weld metal which—for reasons of improved hot cracking resistance—shows low contents of delta ferrite, it must also be considered that the latter is enriched in chromium (see Table 6, page 133), which somewhat accelerates the 475° C embrittlement. Nevertheless, the latter is hardly felt up to about 14 FN. Unstabilized weld metal shows a better performance here than does a niobium stabilized one, since the inhibiting influence of free carbon is not diminished or offset by its stable retention as niobium carbide. According to Malone [281], the impact energy values of unstabilized austenitic weld metal with chromium contents of about 21—23% and delta ferrite contents of 6—20 FN after 100 hours aging at 475° C drop to about 92—94% of the original value and after 3000 hours to 85—92%. Up to 14 FN, niobium stabilized weld metal with about 18—21% chromium behaves practically in the same way as an unstabilized one. At approx. 25 FN, however, the impact energy values drop after 100 hours of aging at 475° C to approx. 70% and after 3000 hours to approx. 57% of the original value.

5 Hot Cracking Resistance During the Welding of Austenitic Stainless Steels

Because of the great variety of possible types of cracks and causes for their formation, the problem of safeguarding against crack formation in welding operations is a highly complex one. A comprehensive survey of the cracking problems associated with the welding of different steel grades, is given by Baker [282] with special emphasis given to the different crack phenomena and the various causes of their formation. A survey of the different types of cracks found and a basic diagram of the temperature for possible crack formation in the welding of steel is also included in DIN 8524, Part 3. Because of their intricate appearance, hot cracks in austenitic welded joints are of particular importance and the literature available on this topic is quite substantial. Borland and Younger [283] compiled a survey of 162 publications up to the year ending 1959 regarding the hot cracking phenomena during the welding of austenitic chromium-nickel steels. In the documentation "Schweißtechnik, Bibliographie zum Thema Heißrisse beim Schweißen" [284] which covers the years from 1968 to 1978, 81 out of a total of 205 papers deal with austenitic materials. Another survey and a classification of hot cracks as they occur in the welding of various steels is supplied by Hemsworth, Boniszewski and Eaton [285]. A comprehensive survey on the present state of knowledge of heat affected zone (HAZ) cracking in thick sections of austenitic stainless steels is given with 91 references by Thomas Jr. [449].

The following important types of hot cracks can be identified in connection with the welding of austenitic stainless steels. The cracks differ in respect of their time and place of origin. They are:

1. Solidification cracks in the weld metal
2. Liquation cracks in the heat affected zone (HAZ)
   a) of the base metal
   b) of the weld metal in multilayer welds.

Solidification cracks occur during solidification of the liquid weld metal, when liquid—i.e. low melting phases remain between the primary, mainly dendritic crystals. Liquation cracks are produced by the enrichment of liquid, low melting phases at the grain boundaries of the crystals formed in the vicinity of the HAZ of a weld bead or, to be more precise, in the high temperature zone which borders immediately upon the liquid weld metal but has not yet been fully fused. This type of crack can be found in both the HAZ of the base metal and the heat affected zones of previously deposited weld metal beneath newly deposited weld beads.

5.1 Formation of Hot Cracks

For crack formation, the deformation potential of the material in question must be locally exceeded. If at first, an initial crack has appeared, widening and propagation of this crack is determined by the level of material stresses and by the deformation potential of the material at the tip of the crack. The complexity of the hot cracking problems associated with welding operations results from the fact that—during cooling—a weldment passes through the full temperature range from melting point to ambient temperature, with both the deformation potential and the deformation forced upon the material by shrinkage stresses being continuously changed as the temperature drops. The latter are additionally influenced by the rigidity of the weldments. If the rigidity is low, e.g. in the welding of thin sheet, deformation becomes transferred from the welding zone into the base metal which may then become distorted more severely. Thus the deformation within the actual welding zone is reduced. However in the case of thickplate and rigid components this is not possible. The deformation by shrinkage is more or less limited to the welding zone itself where the risk of possible crack formation is at its greatest in this area. To produce crack-free weldments, the deformation potential of the welding zone must in all phases of the cooling process be greater than the total deformation forced onto the material by shrinkage stresses.

5.1.1 Formation Mechanics of Solidification Cracks

The formation of solidification cracks has been reported on, amongst other authors, by Aplett and Pellini [286], Prochorow, Jakuschin and Prochorow [287], Probst and Schirmer [288], Klug [289], Arata, Matsuda, Nagakawa, Katayama and Ogata [290, 347], Matsuda, Nagakawa, Ogata and Katayama [291, 429, 430], Thier [292], Homberg and Wellnitz [293], Perteneder, Rabensteiner, Schabereiter and Tösch [294], Kujanpää, Suutala, Takalo and Moisio [150], Masumoto, Tamaki and Kutsana [138], Müsch [295], van Bemst [296], Ritter and Savage [453], Kujanpää [454, 455] and Cieslak and Savage (459).

The main cause for the formation of solidification cracks is to be found in the metallurgically dependent formation of low melting phases which remain during solidification in the form of a liquid film at the grain boundaries of the primary solidified crystals.

During shrinkage, micro fissures form within the weld which, during further cooling, may propagate to the surface where they then appear as macro cracks. In general, they run perpendicular to the direction of the strongest shrinkage stress. Depending, however, upon shrinkage restraint, shape of weld, and type of crystallization structure they may run parallel, transverse or diagonally to the direction of welding. If hot cracks reach as far as the weld surface, the crack surface does not appear metallicly bright but discolored due to the formation of oxides as a result of the exposure of the hot crack surface to the atmosphere.

Fig. 95a shows a schematic representation according to Arata, Matsuda, Katayama et al. [347] of the formation of hot cracks in fully austenitic weld metal with dendritic solidification where the liquid phases settle between the dendritic branches at the moment when the liquid-solid interfaces of the crystals meet (see

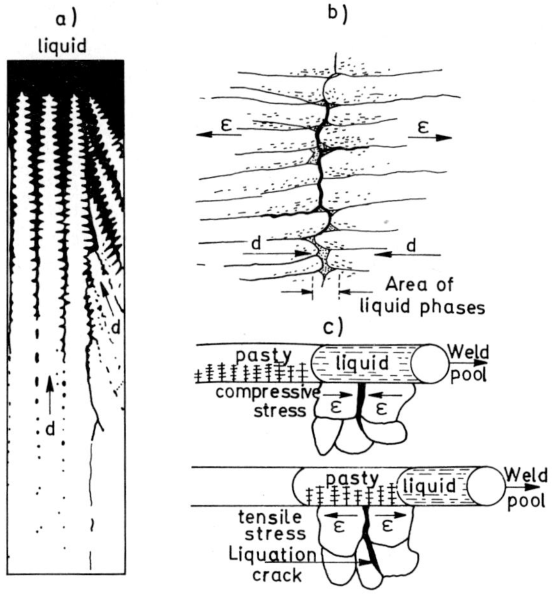

Fig. 95. The formation of hot cracks in austenitic weld metal due to the formation of low melting phases during solidification. $\varepsilon =$ deformation by shrinkage, d = direction of crystallization. *a* formation of solidification cracks at the junction between liquid-solid interfaces with dendritic solidification according to Arata et al. [347], *b* with cellular solidification of the weld metal according to Baker and Newman [297, 347], *c* formation of liquation cracks in the HAZ of an austenitic steel according to Apblett and Pellini [286]

also Fig. 52 and 53, page 59 and 60). If the solidification is cellular, the formation of liquid phases on the liquid-solid interphases is shown in Fig. 95b according to Baker and Newman [297].

Because of the fine branches of the dendritic crystals, it is easier for low melting phases to settle between the grain boundaries here than in a cellular crystal structure. Solidification cracks show a surface structure which is equivalent to the grain boundaries of primary crystals, where low melting phases have been deposited as a thin film. Fig. 96 shows according to Kulmburg [298] the crack surface of a solidification crack in fully austenitic chromium-nickel-molybdenum steel weld metal. The dendritic structure, where the solidification cracks have started at the grain boundaries, is clearly visible. Senda, Matsuda et al. [316] have investigated the disturbance of the coherence of solidifying primary crystals by liquid phases during cooling of the weld from the liquid state. Starting with the liquidus temperature of between 1400–1450° C, there is first a range of very low strength in which the weld metal can be practically separated without any plastic deformation at all. This represents the liquid phase range, termed "brittleness temperature range" (BTR) by the authors, i.e. the temperature range within which falls the minimum deformation potential of the weld metal. In 25/20 chromium-nickel steels, this range may extend down to about 1250° C [316]. Down to this temperature, low melting phases can still be liquid and prevent the coherence of the already solidified crystals. During further cooling to below 1250° C, the strength steeply increases, a sign that the liquid phases too are now solidified and a coherent crystal structure is present.

5.1.2 Formation Mechanics of Liquation Cracks

As mentioned earlier, liquation cracks may appear in the HAZ of the base metal as well as in the HAZ of previously deposited weld metal. They are rather small and normally not externally visible. For this reason, the liquation cracking phenomena is also termed in some publications as "micro cracking, fissuring or HAZ cracking". Liquation cracks may be widened by additional plastic deformation of the welded joint, with the crack length in austenitic weld metal being mainly in the range of 0.1–0.4 mm. Because of their minuteness and location beneath the weld, they can be made more visible e.g. by superficial removal of the weld metal surface followed by a polishing and etching treatment with a solution of 2 parts strong hydrochloric acid, 1 part strong nitric acid and 1 part hydrogen peroxide. The visual examination is made with the aid of a binocular microscope with a magnification of × 20 to × 50. With stronger plastic deformation, it is also possible to make liquation cracks better visible without etching with the aid of the fluorescent-dye penetrant method. This only permits, however, the detection of the coarser liquation cracks. The appearance of liquation cracks in the HAZ of base metals is discussed by Apblett and Pellini [286]. The literature listed below deals with the occurrence and causes of liquation cracking in welded joints made from austenitic stainless steels. It has been compiled by the following authors: Stauffer and Keller [299], Honeycomb and Gooch [300], Tamura and Watanabe [58], Schabereiter, Rabensteiner and Folkhard [301], Lundin

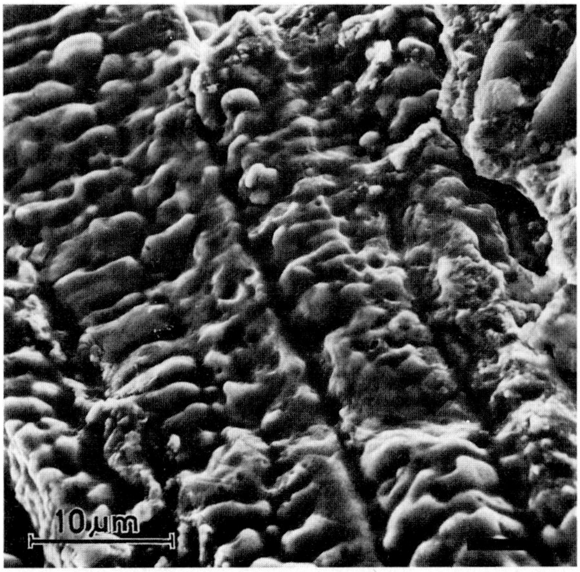

Fig. 96. Surface of solidification cracks in fully austenitic weld metal with a dendritic structure of primary precipitated austenite crystals according to Kulmburg [298]. (Image of crack surface obtained with a scanning electron microscope)

and Spond [302], Gooch and Honeycomb [303], Lundin, Chou and Sullivan [304], Morishige and Okabayashi [450], Lundin and Chou [456] and Nakao, Hiroaki, Oshige, Koga, Nishihara and Sugitani [458]. A comprehensive survey of the liquation cracking phenomena in the welding of thick cross sections of austenitic materials is given with 91 references by Thomas Jr. [449]. The formation of liquation cracks during the welding of low alloyed steels together with a great deal of supplementary literature is given by Klingauf [305].

According to Apblett and Pellini [286], liquation cracks in the HAZ of the base metal are formed by grain boundary segregations which lead to low melting phases forming at the grain boundaries. During deposition of the subsequent weld beads the segregations become locally liquified again and the fused low melting phases will produce material separations during the shrinkage process, as is shown schematically in Fig. 95c (page 146). Initially, the HAZ is exposed to compressive stresses during welding. Tensile stresses are only produced at a later point, normally after solidification of the weld metal. Klug [289] has shown that it is very important at which time and temperature the compressive stresses produced during heating of the weld change into tensile stresses during cooling, as they have a decisive influence on the formation of liquation cracks. If then the low melting phases which are present at the grain boundaries of the HAZ are still in the liquid state, the liquid grain boundary films are widened to form material fissures and liquation cracking.

If the temperature of the change from compressive to tensile stresses is lower, however, and if the liquid phases have already solidified, the crystal structure is no longer separated by the liquid films and there is normally no material fissure which may widen to form liquation cracks. If a weld cools rapidly as happens with thicker materials, the temperature at the time of changeover from compressive to tensile stress will be high and the risk of liquid phases remaining at the grain boundaries will also be greater. According to Klug [289] with thinner materials a stress reversal takes place at lower temperatures, i.e. at a time when the liquid phases have already solidified. This is also the major explanation why thicker materials often show a stronger tendency towards liquation cracking during welding operations than do thinner ones.

The formation of liquation cracks in the HAZ of previously deposited weld metal is considerably more complicated. In the first place, it is difficult to explain why hot cracking does not occur initially in the form of solidification cracks during cooling from the liquid state, but rather later in the heat affected zone produced by depositing subsequent weld layers on top of the previous ones. According to Tamura and Watanabe [58], a possible explanation may be that in a weld metal which has exhibited sufficient hot cracking resistance during solidification to be free from solidification cracks, new, coarser crystal grains may be formed by recrystallization in the heat affected zone which is produced by the deposition of a new bead. In weld metal which solidifies primarily to austenite, the alloying elements chromium and nickel may, because of the similar direction of segregation in fully austenitic alloys (see Fig. 66, page 73), be enriched with these elements at the newly formed grain boundaries. Tamura and Watanabe [58] also found an enrichment of the elements manganese, silicon and niobium at these new grain boundaries. These enrichments lead to a drop in the melting point which means that new lower melting phases are produced in the heat affected zone at the newly formed grain boundaries during the recrystallization which had not been present in the previously deposited weld metal layers. With increasing grain coarsening in the HAZ due to recrystallization, the enrichment of chromium, nickel, manganese and silicon at the grain boundaries will also increase as does the tendency towards liquation cracking in the weld metal.

The degree of grain coarsening in the HAZ influences the tendency of the weld metal towards liquation cracking. In Fig. 61a (page 68) for example, we can see that the formation of liquid phases in the lower weld bead near the boundary of the two weld passes, has not yet led to liquation cracking. Fig. 61b (page 68) shows the secondary grains produced by recrystallization, which are considerably coarser than the primary cellular crystals formed during solidification.

In fully austenitic weld metal which solidifies primarily to austenite, there is normally a much stronger grain growth due to recrystallization than in weld metal which primarily solidified to delta ferrite and experiences immediately afterwards a δ—γ-transformation which represents an effective hindrance for grain growth by recrystallization. According to Tamura and Watanabe [58], this is one of the major reasons why weld metal which primarily solidifies to austenite shows a much stronger tendency towards liquation cracking than austenitic weld metal which primarily solidifies to delta ferrite and which transforms immediately afterwards to austenite by the δ—γ-transformation (see also Fig. 6 and 9, page 8 and 10).

Another theory of HAZ cracking in the reheated regions of weld metal deposits in multipass welds is proposed by Lundin and Chou [456]. In fully austenitic welds the regions near the fusion zone can exhibit low ductility due to grain boundary embrittlement caused by segregations of harmful elements. In this degraded austenite grain boundaries fissuring occurs, provided sufficient restraint is imposed by the structure and the welding conditions employed. The extent of this degraded zone depends primarily on the composition of the weld metal with respect to the Cr/Ni ratio combined with the Cr content, as well as to Mn + Mo levels. The degree of degradation of the grain boundary ductility is dependent on the P, S and Si contents coupled with the number of HAZ exposures.

Chemical elements which have a lower solubility in austenite than in ferrite, e.g. sulphur, phosphorus and boron, and which at the same time form low melting phases and exhibit a tendency towards segregation, considerably increase the tendency in fully austenitic weld metal towards liquation cracking (see also Fig. 44, 45 and 46, page 39, 41 and 43). Furthermore, it must be considered that fully austenitic weld metal shows a much stronger tendency towards constitutional supercooling, towards segregations and towards the formation of low melting metallic phases than weld metal which primarily solidifies to delta ferrite and subsequently experiences a relatively severe δ—γ-transformation (see sections 2.6 and 2.7 and Fig. 69 respectively, page 76).

5.2 Hot Cracking Tests

In order to understand the hot cracking behaviour of austenitic stainless steels a knowledge of the deformation potential of the material in the solidification range is required, i.e. between the actual liquidus and solidus temperature and in the range where liquid phases exist. In order to make assumptions on the probable cracking behaviour of welded joints, this deformation potential which is influenced by material and metallurgical parameters must be compared to the deformation caused by shrinkage stresses in the welding zone. The basic principles of such a procedure are outlined by Prochorow, Jakuschin and Prochorow [287]. Probst and Schirmer [288] developed nomograms for the determination of crackproof welding parameters, but without consideration of the chronology of stress buildup. In the work of Klug [289, 306], the influence of the latter is also considered. It appears that the HAZ of fully austenitic weld metal experiences during solidification an initial upsetting phenomena, produced by compressive stresses which change over to tensile stresses and tensile deformations during further cooling. The temperature at which this reversal takes place during cooling is mainly dependent upon the thickness of the material and the amount of heat input used. If tensile stresses occur in the temperature range where liquid phases still exist, hot cracks are formed because the deformation behaviour of the material is practically zero in this range.

If a material is to be tested for its hot cracking susceptibility, the criteria being used may be the deformation rate at which no hot cracking becomes apparent in the welding zone during solidification and cooling. If we succeed in determining this critical deformation rate with the aid of a suitable hot cracking test, the calculation model developed by Klug [306] permits in the case of simple components predictions as to the materials probable hot cracking behaviour.

5.2.1 Hot Cracking Tests for the Determination of the Critical Deformation Rate in the Welding Zone

A survey of the most common crack testing methods up to the year ending 1961 is given by Granjon [307]. Wilken [308] and Wilken and Schönherr [309] deal in particular with the significance of the hot cracking tests. Hot cracking tests for austenitic weld metals are discussed by Homberg [310], Chène [311], Wilken [318, 451] and Pohle [452]. A compilation of literature prepared by Halkes [312] on behalf of the International Welding Institute regarding the works in this field between 1956 and 1978 lists 23 different testing methods for the determination of the hot cracking susceptibility of weldments. A survey and an evaluation of the significance of the major hot cracking test methods up to 1980 is given by Klug [289].

Hot cracking test methods can be grouped into self-stressing test methods, where the deformation stresses in the welding zone are produced by the use of a rigid fixture which restricts the material shrinkage and test methods where external stress is applied, i.e. where the deformation stresses in the welding zone are produced by an external load. In evaluating the different test methods, Klug [289] arrives at the following result: "When trying to evaluate, whether the hot cracking test methods investigated meet the demands of Wilken [308], it must be noted that in the self-stressing methods no distinction can be made between metallurgical and mechanical influences. In the same way, the double fillet weld specimen, the cylindrical specimen and the annular segment specimen (for specimen shape and test procedure please refer to the supplementary literature [308]) answer only the qualitative question–is there hot cracking?–yes or no. In addition, the test result is strongly influenced by the welding parameters. The Fisco test [308] and the fissure bend test according to Lundin, DeLong and Spond [313] do permit qualitative statements as to the hot cracking susceptibility. However in these tests the level of stress is limited by the rigid fixture and cannot be altered. The advantage of these methods is the rather simple execution of the test. Hot cracking test methods with applied external loads require a more complicated setup. Their great advantage is in the possibility of separating the metallurgical and mechanical criteria. In addition, both criteria can be varied within wide limits. The demand, however, for quantitative transferability and that for quantitative characteristics are still only met to a limited extent. This was the reason why, from the range of available test methods with applied external loads, the newly developed PVR test[8] according to Folkhard, Rabensteiner, Schabereiter, Fuchs and Tösch [314] was chosen for these particular investigations to be able to transfer the test results to simple components. This test produces a favourable situation of the development of a calculation model which permits a prediction to be made on the basis of the test results as to the probable hot cracking behaviour of welded components made from austenitic stainless steels."

Fig. 97 shows the schematic representation of the different hot cracking test methods with external stress loading. In the Murex test (Fig. 97a), two test sheets are joined by welding. Five seconds after the start of welding, one of the two test sheets is twisted with variable speed. In this way, the weld metal is deformed during solidification and cooling. Here, the measure for the hot cracking sensitivity is the crack length which is produced at a certain twisting rate. In the KSLA test (Fig. 97b), the test specimens used are two cylinders rotating in opposite directions. Thus, in contrast to the Murex test, the weld cross section remains the same during the whole welding operation. With a higher speed at the start of welding, an initial hot crack is produced. Then, the speed is reduced gradually until the point is reached where the crack can be stopped. The speed reading is taken at this point and is used as a measure for the hot cracking susceptibility of the weld metal. Homberg [310] uses an apparatus similar to that of the Murex test (Fig. 97a) for the testing of hot cracking susceptibility. Here, the preparation of the sheet specimen has been arranged such that a configuration is obtained, which guarantees that the weld cross section will not change during twisting (Fig. 97e). The twisting rate is adjustable and the bending force can be measured. Similar to the KSLA test, an initial hot crack is produced at the start of the welding operation by a higher twisting rate. Then, the twisting rate is reduced to the point where the crack has been stopped which is indicated by a sudden jump in the bending force. The deformation rate measured at the point of crack arrest is used as a measure for the crack susceptibility of the material. With the three hot cracking test methods described above, it is possible to determine the crack susceptibility of weld metal for solidification cracks but not for liquation cracks.

With the Varestraint test (Fig. 97c) according to Savage and Lundin [315], testing of welded joints for both liquation and solidification cracking is possible. With the Trans-Varestraint test according to Senda, Matsuda, Takano, Watanabe, Kobayashi and Matsuzaka [316], the hot cracking behaviour transverse to the direction of welding can also be tested (Fig. 97d). In this method, a weld is deposited on a sheet specimen, with the specimen being bent over a mandrel during welding or

---

[8] PVR test stands for: Programmierter Verformungs-Riss-Test (program controlled deformation cracking test).

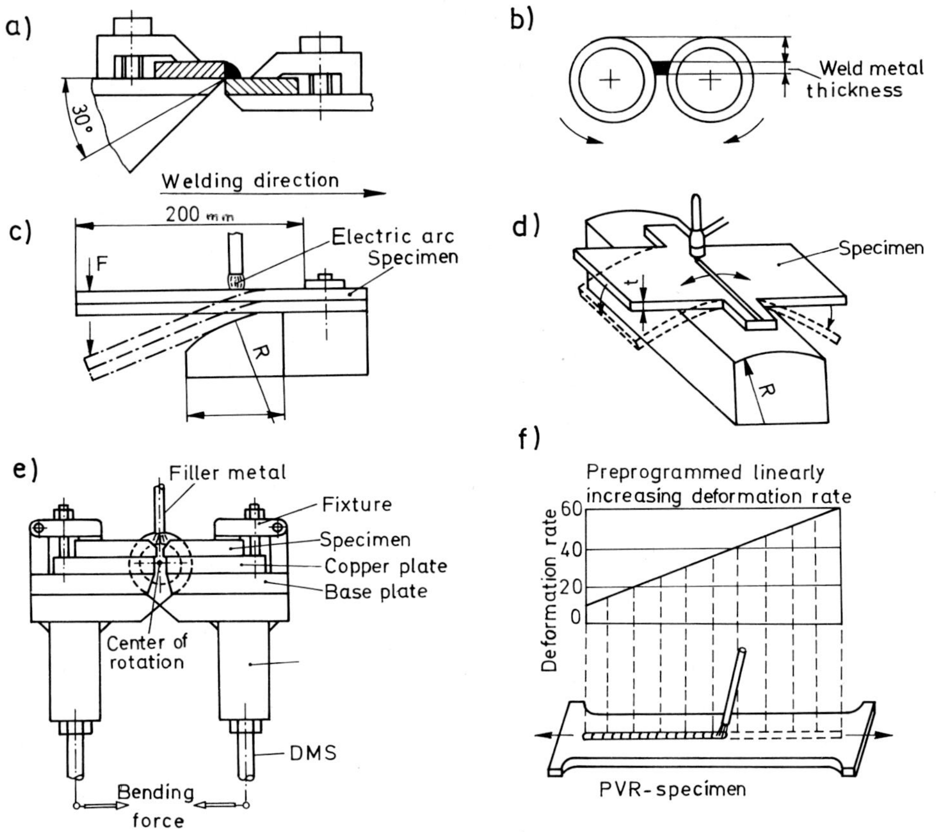

Fig. 97. Schematic representation of the hot cracking test methods with applied external stress. a) Murex test to Wilken and Schönherr [309], b) KSLA test to Wilken and Schönherr [309], c) Varestraint test to Savage and Lundin [315], d) Trans-Varestraint test to Senda, Matsuda et al. [316], e) Hot cracking test machine to Homberg [310], f) PVR test with programmable rate of deformation during testing according to Folkhard et al. [314]

immediately after welding longitudinally or transversely to the direction of welding. By changing the mandrel radius or sheet thickness, the elongation can be increased by up to 5%. The test evaluation takes place by measuring the maximum crack length, the number of cracks and the accumulated total crack length. With the aid of the Slow Bending Varestraint Test according to Arata, Matsuda, Nagakawa, Katayama and Ogata [290], the existence range of liquid phases during the solidification of weld metal can also be measured. Liquation cracks can be detected by the removal of the weld or the sheet surface. The MVT test (modified Varestraint and Trans-Varestraint test) according to Wilken and Kleistner [317] offers a great range of variations. The major modification consists of the continuously variable rate at which the sheet specimen is bent over the mandrel. Individual test parameters such as bending radius, bending rate, welding process, welding parameters, specimen size and material combination can be varied independently of each other. The possibility of transfering the test results of hot tensile tests to the components behaviour is discussed by Wilken [318].

The hot cracking test with a welding simulator (Gleeble test), the principle and test procedure of which are described by Dahl, Düren and Müsch [57] represents a special test in which there is no welding operation but only a simulation of the welding process. In this test, the specimen is quickly heated to a temperature just below the melting point of the material and quickly cooled again in accordance with the temperature gradients as experienced during welding. By simultaneous application of tensile load, both tensile strength and deformation potential can be measured as being dependent upon temperature, with the temperature variation of the smallest deformability being used as criteria for the hot cracking behaviour. The results obtained by the Gleeble test are of limited significance, however, for the hot cracking behaviour of weld metals, because in actual welding operations the material enters the critical temperature range just below the solidus temperature by cooling from the liquid state, in the Gleeble test, however, this happens by reheating the previously cooled material. Because of the resulting processes of recrystallization, diffusion and possible structure changes, the material is tested in a different state to that normally experienced during solidification from the welding heat.

5.2.2 The Program Controlled Deformation Cracking Test (PVR Test)[8]

With the PVR test according to Folkhard, Rabensteiner, Schabereiter, Fuchs and Tösch [314], both solidification and liquation cracks can be tested in a single test where the hot cracking susceptibility of a material or material combination and the weldability of austenitic materials can be numerically evaluated. According to Rabensteiner, Tösch and Schabereiter [319] in this hot cracking test, a flat tensile specimen made from the type of material to be tested for liquation cracking is clamped into a horizontal tensile testing machine, where it is loaded first up to the yield strength of the material (Fig. 97f, page 152). On this prestrained flat specimen a weld is deposited with the filler metal that is to be tested for solidification cracks. At the start of the welding operation, the specimen is subjected to a plastic deformation with a certain programmed rate of elongation. The elongation rate can be kept constant or varied according to a specific program. Normally there is an applied linear increase in the elongation rate from the start of the welding operation from 10 to 60 mm/minute, whilst a weld bead of approx. 220 mm of length is being deposited. Then the specimen is air cooled to ambient temperature. If hot cracking appears from a certain point in the weld onwards, the rate of elongation present at this point can be determined (see standard program in Fig. 97f). This is then used as a measure for the hot cracking susceptibility of the material. For the evaluation of the test, the specimen is first checked visually for macro cracks. Then, the material is removed from the surface of the weld bead and from the specimen surface to a depth of 0.5 mm with subsequent polishing and etching with a hydrochloric acid/nitric acid solution. Now, with $\times$ 40 magnification, micro cracks in the weld metal (solidification cracks) and in the HAZ (liquation cracks) can be detected and allocated-like the initial macro crack—to the pertinent rate of elongation. Fig. 98 shows an example of the testing of the hot cracking behaviour of a fully austenitic

---

[8] See page 151.

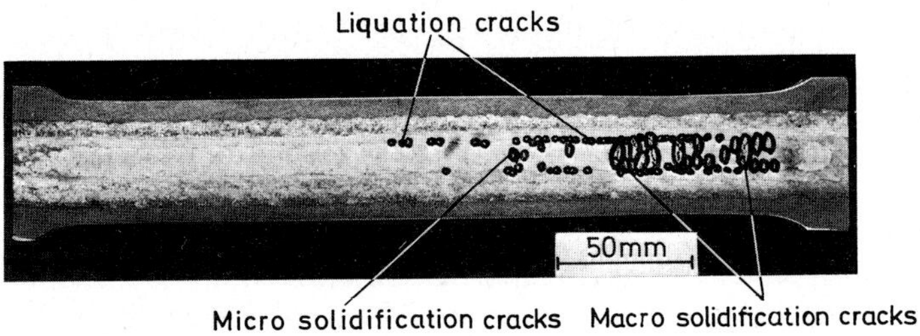

Fig. 98. Example of the PVR specimen used for testing the hot cracking behaviour of fully austenitic weld metal. For better visibility, micro and macro cracks are encircled by black lines. The rate of elongation has been constantly increased from the start of the welding operation from 10–60 mm/min. The test bead was laid on previously deposited weld metal. For specimen shape, please refer to Fig. 99, specimen cross section = 10 × 40 mm

weld metal using the PVR test. When the deformation potential has been exceeded by continuously increasing the rate of elongation, liquation cracks generally occur first in the HAZ of the test weld bead, with their number increasing as deformation rate increases. Solidification cracks in the weld itself can at first only be detected as fine fissures which do not break through to the weld surface. These fissures only become visible at × 40 magnification after removing the weld bead top surface (micro solidification cracks). Macro solidification cracks on the other hand are those cracks which can be detected on the surface with the naked eye without having to remove the weld surface.

The hot cracking susceptibility of a filler metal or a combination of base metal and filler metal can be defined by the following characteristics:

- Critical rate of elongation at the first macro solidification cracks which can be detected at the surface of the weld with the naked eye.
- Critical rate of elongation at the first micro solidification crack visible at × 40 magnification inside the test weld after the removal of the surface layer to a depth of 0.5 mm below the original specimen surface.
- Critical rate of elongation at which for the first time three liquation cracks per 10 mm of weld length appear at a × 40 magnification in the HAZ near the test weld after removal of the specimen surface.
- Critical rate of elongation at the beginning of an increase in the occurrence of liquation cracks in the HAZ near the weld bead with the criteria that 9 cracks per 10 mm bead length are detected at × 40 magnification.

Fig. 99 shows an example of these four hot cracking characteristics for six different weld metal grades according to results obtained by Rabensteiner and Tösch [486]. The chemical composition, the type of coating and the delta ferrite content can be taken from Table 10 (page 155). The ferrite containing weld metal 1 shows a very high deformation potential which means high hot cracking resistance. In weld metal 1 with a ferrite content of 9.3 FN there are no macro cracks at all and there is no accumulation of liquation cracks. The behaviour of fully austenitic weld metal is much less favourable. Even unstabilized niobium free weld metal shows rather poor deformation values. By alloying with molybdenum, these values can be slightly improved (weld metal 2). With the addition of about 4–5% manganese, the hot

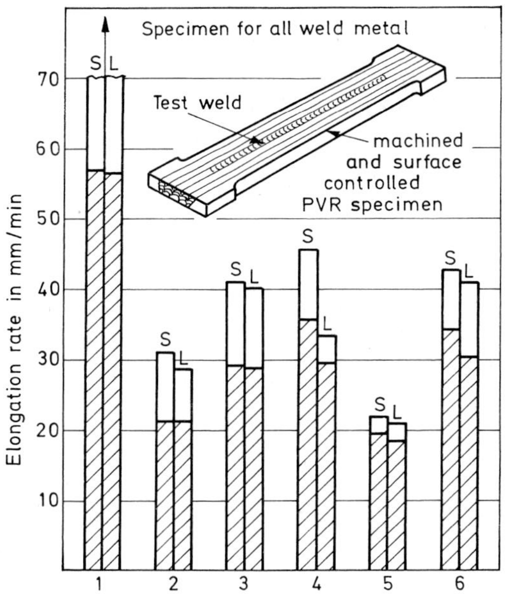

No.	Weld metal type
1	E 316L-16 FN = 9,3
2	18Cr/16Ni/3,5Mo/NL FN = 0
3	20Cr/25Ni/6Mo/4Mn/CuNL FN = 0
4	25Cr/20Ni/2Mo/5Mn/NL FN = 0-1
5	E 320 - 15 FN = 0
6	17Cr/16Ni/2Mo/5Mn/L FN = 0

Fig. 99. The hot cracking behaviour of different austenitic weld metal alloys during the PVR test according to results by Rabensteiner and Tösch [486]. For chemical compositions see Table 10 below. S = solidification cracks, L = liquation cracks, hatched area = rate of elongation to first micro crack, upper boundary = rate of elongation to first macro solidification crack and first accumulation of liquation cracks

Table 10. *Chemical composition of all weld metal, the hot cracking behaviour of which has been shown in Fig. 99 with the results of the PVR test according to Rabensteiner and Tösch [486]*

		Chemical Composition of All Weld Metal, %									
Designation	Types of covered Electrode	C	Si	Mn	Cr	Mo	Ni	N	Other Elements	Ferrite number FN	Remarks
1	Rutile	0.025	0.85	0.80	18.50	2.50	12.20	0.06	–	9.3	Type E 316L-16
2	Rutile	0.020	0.70	0.80	17.60	3.80	16.50	0.12	–	0	standard: chemical plants
3	Basic	0.035	0.45	4.30	20.00	6.40	25.00	0.14	Cu = 2.30	0	high corrosion resistance
4	Basic	0.035	0.32	5.50	25.00	2.00	19.60	0.13	–	0–1	reactors for fertilizer production
5	Basic	0.035	0.30	2.20	20.00	2.60	34.50	0.05	Cu = 3.10 Nb = 0.40	0	Type E320-15
6	Basic	0.035	0.20	4.50	17.00	2.30	16.30	0.05	–	0	cryogenic application

cracking tendency is reduced in particular with regard to the range of micro cracks. Weld metal 3 with about 6.5% molybdenum and 2.3% copper performs relatively well in the range of macro cracks. With regard to analysis, the weld metal 4 is located at the boundary between primary austenite and delta ferrite solidification. It shows a delta ferrite content in the range of 0–1 FN which has a positive effect with regard to the appearance of the first micro cracks. The most unfavourable behaviour is exhibited by the fully austenitic niobium containing weld metal 5, which already shows at only low elongation rates a strong tendency towards the formation of solidification and liquation cracks. Weld metal 6 shows a reduced hot cracking tendency due to the increased manganese content as compared for example with steel 2. The accumulated occurrence of liquation cracks is shifted to relatively high elongation rates. This means that the tendency towards the formation of liquation cracks rises only slowly with an increasing elongation rate.

The PVR test also offers the possibility to investigate and evaluate the weldability of austenitic steels, both regard to their susceptibility towards solidification and liquation cracking during welding without filler metals. Fig. 100 shows, for example, the behaviour of two fully austenitic steels of the type 20 Cr/18 Ni/6 Mo/0.12 NL. Both specimens have been welded in the same way by the GTAW process without filler metals. The loading in the PVR test was carried out with an increasing elongation rate, i.e. from 10 mm/min at the beginning to 60 mm/min at the end of the weld (see Fig. 97f, page 155). Because of the absence of a filler metal, only the cracking behaviour of the steel during welding could be tested. As shown in Fig. 100, liquation and solidification cracks appeared in both steel grades. Both types of crack occur immediately at the boundary between the non fused HAZ and the fused material. Since we are dealing here with rather small micro cracks they have been made visible by encircling them with black coloured lines. The upper specimen in Fig. 100 (designated A 6/3) shows the first signs of cracking at an elongation rate of approx. 48 mm/min without an accumulation of cracks being visible up to an elongation rate of 60 mm/min. The lower specimen (designated B 6/3) shows liquation cracking already beginning at about 24 mm/min elongation rate with an accumula-

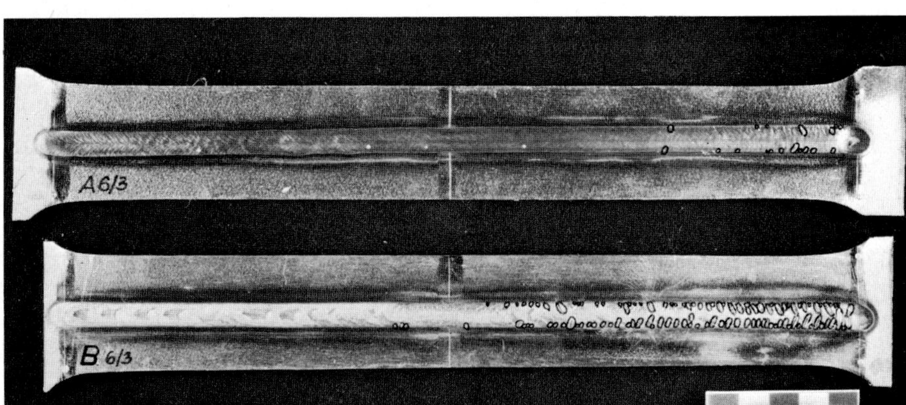

Fig. 100. The behaviour of two steel grades of the Type 20% Cr/18% Ni/6% Mo/0.12% NL in the PVR test.

tion of cracks being found from about 30 mm/min elongation rate onwards. In other words, the latter grade shows a much more unfavourable performance toward cracking than does the former specimen. In both instances the steel grades were welded by the GTAW process without filler metal under the same conditions. Experience showed that many cracks appeared in steel B (lower specimen) which were similar to those experienced during the PVR test, whereas steel A (upper specimen) could be welded crackfree.

From practical experience gained since 1976, with the beginning of systematic testing using the PVR test, it can be concluded that with regard to the actual cracking behaviour of austenitic stainless steels and austenitic stainless filler metals during welding operations, hot cracking will not occur, even in thick components, if they are properly welded and if the materials or material combinations used show no crack formation during the PVR test at elongation rates below 30 mm/min. In the case of thin-walled weldments and components of less rigidity, this limit is shifted to lower values. The weld metal 2, for example, the behaviour of which during the PVR test can be seen from Fig. 99 (page 155) can still be welded without danger of hot cracking in thicknesses of up to about 15 mm.

In the PVR test it is possible to use the conventional electric arc welding processes. With specimen cross sections of 40 × 10 mm, a 300 kN tensile testing machine will be sufficient for austenitic materials. For the heavy duty high current welding processes larger test specimen and more powerful tensile testing machines must be used. If only the hot cracking behaviour of filler metals is to be tested, an all weld metal specimen according to Fig. 99 (page 155) is used, and the test weld is laid directly upon a similar weld metal specimen which has been previously deposited. Prior to the hot cracking test, the specimen must be checked for freedom from cracks. If a certain material combination is to be tested, the test weld is executed with a specific filler metal on the desired base metal. If only the base metal is to be tested, the test weld is carried out with a GTAW torch without filler metal (see Fig. 100, page 156). Another phenomenon of the PVR test should also be mentioned here. Austenitic materials which present difficulties in hot forming operations (forgeability) may show a reduced deformability in the temperature range between 600–1200° C, often in connection with phase precipitations, impurities and two-phase structures. They show a tendency towards the formation of forging cracks, with the latter also showing up in the PVR test in the form of cracking. The same behaviour has also been identified by Arata, Matsuda and Katayama [320] in the Trans-Varestrain test. Such cracks are of a different appearance, however, to those of hot cracks. With the aid of a scanning electron microscope, they can be distinguished from the actual hot cracks which are produced during welding operations.

5.3 Criteria which Influence the Hot Cracking Susceptibility of Austenitic Stainless Steel Weldments

According to Klug [289], hot cracks are formed by an interaction between metallurgical and mechanical criteria. Amongst the mechanical criteria, of particular importance is the level of stresses and deformations that occur, their distribution and chronological growth. They originate from the shrinkage produced during cooling, with the shape and rigidity of the component also being of decisive influence. The chronological development of the deformations can be determined by a calculation model supplied by Klug [289] and the characteristics obtained in the PVR test make it possible to make predictions on the probable hot cracking behaviour of simple components. Amongst the metallurgical criteria first of all is the presence of low melting phases at the grain boundaries which, in the temperature range where they exist, may greatly reduce the deformation potential of the material. Any process which enlarges the existence range of the liquid phases will normally also increase the hot cracking susceptibility during welding.

In austenitic stainless steel materials, the hot cracking resistance is mainly determined by the following criteria:

— Type of primary crystallization and content of residual delta ferrite.
— Influence of alloying elements and impurities.
— Influence of segregations.
— Influence of deoxidation.
— Influence of material thickness and welding parameters.
— Influence of design features.

The first four criteria are metallurgical ones, the influence of the design features is a mechanical one. With regard to material thickness and welding parameters, it is important to know how the metallurgical and mechanical influences will interact since they influence not only the type of solidification, crystallization and supercooling but also the internal stresses and the deformation by shrinkage.

5.3.1 Influence of Primary Solidification on the Hot Cracking Behaviour of Austenitic Stainless Steel Weld Metal

Many authors point out that the hot cracking susceptibility of austenitic stainless steel weld metal is not determined by the amount of delta ferrite which is present at ambient temperature but by the type of primary crystallization during solidification (13, 14, 15, 58, 138, 143, 145, 150, 290, 292, 294, 325, 334, 443, 446, 449, 453, 454, 455). If the eutectic groove of the iron-chromium-nickel constitution diagram (Figs. 4 and 5, page 6 and 7) which separates the areas of primary solidification into primary delta and gamma crystals is transferred to the Schaeffler diagram, as was done by Masumoto, Tamaki and Kutsuna [138] and by Thier [292], we note that the eutectic groove does not coincide with the zero ferrite line. Greater deviations occur especially at higher chromium and nickel equivalents. This discrepancy is understandable, however, if we consider that the iron-chromium-nickel constitution diagram represents an equilibrium diagram whereas the Schaeffler diagram is a practical diagram, representing the actual conditions during electric arc welding with coated electrodes. In addition, the delta ferrite content can also be influenced by welding parameters. Respective investigations have been carried out by Tösch and Schabereiter [323], Gooch and Honeycomb [303], Lippold and Savage [157], Asakura, Wachi and Watanabe [324], Mundt and Hoffmeister [161] and Suutala [185]. In addition, the dilution with the base metal will also influence the delta ferrite content of the weld metal. The reasons why chromium-nickel weld metal which solidifies primarily to austenite is considerably more susceptible to hot chracking than weld metal solidifying primarily to delta ferrite are of a highly complex nature. There are normally several interacting influences of which the major ones are listed below:

— For constitutional reasons, chromium-nickel weld metal which solidifies to austenite (peritectic solidification with a relatively steeply sloping solidus area and a rather flat liquidus area) shows a much stronger tendency towards the formation of low melting phases and supercooling than weld metal which solidifies to ferrite, where supercooling is blocked by delta-gamma transformation which sets in with relatively high activating energy (see also sections 2.6 and 2.7 and supplementary literature [326] and [327].
— Austenite crystals have a much lower solubility for hot crack promoting elements than do ferrite crystals (see Table 11, page 162). This means that during the solidification to austenite crystals, these elements or their components are precipitated in the form of low melting phases at much lower levels than they would be during the solidification to delta ferrite crystals.
— The rather poor diffusability of many elements in austenite (see Table 1, page 50) prevents a stronger equalization by diffusion of the precipitated phases. Thus, the precipitation of low melting phases may take place at considerably lower contents, as in fact it should be according to the metallurgical equilibrium (see section 2.7).
— According to Lippold and Savage [325], the tendency towards segregation is much greater with austenitic solidification in combination with the effects mentioned above than with solidification to delta ferrite. This applies in particular to hot crack promoting elements which, in accordance with their constitution diagrams with iron, already show a stronger constitutional tendency towards segregations (see also section 5.3.3).
— The coarse grain formation in the HAZ due to recrystallization takes place according to Tamura and Watanabe [58] much more intensively in a fully austenitic structure than in an austenitic structure with small contents of delta ferrite, where it is impeded by the latter. Therefore, low melting phases can accumulate at the newly formed grain boundaries easier in a fully austenitic structure than in a ferrite containing one, a fact which increases the hot cracking susceptibility.
— According to Elmer, Olson and Matlock [328] and investigations by Kulmburg [298], the coefficient of thermal expansion at elevated temperatures of austenite is greater than that of ferrite. It must be considered here, that with primary delta ferrite solidification, the ferrite content at elevated temperatures in the range of possible hot cracking phenomena will be much higher, i.e. the difference in thermal expansion will be felt much stronger there. Thus, the plastic deformations caused by shrinkage will be more extensive in weld metals with primary austenitic solidification than in weld metal with primary ferritic solidification.

The interaction of these criteria, in which weld metal with austenitic solidification performs less favourably than weld metal with ferritic solidification, is the cause for the sudden drop in the hot cracking resistance which occurs at the transition from primary delta ferrite solidification to primary austenite solidification.

5.3.2 Influence of the Residual Delta Ferrite Content[9] on the Hot Cracking Behaviour

If similar austenitic stainless steel weld metals with low delta ferrite contents are tested for hot cracking sensitivity under similar welding and testing conditions, a distinct dependence upon delta ferrite content can be established, provided the ferrite content can be measured by a method which guarantees little scattering (see section 3.3). Fig. 101 a shows by way of an example the marked increase in the deformation potential of different weld metals with increasing delta ferrite contents as obtained from the PVR test. In austenitic weld metals, the critical elongation rate at the first crack quickly increases with increasing delta ferrite contents, with different weld metals behaving differently however. If we assume as a measure for high hot cracking resistance the empirical elongation rate value of 30 mm/min as obtained from the PVR test, it is possible to establish from Fig. 101 a the theoretical minimum contents of delta ferrite F 1 to F 4, at which weld metals will most likely produce welds free from hot cracking when welding components even under unfavourable restraint conditions. Similar results were obtained by Lundin, DeLong and Spond [322] in the testing of weld metal with the fissure bend test [313] which indicates that different ferrite contents are required for different weld metals to achieve crack free welds (Fig. 101 b).

These results seem to contradict the findings given in the previous section 5.3.1 which stated that it is primarily the type of solidification which is of decisive im-

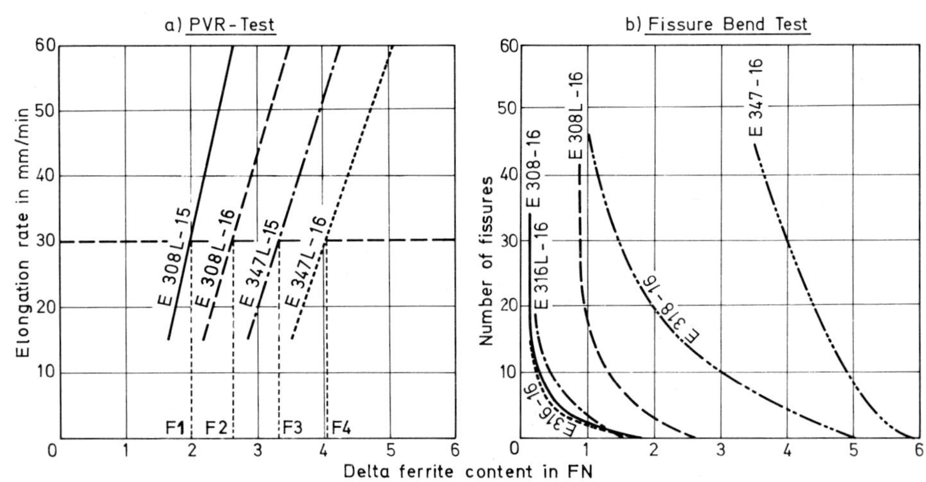

Fig. 101. Examples of the influence of the delta ferrite content of austenitic electrodes with basic and rutile coatings on the critical elongation rate in the PVR test (a) and (b) on the behaviour in the fissure bend test according to Lundin, DeLong and Spond [322]. The definition of the filler metals is in accordance with US Standard ANSI/AWS A 5.4–81

---

9 With residual delta ferrite, we mean that amount of delta ferrite in the austenitic matrix which is present at ambient temperature in the weld metal after δ-γ transformation in the case of primary delta ferrite solidification. In the following pages it shall be called delta ferrite content or ferrite content for short, in accordance with the pertinent literature.

portance and of secondary importance the amount of delta ferrite leftover at ambient temperature which is also influenced by other criteria not connected with the hot cracking resistance. In practical terms however, the determination of the delta ferrite content at ambient temperature has proven to be a very simple means of making predictions regarding the hot cracking behaviour of austenitic weld metal. Polgary [329] states that the lowest delta ferrite level permissible to achieve sufficient hot cracking resistance is 4%. In [330] he points out, however, that possible scattering must be considered when operating with such minimum contents. Ebert [331] stresses that contents lower than 4% ferrite are difficult to set with any reliability. Hull [332] recommends a range of 5—10% delta ferrite to achieve weld metals free from hot cracking. Kujanpää, Suutala, Takalo and Moisio [150] state that the highest hot cracking resistance is found within the range of 5—20% ferrite. Matsuda et al. [327] recommend a minimum ferrite content of 4—5% to achieve good resistance against solidification cracking in austenitic chromium-nickel weld metal containing 17—25% chromium and 8—20% nickel.

From Fig. 101 and investigations by Lundin, DeLong and Spond [322] it appears that higher ferrite contents are required for niobium stabilized weld metals to obtain the same sufficient cracking resistance as, for example, niobium free 20/10 chromium-nickel weld metal. The same applies to weld metals with increased silicon, phosphorus and sulphur contents. By increasing the ferrite contents, however, the negative effect of these elements can be largely compensated for, this is stated, for example, by Polgary [333] for silicon and Brooks and Lambert [69] for phosphorus and sulphur. In other words, to obtain sufficient hot cracking resistance it is necessary to modify the minimum ferrite content in accordance with the weld metal composition. In addition, possible scattering resulting from the method of ferrite measurement and from the influence of welding parameters, must also be considered. To obtain good hot cracking resistance in austenitic stainless steel weld metals with silicon contents below 1.2%, sulphur contents below 0.020% and phosphorus contents below 0.025%, the following recommendations with regard to delta ferrite content have been established:

— Fig. 101 shows examples of the theoretical minimum delta ferrite contents necessary to obtain high hot cracking resistance in different weld metal grades.
— If delta ferrite contents are measured with the aid of calibrated ferrite standards (see section 3.3), scattering of ± 10 relative percent must be expected (at 10 FN ± 1 FN). If the ferrite content is calculated from chemical composition, this value rises to $\pm 25\%$ (at 10 FN ± 2.5 FN).
— If welding parameters deviate considerably from the standard values, a further addition of 1 FN is required, in the case of very high welding speeds this will be increased to 1.5—2 FN.
— The dilution with the base metal must be considered such that the base metals delta ferrite content which is established during welding is first determined from its chemical composition or by running a GTA weld on top of it, followed by subsequent ferrite measurement. Then, the ferrite content of all weld metal and base metal must be plotted in the DeLong diagram and connected by a straight line. The ferrite content to be expected in the weld metal is determined by the degree of dilution. This can be calculated from the ratio of base metal to

weld metal along the connecting line (see also Fig. 112, page 232). In SMA welding with coated electrodes, a degree of dilution (i.e. the share of base metal) of 25–35% must normally be expected. With GTAW, GMAW and SAW processes it may be higher. In the fused weld metal, a ferrite content must be achieved which is located above the theoretical minimum value given in Fig. 101 taking also into account the possible scattering of the applied ferrite measuring method.

If this ferrite content is not reached due to the dilution effect with the base metal, an intermediate layer with an over-alloyed filler metal with an increased ferrite content of 15–20 FN is deposited on the base metal prior to carrying out the actual joint welding operation [e.g. 23/12 Cr-Ni (E 309 L) or 23/12/3 Cr-Ni-Mo weld metal (E 309 Mo L)—see also Table 24, Nos. 2 and 3, page 227].

5.3.3 Influence of Alloying Elements and Impurities which Promote Hot Cracking

This section deals with the influence of chemical elements which may form low melting phases in austenitic weld metal and thus reduce its hot cracking resistance. In ferrite containing weld metal which primarily solidifies to delta ferrite, the hot crack promoting effect of these elements is compensated for by the positive effect of the delta ferrite solidification which counteracts the hot cracking tendency. Thus the effect of the hot crack promoting elements is mainly felt in fully austenitic weld

Table 11. *Solubility of hot crack promoting chemical elements in austenite and ferrite; structure and melting points of low melting phases*

	Solubility in pure iron				Low melting phases	
	in austenite		in ferrite			
Constituent	%	Temperature °C	%	Temperature °C	Structure	Melting point °C
Sulphur	0.05	1365	0.14	1365	Eutectic Fe-FeS Eutectic Ni-NiS	988 630
Phosphorus	0.20	1250	1.6	1250	Eutectic $Fe\text{-}Fe_3P$ Eutectic $Ni\text{-}Ni_3P$	1048 875
Boron	0.005	1381	0.5	1381	Eutectic $Fe\text{-}Fe_2B$ Eutectic $Ni\text{-}Ni_2B$ Eutectic $(Fe, Cr)_2B$-Austenite	1177 1140 1180
Niobium	1.0	1300	4.1	1300	Eutectic $Fe\text{-}Fe_2Nb$ Eutectic NbC-Austenite Nb-Ni-rich phases	1370 1315 1160
Titanium	0.36	1300	8.1	1300	Eutectic $Fe\text{-}Fe_2Ti$ Eutectic TiC-Austenite	1290 1320
Silicon	1.15	1300	10.5	1300	Eutectic $Fe\text{-}Fe_2Si$ Eutectic $NiSi\text{-}Ni_3Si_2$ NiSi	1212 964 996

metal with primary austenite solidification. A general survey of the influence of alloying elements on the hot cracking sensitivity of austenitic steels is given by Duhl, Düren and Müsch [56], Hull [71], Borland and Younger [283], Gueussier and Castro [335] and by Thomas [449]. In fully austenitic weld metals, the main elements that increas the hot cracking sensitivity are sulphur, phosphorus, boron, niobium, silicon and titanium whereas manganese and molybdenum reduce it. Table 11, page 162, gives a survey of the solubility of major hot crack promoting elements in austenite and ferrite at elevated temperatures, i.e. in the range of possible hot crack formation. In addition, the low melting phases which are possibly formed are listed together with their melting points. The solubility limits have been taken from the respective constitution diagrams (see Figs. 38, 40, 41, 44, 45 and 46). In all cases, austenite shows a lower solubility for these elements than does ferrite. With the presence of nickel this solubility is diminished even further, because the solubility of sulphur and phosphorus in pure nickel, for example, is practically zero. The situation is aggrevated by the fact that in the case of sulphur, for example, a low melting eutectic with 31% sulphur may appear, a content that is very far away from the theoretical solubility limit of sulphur in austenite of 0.05% at 1365° C (see also iron-sulphur constitution diagram in Fig. 44, page 39). Because of the great difference in composition between precipitated crystals and residual melt an equalization of concentration by diffusion is impeded in view of the high cooling rates during welding and the low diffusability in austenite. This means that during welding, the solubility limit may still shift considerably below the values listed in Table 11, page 162 (see also section 2.7 "segregations"). From practical experience it can be assumed that in fully austenitic chromium-nickel weld metal the solubility for sulphur has dropped to the range of 0.005—0.010% and that of phosphorus to the range of 0.010—0.015%. In addition, the lack of equalization by diffusion leads to a concentration of low melting phases at the grain boundaries of the precipitating austenite crystals, where they may become locked in as liquid grain boundary films between the dendrite branches (see also Fig. 95 a, page 146).

The existing literature on the influence of various elements on the hot cracking sensitivity of austenitic stainless steels is quite substantial. Therefore, we shall deal here only with publications which are of importance for hot cracking phenomena in austenitic weld metal. Nakagawa, Matsuda, Senda et al. [336] have found grain boundary films consisting of an iron-iron sulphide eutectic which induced hot cracking. Nickel increases the tendency towards segregation and solidification cracking and at the same time reduces the eutectic temperature. With the addition of 2—3% manganese, the sulphide inclusions will coagulate and the grain boundary films are made to disappear. Masumoto, Tamaki and Kutsuna [337] measured the free energy of sulphide formation for the pure elements and have established the following order of sulphide formation:

$$\mathrm{Ce} > \mathrm{Ca} > \mathrm{Cs} > \mathrm{Mg} > \mathrm{Al} > \mathrm{Ti} > \mathrm{Mn} > \mathrm{Mo} > \mathrm{Ni} > \mathrm{W} > \mathrm{Fe} > \mathrm{Si}.$$

According to the same authors, however, a different order applies for iron alloys due to the influence of other elements. This is listed below, with elements to the left of (Fe) showing a higher affinity towards sulphur than those to the right:

$$\mathrm{Zr} > \mathrm{Ti} > \mathrm{Mn} > \mathrm{Nb} > \mathrm{V} > \mathrm{Cr} > \mathrm{Al} > \mathrm{Mo} > \mathrm{W} > (\mathrm{Fe}) > \mathrm{Ni} > \mathrm{Co} > \mathrm{Si}.$$

Morgenfield, Solari and Ovejero-Garcia [338] report on a method by which it is possible to make the sulphur distribution and sulphur segregation and their correlation to the micro structure visible and which is significant for investigating the influence of sulphur on the hot cracking sensitivity. Cieslak and Savage [339] verify that the unfavourable influence of phosphorus on the hot cracking sensitivity of austenitic cast steel is considerably stronger with primary austenite solidification than with primary delta ferrite solidification. Matsuda, Nakagawa et al. [291] have investigated the influence of phosphorus in 25/20 chromium-nickel fully austenitic steels on that temperature range where low melting phases are still liquid. They state the following lower temperature limits for this range:

Phosphorus content	Lower temperature limit in °C
0.003%	1300
0.013%	1270
0.022%	1240
0.032%	1200

In evaluating the combined influence of phosphorus and sulphur, Bernstein, Carlen and Rick [70] found that for 25/20 chromium-nickel fully austenitic weld metal sulphur will increase the hot cracking sensitivity more strongly than will phosphorus and that each element should be limited to 0.015% max. Lippold and Savage [325] found high enrichments of chromium, phosphorus and sulphur in the liquation cracks of fully austenitic 25/20 chromium-nickel weld metal. Brooks and Lambert [69] recommended that (P + S) be limited to 0.02% max. in fully austenitic weld metal. If the weld metal contains ferrite, then every increase above the level of (P + S) = 0.02% max. by 0.01% requires an increase in the delta ferrite content by 4 FN. Therefore with (P + S) = 0.03%, the weld metal should show 4 FN, with (P + S) = 0.04% 8 FN and with (P + S) = 0.05% 12 FN, in order to achieve the highest hot cracking resistance.

The unfavourable influence of niobium on the hot cracking resistance of fully austenitic weld metal is clearly apparent by the behaviour of alloy 5 in Fig. 99 (page 155). According to Jolley and Geraghaty [340], the large quantity of niobium containing liquid phases plays an important role here. Whereas in the ferrite containing niobium alloyed weld metal with 5 FN niobium containing liquid phases amount to only to 0.3%, this content increases in fully austenitic weld metal with 0 FN to considerably higher values of around 1.5%. This increase can be traced to the low solubility of niobium in Austenite (see Table 11, page 162). The low melting grain boundary phases represent niobium rich eutectics which promote hot cracking. Additions of magnesium in contents above 0.1% excert a positive influence in reducing the hot cracking tendency. According to Dahl, Düren and Müsch [56, 57] niobium together with phosphorus, chromium and manganese forms low melting phosphides and together with silicon, chromium and manganese low melting sulphide-oxide inclusions. Niobium concentrations at the grain boundaries may lead to the formation of low melting niobium and nickel rich phases which even partly solidify below 1160° C. Tamura and Watanabe [58] and Schüller [59] report that the formation of hot cracks in the heat affected zone of niobium containing austenitic steels are probably caused by a low melting eutectic of niobium carbide or niobium carbonitride with austenite. Similar results were obtained by Ogawa, Zaizen and

Tsumetomi [457]. Folkhard, Fladischer, Rabensteiner, Schabereiter and Tösch [321] show that the unfavourable influence of niobium on the hot cracking resistance of low carbon weld metal with niobium contents below 1% can be almost completely compensated for by primary delta ferrite solidification. The necessary minimum delta ferrite contents F 3 and F 4 can be taken from Fig. 101 (page 160) for the niobium containing weld metals 3 and 4. For practical applications, the possible scattering effect must be considered, as previously discussed at the end of section 5.3.2.

The possibility of the formation of low melting grain boundary phases by the element titanium has been investigated by Schüller [59]. The formation of the first low melting phases in which titanium carbo-nitrides are mainly involved, takes place at about 1340° C. The formation of liquid phases by titanium and their influence on hot cracking resistance is less pronounced, however, than in the case of niobium. Because of its high affinity towards oxygen titanium is not normally used for the stabilization of weld metal but only for that of steel, and the effect of titanium is mainly of significance for the base metal and the formation of liquation cracks in the HAZ.

The unfavourable influence of the element silicon on hot cracking sensitivity is practically only felt in fully austenitic weld metal with primary austenite solidification. According to Table 11 (page 162), silicon possesses at 1300° C a solubility in austenite which is almost 10 times lower than that in ferrite. At the same time, silicon favours according to Düren, Dahl and Müsch [56] and Ogawa and Tsunetomi [65] the formation of low melting phases by silicon enrichment at the grain boundaries of primary precipitated austenite crystals. Due to the high silicon solubility of ferrite, however, they do not occur with primary delta ferrite solidification. According to Polgary [333], a silicon content of 0.6–0.7% should not be exceeded in weld metal with primary solidification to austenite.

The combined influence of the elements phosphorus, sulphur, silicon and niobium on the hot cracking susceptibility of fully austenitic 16/16 chromium-nickel steels and weld metals has been investigated by Dahl, Düren and Müsch [56]. Arata, Matsuda, and Katayama [320] and Ogawa and Tsunetomi [65] have investigated the influence of these constituents together with those of carbon and nitrogen on the hot cracking sensitivity of various steel grades, with special emphasis on 25/20 chromium-nickel steels and weld metals. In general, it was found, that the combined effect of these elements was the same as the influence of individual elements, with certain deviations, however, due to the interaction between the elements and the sometimes diverging effect on differently alloyed steels and weld metals.

The effect of carbon on the hot cracking sensitivity is only felt in fully austenitic weld metal with primary austenite solidification. It is not uniform as it is dependent upon the overall alloying composition. An increase in the carbon content from 0.05 to 0.10% in unstabilized 25/20 chromium-nickel weld metal, for example, has a positive effect on the latters hot cracking resistance. In niobium stabilized weld metal, on the other hand, carbon may increase the hot cracking sensitivity by the formation of low melting carbide eutectics.

The effect of nitrogen on the hot cracking sensitivity of weld metal is practically neutral. This conclusion is drawn from the many hot cracking tests performed and many years of experience gained with nitrogen alloyed weld metals. The indirect effect of nitrogen pickup from the surrounding air which may produce a reversal from primary delta ferrite solidification to primary austenite solidification will be discussed in section 5.3.5 (see also Fig. 102, page 169).

Boron is described by Hull [71] as the element with the most unfavourable influence on hot cracking resistance. The solubility of boron in austenite at elevated temperatures is very low at 0.005% (see Table 11, page 162). In addition, boron forms low melting eutectics with both iron and nickel (see also Fig. 46, page 43). Donati, Guttman and Zacharie [77] have found deposits of a eutectic of iron-chromium-boride $(FeCr)_2B$ and austenite with a melting point of 1180° C at the grain boundaries of boron containing steels. They recommend limiting the boron content in 18/10 chromium-nickel steels to 0.0035% (35 ppm). If we are dealing with titanium alloyed grades, however, a boron limitation of up to 0.0050% (50 ppm) is sufficient. According to Donati, Lehuede and Zacharie [345], the presence of delta ferrite in weld metal has normally a positive effect. This positive effect is lost, however, if the tendency towards the formation of liquid phases becomes more pronounced. Therefore, in the case of boron alloyed steels, primary delta ferrite solidification does not give a complete protection against hot cracking. According to Gerlach and Schmidtmann [39], boron exerts a strong influence in fully austenitic niobium stabilized chromium-nickel-molybdenum steels in lowering the solidus temperature which increases the hot cracking sensitivity. This drop can be counteracted by the addition of nitrogen. If the boron content is limited to 0.005% max., no cracks will occur during welding according to [39]. According to the experience of the authors, an increase in the hot cracking sensitivity is to be expected if the boron content is raised above 0.005%. If in the case of a weld metal which due to its chemical composition is already at the limit of its hot cracking resistance then the maximum boron content may have to be reduced further.

5.3.4 Hot Cracking Sensitivity of Fully Austenitic Weld Metal

If no specific measures for the improvement of the hot cracking resistance are taken when using fully austenitic weld metal with primary austenite solidification, it becomes difficult when welding heavy wall thicknesses to avoid the formation of cracks in the weld metal and the HAZ during the welding operations. Alloy 2 in Fig. 99, for example, shows the behaviour of a crack sensitive weld metal in the PVR test. Liquation cracks already show up at low elongation rates in the range of 20 mm/min. In addition, Fig. 99 also shows that the first liquation cracks in fully austenitic weld metal alloys often begin to appear at lower elongation rates than the first solidification micro and macro cracks.

This means that liquation cracks which are not externally visible represent a critical cracking problem in fully austenitic weld metal. The influence of harmful elements on the cracking sensitivity of fully austenitic steels is documented by numerous publications which have been discussed already in the previous sections. New publications on this subject are listed in the following literature [454, 455, 456, 457, 458 and 459]. They also verify that fully austenitic grades show a much higher crack sensitivity than austenitic grades which contain more than 4% delta ferrite (see also Fig. 101, page 160). The elements which are considered particularly harmful are sulphur, phosphorus, silicon and niobium.

At the end of section 5.3.1, the major reasons were listed why weld metals with primary austenitic solidification show a strong hot cracking sensitivity. It is essentially the result of constitutional criteria of primary solidification, the tendency towards segregation and the properties of primary solidified austenite crystals, i.e. of processes which are difficult to influence. The measures to be taken to ensure an improvement in the hot crack sensitivity of an alloy must focus, therefore, on a reduction in the alloying impurities which promote hot cracking.

Sulphur is an element which is practically always present in steel in the form of impurities and which may already induce hot cracking by the formation of low melting phases even at relatively low contents. It is common knowledge that the hot crack promoting effect of sulphur can be reduced by alloying weld metals with manganese. Back in the thirties, crack resistant 18/8 chromium-nickel welding rods with 6–7% manganese were marketed which had a good performance record. Manganese has a higher affinity to sulphur than iron and is capable of retaining the sulphur which is present in the weld metal in the form of manganese sulphide, with the melting point of this manganese sulphide eutectic being 1580° C, it is considerably higher than that of the iron-sulphur eutectic at 988° C. During welding, the higher melting manganese sulphide is formed in preference to the low melting iron sulphide. This manganese sulphide precipitates as spherical inclusions in the still liquid weld metal before the metallic matrix has solidified. Nakagawa et al. [336] found that when the material is alloyed with manganese, the grain boundary films of the iron sulphur eutectic from about 2% manganese onwards become increasingly transformed into spherical sulphide inclusions with high manganese contents.

The positive influence of manganese on the hot cracking resistance of fully austenitic weld metal is verified by many different authors (see literature [67, 68, 71, 265, 266, 283, 300, 304, 320, 336, 341, 342]). According to Honeycombe and Gooch [346] the positive effect of manganese in fully austenitic 25/20 chromium-nickel weld metal starts at about 2% and produces optimum results between 4–6%. Lundin, Chou and Sullivan [304] report that apart from manganese, increasing molybdenum contents will also diminish the sensitivity towards liquation cracking which coincides well with the experiences of the authors. Arata, Matsuda, Katayama et al. [347] deal very thoroughly in six individual reports on the influence of sulphur on the hot cracking sensitivity of fully austenitic weld metal. They verify the positive effect of manganese which can be supported by the addition of rare earths, in particular lanthanum. Lanthanum increases the melting point of manganese sulphide and diminishes at the same time the crack promoting effect of phosphorus. From microscopic investigations they conclude that the sulphur content of fully austenitic weld metal should be limited to values of between 0.005–0.010%. When the sulphur content is kept below 0.005% and the phosphorus content below 0.006% sulphides and phosphides are undetectable under the scanning electron microscope. But even with sulphur contents below 0.003% and phosphorus contents below 0.001%, fully austenitic weld metal is still considerably more prone to hot cracking than delta ferrite containing weld metal with higher contents of S + P, e.g. 0.025–0.052% phosphorus and 0.01% sulphur. In other words, even with the extreme lowering of the phosphorus and sulphur contents, fully austenitic weld metal cannot be made as hot crack resistant as ferrite containing weld metal with primary solidification to delta ferrite. The latter is able to handle much higher contents of impurities without becoming prone to hot cracking.

Rabensteiner, Tösch and Schabereiter [348 and 486] report on the investigation of the hot cracking phenomena in different fully austenitic weld metals with the aid of the PVR test. The alloying elements manganese, in contents of 4–6%, and molybdenum, in contents above 2%, reduce the cracking sensitivity. Silicon and in particular niobium increase the risk of hot cracking. In addition, the hot cracking sensitivity of an alloy is influenced by the position of the weld metal alloy within the DeLong diagram. It is reduced as the fully austenitic alloys get gradually closer to the zero ferrite line. With the present state of technology it is difficult to produce a fully austenitic weld metal that shows no micro cracks at all when welding thick walled components made from fully austenitic stainless steels. By adopting the correct alloying technology, however, measures can be taken to substantielly reduce the tendency towards hot crack accumulations. Alloys 4 and 6, for example, in Fig. 99 (page 155) already show a relatively high critical elongation rate with the first appearance of liquation cracks. But contrary to alloy 5, an increase in the accumulation of liquation cracks occurs only with higher elongation rates in the range of about 40 mm/min, a value which is considerably above the critical limit of 30 mm/min. Therefore, it can be assumed that for welded components with heavy wall thicknesses liquation cracks will only occur sporadicly and not in an accumulation form.

Many recommendations are given in the pertinent literature as to the limitation on alloying elements and impurities to avoid hot cracking in fully austenitic weld metals. Silicon, for example, should be limited according to Polgary [333] to between 0.6–0.7%, according to Dahl, Düren and Müsch [56] to 0.5% max. and according to Gooch and Honeycomb [350] to 0.3% max. A very unfavourable behaviour is exhibited by niobium in fully austenitic weld metals as indicated by the performance of alloy 5 (Fig. 99) in the PVR test. Apart from the effects of niobium, other harmful effects such as a low manganese content, high nickel contents and increased contents of sulphur and phosphorus also produce a negative influence.

By lowering the sulphur content to 0.011%, the performance of alloy 5 in the PVR test can be improved, without the possibility however of completely eliminating the strong hot cracking sensitivity of niobium stabilized fully austenitic weld metals. Brown and Koch [67] recommend phosphorus to be limited to a maximum of 0.010% and sulphur to be limited to a maximum of 0.005%, whereas Matsuda et al. [349] call for a limitation of 0.010% max. on the total amount of sulphur and phosphorus. According to Ogawa and Tsunetomi [351], the contents of phosphorus and sulphur should even be limited to 0.002% max. each. Husemann [352] and Lorenz, Fabritius and Kranz [353] recommend the use of unstabilized fully austenitic steels for thick walled components which entail less problems with regard to hot cracking. If, however, for reasons of extra high corrosion resistance niobium stabilization together with a fully austenitic structure cannot be avoided, every attempt must be made from a design and manufacturing point of view to keep the hot cracking sensitivity inside the component as low as possible. At the same time more stringent limitations on the crack promoting elements and impurities both in a stabilized steel and in a niobium containing weld metal must be provided than would normally be required for unstabilized fully austenitic materials. Which limits are to be followed or whether niobium free fully austenitic steel grades and weld metal grades with low carbon content (L grades, see also ANSI/AWS A 5.4—81) which are not so hot cracking sensitive, can be used, must be determined individually for each specific case.

5.3.5 Influence of Welding Parameters and Material Thickness on the Hot Cracking Sensitivity of Austenitic Stainless Steel Weldments

As mentioned earlier, the hot cracking sensitivity of weldments and welded components is determined by the interaction of mechanical and metallurgical criteria. Amongst the metallurgical criteria, primary delta ferrite solidification of the weld metal is by far the most effective measure for a substantial improvement in the hot cracking resistance of an austenitic weld metal (see sections 5.3.1 and 5.3.2). However, the primary delta ferrite solidification can be influenced by welding parameters, for instance nitrogen pickup from the surrounding atmosphere during welding may cause a transition from primary delta ferrite to primary austenite solidification. The reason for such a transition is found in the strong austenitizing effect of nitrogen. This effect during welding with coated electrodes is discussed by Tösch and Schabereiter [34, 323], Long and DeLong [184], Kobayashi, Kuwana and Kikuchi [354], Asakura, Wachi and Watanabe [355], Okagawa, Dixon and Olson [432] and Bekkers and Berkhout [356]. Due to the high nitrogen solubility of austenitic chromium-nickel weld metal, welding with excessive arc lengths may produce increases in the nitrogen content by approx. 0.10—0.12%. With a typical nitrogen content of 0.05% in a 20/10 chromium-nickel weld metal (AWS E 308), this means an increase in the nitrogen content within the weld metal of 0.05—0.07%

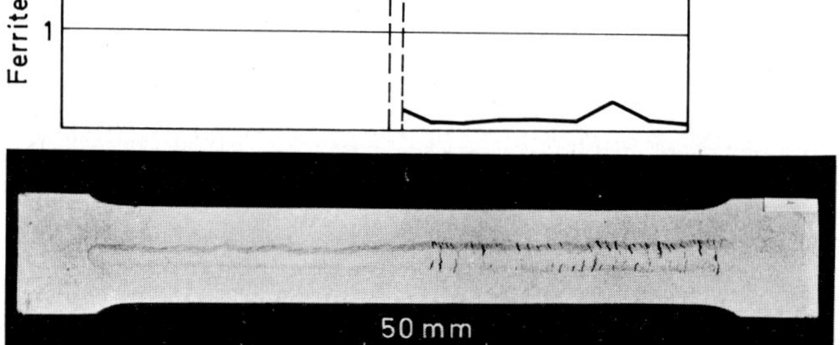

Fig. 102. Influence of the arc length on nitrogen content, delta ferrite content and hot cracking sensitivity of weld metal when investigated by the PVR test with a constant elongation rate of 30 mm/min. The weld was deposited with a 4 mm rutile covered electrode with a welding current of 135 A. Chemical composition of diluted weld metal in %: Short arc—C = 0.011, Si = 1.20, Mn = 1.10, Cr = 17.05, Ni = 10.80, N = 0.065; long arc—C = 0.011, Si = 1.10, Mn = 1.0, Cr = 16.60, Ni = 10.80, N = 0.110. According to Tösch and Schabereiter [34]

which, according to the DeLong diagram results in an increase of the nickel equivalent of 1.5–2.1 and a decrease in delta ferrite content by about 3–4 FN. Thus, weld metal may solidify fully austenitic and become sensitive to hot cracking. According to Tösch and Schabereiter [34] Fig. 102 shows very distinctly the influence that the arc length and the corrresponding nitrogen pick up has on the delta ferrite content and on the hot cracking behaviour of austenitic weld metal during the PVR test. The PVR specimen shown in Fig. 102 has been deformed during test welding with a constant elongation rate of 30 mm/min with the left half of the weld being deposited with a short arc and the right half with a long arc. The right hand half of the weld shows primary solidification to austenite and numerous solidification cracks whereas the left half with primary delta ferrite solidification is free of cracks. This welding error may occur more often with rutile coated electrodes than with basic ones, because with rutile electrodes the coating formulation makes it easier to draw a longer arc than it is possible with basic coated electrodes which due to their coating design and welding characteristic cannot be welded successfully with a long arc.

But even with similar arc lengths the weld metal from basic coated electrodes will normally pick up a little less nitrogen than the weld metal from rutile coated ones. In submerged arc welding according to Killing and Thier [357] and in GMA welding according to Kotecki [358] nitrogen can also be picked up from the surrounding atmosphere, especially in the case of excessive arc voltages and when welding with an electrode extension that is too long. Correct protection of the welding zone by the shielding gas is also of importance here, with the pickup of nitrogen being considerably promoted by a draft or wind. The use of a shielding gas with a higher CO_2 content may also have a negative effect, due to the carburizing effect of CO_2, the carbon content in the weld metal will be increased. Consequently, the strong austenitizing effect of carbon may reduce the delta ferrite content or even promote solidification to primary austenite instead of primary delta ferrite crystals.

When welding austenitic steels which due to their chemical compositions are located well inside the fully austenitic region of the DeLong diagram, with welding parameters which cause relatively high amounts of dilution between the weld metal and the base metal, the delta ferrite content may be decreased to such an extent that hot cracking will occur. All welding processes that produce a high degree of dilution will promote this effect. If such steels must be welded, it is useful first to measure the delta ferrite content of the diluted weld metal by preliminary bead on plate tests, in order to find out if the chosen welding parameters and welding processes will still guarantee sufficient delta ferrite contents in the diluted weld metal.

Austenitic chromium-nickel stainless steel weld metal with primary solidification to delta ferrite generally shows such a low hot cracking susceptibility that no further measures for the improvement of the hot cracking resistance are normally required unless they are demanded from the base metal point of view. The latter may happen if the steel grade in question shows insufficient cleanliness and strong segregations or is located with regard to its chemical composition well inside the fully austenitic region of the DeLong diagram so that liquation cracks will appear in the HAZ. In such cases measures must be taken as discussed in section 9.3 for fully austenitic weldments.

Fully austenitic weld metal, on the other hand, shows such a high crack sensitivity that it is normally necessary to take special preventive measures when carrying out welding operations. The influence of the material thickness is manifold. With increasing wall thickness, crack sensitivity also increases. In contrast to low alloyed steels which show a tendency towards martensite formation in the HAZ during cooling, the influences which determine crack sensitivity in austenitic steels are mainly mechanical. With increasing material thickness it becomes less likely that the welding stresses will become relieved by the deformation of the whole component. In thicker materials, stress concentration and internal strain caused by material shrinkage become stronger and are more concentrated in the welding zones. The mechanical effect on the crack formation becomes important when welding heavy material sections. According to Klug [306], the change from compressive to tensile stresses in the welding zone take place more quickly and at higher temperatures in thick materials than in thin ones. Thus, both the weld metal and the HAZ are subjected more quickly and at higher temperatures to tensile stresses, a fact which promotes a high tendency towards the formation of hot cracks. The practical measures with regard to welding parameters and welding technology which must be employed in order to avoid hot cracking during the welding of fully austenitic steels and filler metals are discussed in detail in section 9.3.1.

6 Welding Metallurgy of Ferritic Stainless Chromium Steels with Carbon Contents Below 0.15%

Depending upon the alloying contents of chromium, nickel and carbon which are present in stainless steels, the structure after cooling to ambient temperature will consist of varying amounts of delta ferrite, austenite and martensite. Table 12 shows average values of the typical analysis of some weldable chromium steels, the carbon content of which has been set below 0.15% in order to improve the weldability. There are two main groups of steel, one with 12% and the other with 17% chromium, which differ mainly with regard to their corrosion resistance. With the addition of 1—2% molybdenum, the pitting resistance and retention of hardness can be improved. The corrosion behaviour of these grades, with special consideration towards welding is discussed by Bäumel [206, 207, 227] and by Honeycombe and Gooch [361].

The iron-chromium constitution diagram (Fig. 1) as well as the concentration profiles in the iron-chromium-carbon system (Fig. 10 and 16) indicate that steels with 12% chromium are at about 1050° C already located near the borderline of the γ area and in the case of lower carbon contents already inside the $(\delta + \gamma)$ area. Steels with 17% chromium are located in the $(\delta + \gamma)$ area with relatively high proportions of delta ferrite (see also section 3.2: $\delta - \gamma$ transformation). During air cooling austenite is always transformed to martensite at the subsequent γ-α transformation because the formation of pearlite and bainite will due to the higher chromium contents present be considerably extended to longer periods of time, as indicated by the TTT diagrams for ferritic chromium steels (see also sections 3.4 and 3.4.2). Fig. 81 shows the TTT diagram of a 12% chromium steel upon which the cooling curves for

Table 12. *Chemical composition (average values) of stainless ferritic steels with carbon contents below 0.15%*

Steel grade	AISI Standard	Chemical composition in %			
		C_{max}	Cr	Mo	Others
12 Cr	410 S	0.08	12.3	—	—
12 CrAl	405	0.08	12.3	—	Al = 0.2
12 CrTi	409	0.08	11.2	—	Ti = 6 × % C
12 Cr	410	0.15	12.3	—	—
17 Cr	430	0.12	17.0	—	—
17 CrTi	430 Ti	0.10	17.3	—	Ti = 5 × % C
17 ClrMo	434	0.12	17.0	1.0	—

single pass welds have been plotted. Even with relatively high preheating temperatures and heat inputs, these curves are outside the area of pearlite transformation which means that, after welding and cooling to ambient temperature, the HAZ and the weld metal will always show a martensitic structure. The transformation behaviour of different steels with 12—14% chromium is discussed by Peter and Matz [364]. The beginning of martensite transformation is located within the range of 200—300° C and the martensite hardness is, depending upon the carbon content in the range of 450—720 Vickers. In the case of air cooling, the structure will consist of martensite and delta ferrite, with the latter's content being in the range of 20—30% for 12% chromium steels and in the range of 50—80% for 17% chromium steels. In steels with a carbon content of about 0.1% the martensite is relatively hard, brittle and susceptible to corrosion. The 12% chromium steels are always heat treated. The heat treatment consisting of an air or oil hardening process followed by tempering in the range of 700—750° C. The 17% chromium steels are quenched and tempered or annealed. After tempering, the structure of these chromium steels consists of tempered martensite with finely dispersed carbides and delta ferrite. The influence of other alloying elements is discussed by Briggs and Parker [362]. With regard to the general properties of stainless ferritic chromium steels, please refer to the pertinent literature.

The structure of similar weld metal in the as-welded-condition consists of delta ferrite and martensite with small quantities of residual austenite. Segregations resulting from solidification are relatively rare because liquidus and solidus areas are located close to each other and incline rather steeply (see Fig. 7).

If the typical chemical analysis of grades 12 Cr and 17 Cr are plotted in the modified Schaeffler diagram (see Fig. 103), rectangular fields are obtained which

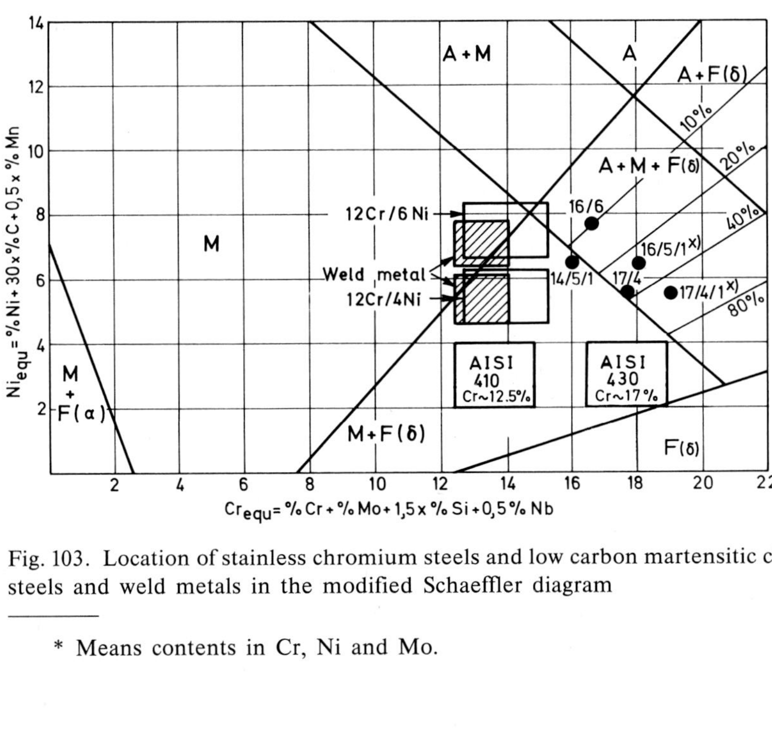

Fig. 103. Location of stainless chromium steels and low carbon martensitic chromium-nickel steels and weld metals in the modified Schaeffler diagram

---

* Means contents in Cr, Ni and Mo.

indicate a relatively high delta ferrite content. In chromium alloyed weld metal, however, delta ferrite does not represent a desirable structural constituent because of its unfavourable influence on impact energy values. Therefore its content should be kept as low as possible. For this reason the carbon content in the weld metal is often increased to 0.1% which coincides with the 12 Cr and 17 Cr rectangles in Fig. 103. In 12% chromium weld metal, the delta ferrite content is normally in the range of 20–30% and in the 17% weld metal it is between 50 to 80%. The distribution of delta ferrite within a weld is not uniform. The highest contents being normally found in those areas of the weld which cool down most quickly, such as the underside of the weld beads.

Because of the high content of brittle martensite, weld metal with 12% chromium shows in the as-welded condition low elongation and low impact toughness. For this reason, it is generally annealed for longer periods of time at 700–750° C. In this way, it is possible to obtain higher elongation values, but the impact energy values however, remain rather low, because of the delta ferrite content of 20–30%. In steels with 17% chromium too, elongation values can be improved by annealing at 750° C, but the impact toughness is low because of the high delta ferrite content. The major reason for this phenomenon is the tendency of delta ferrite to form a coarse structure by recrystallization. According to Woltron [193], coarse grain formation during welding can be reduced by the addition of titanium and molybdenum because these elements have a nucleating effect on the structure. But even with such additions, the increase in impact energy values is rather small. Therefore, 17% chromium steels are mainly used for weldments which do not require increased toughness properties.

The development of ferritic steels with ultrahigh purity (superferrite) possessing high resistance to chloride induced stress corrosion cracking did not lead to the progress expected in this field, because of their poor weldability and low HAZ impact toughness. Such steels show very low contents of the interstitially dissolved elements such as carbon and nitrogen with the total amount of (C + N) being normally limited to 150 ppm max. A thorough discussion of these steels is given by Lennartz and Kiesheyer [234] as well as in the compendium "Stainless Steel 77" [235]. A survey of the present state of knowledge and the properties that can be obtained from the interstitial ferritic stainless steels is supplied by Krysiak [460]. The author concludes that further research in this area is necessary and suggests some basic research projects which are required in his opinion for an improvement in this group of steels.

6.1 Practical Welding of Stainless Chromium Steels

The weldability of chromium stainless steels and their weld metals is reduced by the following:

- The formation of brittle martensite in the HAZ and the weld metal, particularly with 12% chromium materials because of their high content of brittle martensite.
- The relatively low toughness of the delta ferrite and the strong tendency of this structural constituent towards coarse grain formation in the HAZ and in the

weld metal. This effect is particularly felt with 17% chromium steels because of their high delta ferrite content of 50–80%.
— The strong hydrogen sensitivity of the martensitic structure of both HAZ of the base metal and weld metal.

The formation of brittle martensite produces a strong tendency towards the cold cracking of weldments which may be increased still further by high hydrogen contents giving rise to hydrogen induced crack formation. The strong tendency of delta ferrite towards coarse grain formation causes embrittlement in the HAZ and in the weld metal. The unfavourable properties mentioned above are the reasons why the weldability of ferritic chromium stainless steels is considered to be relatively poor. When welding 12% and 17% chromium steels with matching filler metals both HAZ and weld metal show these unfavourable properties. In non-matching weldments, e.g. welding with austenitic filler metals, the negative influences are limited to the HAZ. In practical applications, both methods are used. Gut and Werner [365] and Gysel and Mayer [366] report on the preheating and post weld heat treatment required for the welding of these 12% chromium steels.

The following procedures have been established for the correct welding of 12% chromium steels with carbon contents below 0.15%:

— The use of matching filler metals containing 12% Cr should be employed if similar properties are required from both the steel and the weld metal. This applies in particular to similar colour, tensile properties and fatigue strength, the latter is particularly important for components subjected to severe cyclic or alternating stresses.
— The use of non matching austenitic filler metals is recommended in all cases where it is not possible to carry out a postweld heat treatment (e.g. the repair welding of assembled components).
— Coated electrodes and submerged arc fluxes should always be rebaked immediately prior to welding in accordance with the manufacturer's instructions or at least to 300° C for 2 hours min in order to keep the content of diffusible hydrogen in the weld metal to a minimum.
— A preheating temperature of 200–300° C is highly recommended. The interpass temperature should be as high as the preheating temperature, with the upper temperature applicable to thick materials and the lower temperature to thin materials.
— When welding thick materials with high interpass temperatures a transformation of the welding zone from austenite to martensite does not take place due to the high interpass temperatures. For this reason, the welded component must be slowly cooled immediately after welding to 100–150° C and held at this temperature for at least one hour–if possible inside the furnace–to allow for martensite transformation. Such transformation treatment should also be carried out when joining ferritic steels using austenitic filler metals where subsequent post weld heat treatments are not possible due to practical reasons. When the component has been welded using matching filler metals, annealing in the range of 700–750° C should take place immediately from the martensite transformation temperature without intermediate cooling to ambient temperature. When austenitic filler metals are used and annealing is necessary due to

manufacturing regulations, it should be carried out at a temperature below the sigma phase precipitation range of the weld metal (see also section 4.2, page 125).

The following procedures have been established for the welding of 17% chromium stainless steels using matching filler metals:

— A preheating and interpass temperature of 200—300° C reduces the danger of crack formation and also reduces the delta ferrite content thus leading to higher elongation values in the tempered weld metal to be obtained.
— A tempering treatment at 700—750° C carried out immediately after welding without intermediate cooling to ambient temperature is necessary to produce weld metals with better ductility, in this way the elongation can be increased to about 10%. In the impact toughness test, however, the HAZ and the weld metal show low values because of the high delta ferrite content.
— The heat input applied during welding should be limited to keep coarse grain formation in the HAZ of the base metal and the weld metal to a minimum.
— Coated electrodes and submerged arc fluxes should be rebaked according to the manufacturer's instructions or at least at 300° C for two hours minimum.
— The chromium content of the weld metal should be kept below 17.5% to ensure that the delta ferrite content will not rise above 70%.

6.2 Mechanical Properties of Stainless Steel Weld Metals with 12 and 17% Chromium

Table 13 lists some typical mechanical properties for stainless 12% and 17% chromium alloyed weld metals. In the as-welded-condition, 12% chromium weld metal shows low elongation values. After annealing about 20% elongation can be obtained, but impact toughness remains low. The properties of the weld metal in the quenched and tempered condition show no marked differences from those of the parent steel grade.

17% chromium weld metal is brittle in the as-welded condition. During tensile testing it often fractures prematurely at low elongation values. By annealing, the elongation values can be substantially improved, due to the high delta ferrite content, however, the impact toughness remains low. With chromium contents in the weld metal above 17.5%, elongation values in the annealed state decrease. Because of the high delta ferrite content, weld metal with 17% chromium shows lower yield strength and tensile strength values than weld metal with 12% chromium. Weld metal with 17% chromium and 1—2% Molybdenum shows similar mechanical properties. In order to limit the delta ferrite content, the chromium content can be reduced or the carbon content increased, if this is permitted by the corrosion resistance requirements.

Table 13. *Mechanical properties and chemical composition of ferritic stainless steel weld metals with 12 and 17% chromium*

			Mechanical properties					Chemical composition in %				
Grade °C	PHT/IPT*	Condition	0.2 *YS* N/mm²	1.0 *YS* N/mm²	*TS* N/mm²	*El.* %	*IE* CVN J	C	Si	Mn	Cr	Ni
12 Cr	250/250	as welded	793	927	1042	4.8	–					
12 Cr	250/250	annealed 750° C/2h	608	634	769	20.8	12					
12 Cr	250/250	annealed 850° C/2h	444	468	759	21.2	15	0.071	0.60	0.70	12.27	0.36
12 Cr	250/250	quenched and tempered 950° C/air + + 750° C/2h	458	505	627	22.1	65					
17 Cr	250/250	as welded	482	–	536	2.1	–					
17 Cr	250/250	annealed 750° C/2h	409	497	593	20.1	10	0.092	0.45	0.31	16.85	0.23

* PHT = preheating temperature, IPT = interpass temperature.

6.3 Precipitation Phenomena in Ferritic Stainless Chromium Steels and Weld Metals

Chromium steels show a marked tendency towards the precipitation of the carbide $M_{23}C_6$ (see sections 4.1, 4.1.3 and 4.1.4). Fig. 85a (page 108) shows the TTC diagram of a 17% chromium steel. The region of corrosive attack extends into the range of the cooling rates normally experienced during welding operations, especially if preheating prior to welding is used which is generally the case with ferritic chromium steels. Regarding the influence of $M_{23}C_6$ precipitation on corrosion resistance, please refer to section 4.1.3.1 as well as the pertinent literature listed there. In chromium steels with 17% chromium, sigma phase precipitation starts only after rather long periods of time in the range of about 1000 hours at temperatures of about 550° C. With the addition of 2% molybdenum, the beginning of sigma phase precipitation is shifted to shorter times of about 200 hours at temperatures of around 600° C. The precipitation phenomena are described in detail in section 4.2.2 with references to the pertinent literature. In steels with about 14% chromium, 475° C embrittlement starts after about 700 hours. In steels with 18% chromium, this time is reduced to about 1 hour (see Fig. 94, page 142). The processes occurring during 475° C embrittlement are discussed in section 4.3.

7 Welding Metallurgy of Low Carbon Chromium-Nickel Martensitic Stainless Steels (Soft Martensitic Steels)

According to Gysel, Gerber and Trautwein [369] the relatively poor weldability of chromium stainless steels, their cold cracking sensitivity and the often unsatisfactory mechanical properties obtained from welded joints led at the end of the fifties to the development of low carbon martensitic chromium-nickel steels. The basic idea behind this development was first of all the lowering of the carbon content in order to improve the toughness of the martensitic structure, to diminish the cold cracking sensitivity and to achieve a structure by adding 4—6% nickel which was as near as possible free from delta ferrite.

From the concentration profiles of the iron-chromium-nickel constitution diagram in Fig. 11, the strong influence of nickel on the area of austenite (γ-crystals) is quite apparent. Fig. 9a which represents a profile at a chromium-nickel ratio of 3 : 1, illustrates the course of solidification and cooling of a 12/4 or 15/5 chromium-nickel alloy. Initially, these alloys solidify to δ crystals. At around 1300° C, transformation of δ into γ crystals starts which ends, in the case of equilibrium conditions, at around 1200° C. With the actual cooling rates experienced during a welding operation, small amounts of delta ferrite are supercooled during $\delta - \gamma$ transformation. Similar to pure chromium steels, the $\gamma - \alpha$ transformation which is discussed in detail in section 3.4 leads to a martensitic structure containing small quantities of supercooled residual austenite which, depending upon the alloy type, may be in the range of 1—10%. Contrary to chromium steels, the soft martensitic steels show because of their low carbon content and the addition of nickel, good toughness values with a corresponding material hardness of 350—400 HV. At 200—250° C, the martensite transformation temperature (M_s) is lower than that of the pure chromium steels (see Fig. 82). Thus, after cooling has been completed, the structure consists of "soft" martensite with small amounts of supercooled delta ferrite and austenite. With a subsequent annealing treatment, the toughness can be increased with a corresponding decrease in the hardness and tensile strength values.

Table 14 shows the average chemical compositions of some low carbon martensitic steels according to Brezina [93]. The most important alloys are the 13/4 and the 13/6 chromium-nickel steels. Their properties are reported on by Baggström [370], Cristina and Gysel [371], Souresny and Sauer [372], Gysel [373], Gysel and Gerber [374], Niederau [199], Brezina [93, 437], Kulmburg et al. [194], Heimann and Strom [375] and Gümpel, Hook and Strom [461]. Their corrosion behaviour is discussed by Süry and Brezina [239] and the phase precipitations by Brezina et al.

Table 14. *Typical chemical compositions of low carbon martensitic chromium-nickel-molybdenum stainless steels and precipitation hardening grades containing copper and niobium to Brezina [93]*

Steel grade	AISI Standard	Chemical composition in % average values					
		C_{max}	Cr	Mo	Ni	Cu	Nb
13 Cr 1.5 Ni	414	0.05	13	0–0.4	1–2	–	–
13 Cr 4 Ni	–	0.05	13	0.4	4	–	–
13 Cr 4 Ni 1.5 Mo	–	0.05	13	1.5	4	–	–
13 Cr 6 Ni	–	0.05	13	0.4	6	–	–
13 Cr 6 Ni 1.5 Mo	–	0.05	13	1.5	6	–	–
16 Cr 6 Ni	–	0.05	16	–	6	–	–
16 Cr 5 Ni 1.5 Mo	–	0.05	16	1.5	5	–	–
17 Cr 4 Ni	–	0.05	17	–	4	–	–
17 Cr 4 Ni 1.5 Mo	–	0.05	17	1.5	4	–	–
14 Cr 5 Ni 1.5 Mo + + Cu + Nb	–	0.05	14	1.5	5	1.5	0.2
17 Cr 4 Ni 3 Cu + Nb	630	0.05	17	–	4	3	0.3

[236]. The 475° C embrittlement is thoroughly discussed in section 4.3 (see also Fig. 94, page 142), together with a number of references to pertinent literature on low carbon martensitic steels [238, 278, 279, 280, 461].

Low carbon martensitic steels and steel castings are always quenched and tempered. Depending upon the grade, quenching normally takes place in the range of 950–1050° C with subsequent tempering at 600° C. The special effect of the formation of finely dispersed stable austenite during tempering is discussed in section 3.4.3 (see also Fig. 83). Some of the low carbon martensitic steels with chromium contents above 14% can also be precipitation hardened. These are solution annealed at temperatures of 1000–1050° C and then precipitation hardened, depending upon the grade, in the range between 450–580° C. A comprehensive discussion on the heat treatment and the properties of precipitation hardening low carbon martensitic steels is given by Brezina [93], Guiraldeng et al. [243], Mende and Oppenheim [241], Püber and Auer [280], Koren and Diebold [376], Brezina and Sonderegger [240], Lindscheid and Mayer [242] and Schmidtmann and Kieselbach [377].

If the typical chemical compositions of 13/4 and 13/6 chromium-nickel steels are plotted in the modified Schaeffler diagram, the rectangular regions shown in Fig. 103, page 173, for the grades 13 Cr 4 Ni and 13 Cr 6 Ni are obtained. It is apparent that the addition of 4–6% nickel produces a marked increase in the nickel equivalent especially when compared with the values for ferritic chromium steels. This brings about a substantial reduction in the delta ferrite content and an increase in the martensite content. The 13/4 and 13/6 chromium-nickel weld metals require a restriction of the manufacturing analysis limits and a further reduction in the delta ferrite content because in most cases the weld metal is not quenched and tempered like the parent steel but only tempered. The restricted analysis limits for 13/4 and 13/6 chromium-nickel weld metals are shown in Fig. 103 by the hatched areas.

Similarly in steel castings too the analysis limits are often restricted to limit the delta ferrite content. Steels with higher chromium contents are mainly located in the Schaeffler diagram in the three-phase austenite + martensite + delta ferrite sector. In Fig. 103, they are plotted according to the average analysis given in Table 14 (page 180) and characterized by the contents of chromium, nickel and molybdenum.

7.1 Practical Welding of Low Carbon Martensitic Stainless Steels

A major advantage of the low carbon chromium-nickel martensitic steels when compared with ferritic chromium steels is their good weldability, with particular emphasis on the 13 Cr/4 Ni and 13 Cr/6 Ni alloys which have been investigated by Cristina and Gysel [371], Gysel [378], Howard [379] and Brezina [93], Asakura and Nihei [462]. The properties of weld metals containing 12—14% chromium and 1—6% nickel together with the most suitable welding technology necessary for the production of crack-free weldments are given by Tösch, Perteneder and Rabensteiner [94] and for the improvement of 13/4 weld metal by Rabensteiner, Perteneder and Tösch [191]. The weldability of low carbon martensitic steels and weld metals can be characterized by three particular features:

— The formation of tough low carbon martensite in the HAZ and the weld metal which greatly reduces the cold cracking sensitivity.
— The low delta ferrite contents which are normally for 13/4 weld metal in the range of 1—4% and for 13/6 weld metal below 1%. Thus, the tendency towards coarse grain formation is largely reduced.
— The relatively strong hydrogen sensitivity of the martensitic structure in the HAZ and the weld metal. Hydrogen induced cracks may occur with contents of diffusible hydrogen in the weld metal of $H_{DM} > 5$ ml/100 g.

The following welding technology has been successfully applied for the crack-free welding of low carbon martensitic 13/4 and 13/6 chromium-nickel steels:

— Welding should only be carried out with a matching filler metal containing a maximum of 0.040% carbon, approx. 12% chromium and approx. 4 or 6% nickel giving a maximum content of 5% delta ferrite in the weld metal.
— The content of diffusible hydrogen (H_{DM} according to ISO 3690 and AWS A 4.3-86) deposited in the weld metal should be below 5 ml/100 g. Coated electrodes and submerged arc welding fluxes must be rebaked according to manufacturer's instruction or at least at 300° C for two hours minimum to obtain the required hydrogen level.
— Workpieces with heavy wall thicknesses (above 20 mm) should be preheated prior to welding to about 100° C.
— The best resistance against cracking is achieved when the welding operations are carried out below the martensite transformation temperature of the steel used for the workpiece with an interpass temperature of between 100—150° C (see also Fig. 82, page 99 and section 3.4.2). Thus the structure of each weld

bead becomes largely transformed from austenite to martensite and is then tempered during welding by the subsequent weld passes. The martensite transformation of a weld bead can be followed with the aid of a ferrite measuring device or more simply by using a small bar magnet (martensite is ferromagnetic but austenite is not, see also Fig. 82 [page 99] and supplementary literature [94] and [191]). This method of welding should generally be used for all welded components that are not going to be tempered after welding.

— If higher impact values are required, a post weld heat treatment such as tempering or quenching and tempering is necessary (see section 7.2 and heat treatment temperatures in Fig. 104, page 184 and Table 15, page 183).
— In practical applications, e.g. when using high deposition welding processes (SAW or GMAW high current processes) or in the case of the repair welding of thick-walled steel castings, it is sometimes common practice to weld with a high interpass temperature in the range of 250–300° C, i.e. above the martensite transformation temperature of the steel. In such cases, it is absolutely necessary to transform the austenitic structure of the HAZ and of the weld metal completely to martensite again by controlled cooling of the workpiece down to about 100° C prior to starting any post weld heat treatment. However the complete cooling of the welded components down to ambient temperature should be avoided, particularly in the case of thick walled components or where the ambient temperatures are very low. The best procedure is to start the post weld heat treatment immediately after completion of the martensitic transformation.

Precipitation hardening low carbon martensitic steels are only rarely welded. In order to achieve the same properties in the weld metal and the base metal, it may be sometimes necessary to subject the welded component to a full heat treatment consisting of solution annealing, tempering and aging (see Table 15, page 183). Welding should be carried out with matching filler metals. In general, the same recommendations apply for the welding of precipitation hardening steels as those stated above, but no general rules can be established. It is sensible to consider each individual case separately and decide which welding procedure and heat treatment to employ to suit the specific circumstances.

7.2 Mechanical Properties of Low Carbon Martensitic Stainless Steel Weld Metal

Low carbon martensitic materials with nickel contents above 3.5% show a metallurgical peculiarity, i.e. the formation of finely dispersed austenite after tempering at temperatures above 580° C (see also section 3.4.3). This effect produces an increase in the impact toughness of the 13/4 weld metals, with the highest values being obtained after tempering at temperatures, between 600 and 620° C. At higher tempering temperatures impact toughness decreases again because of the transformation of tempered austenite to martensite during cooling (see Fig. 83, page 101). Table 15 represents a survey of the weld metal properties after various heat treatments including mechanical properties, chemical compositions and structural proportions of delta ferrite and austenite in the martensitic matrix as

Table 15. *Properties and chemical compositions for different grades of low carbon soft martensitic Cr-Ni-Mo stainless steel weld metals*

Grade	AWS A5.4-81	Welding	IPT	Condition	Mechanical properties				Chemical composition in %					Structural proportion in %		Martensite trans-
				heat treatment	0.2 *YS*	*TS*	*El.*	*IE/J*						Delta		formation
Short designation	resp. 9-81	process	°C	temperature °C/time	N/mm²	N/mm²	%	CVN	C	Cr	Mo	Ni	Others	ferrite	Austenite	M_S point°C
12 Cr/1 Ni L	E 410 Ni	SMAW	230	as welded	755	994	12.0	23	0.032	12.01	–	1.45	–	12–18	0–2	290
			230	annealed 700/2h	554	678	21.8	85								
			150	annealed 700/2h	541	673	21.0	68								
13 Cr/4 Ni/Mo L	E 410 NiMo	SMAW	130	as welded	891	1073	13.0	36	0.031	12.28	0.50	4.58	–	1–3	2–6	240
				tempered 600/2h	720	891	19.0	73								
				quenched + tempered 950/1h/air + 600/2h	710	864	19.5	95								
13 Cr/4 Ni/Mo L	ER 410 NiMo	GMAW	130	tempered 600/2h	775	885	18.9	88	0.022	12.82	0.48	3.96	–	1–5	2–5	240
13 Cr/4 Ni/Mo L	ER 410 NiMo	GTAW	130	tempered 600/2h	742	855	21.7	132	0.018	12.48	0.51	4.44	–	1–3	2–5	240
13 Cr/6 Ni/Mo L	–	SMAW	120	as welded	834	1062	12.8	31	0.028	12.36	0.58	5.76	–	0–2	8–10	220
				tempered 600/2h	621	879	18.0	58								
				quenched + tempered 950/1h/air + 600/2h	637	825	20.8	68								
16 Cr/6 Ni/Mo L	–	SMAW	80	as welded	480	1115	7.4	27	0.037	16.11	0.57	5.68	–	7–10	~ 12	120
				tempered 540/10h	785	984	16.4	42								
16 Cr/5 Ni/ 1 Mo L	–	SMAW	70	as welded	432	1080	8.8	33	0.044	15.97	1.36	4.93	–	9–12	25–33	120
				tempered 540/10h	743	1045	15.7	38						9–12	16–26	
				tempered 580/10h	708	1038	14.2	41						9–12	14	
14 Cr/6 Ni/ 2 Mo L + Cu + Nb	–	SMAW	70	as welded	363	1054	13.2	32	0.035	13.53	1.95	6.18	Cu = 2.7	1–3	> 50	110
				quenched + tempered 1050/2h/air + 580/4h	787	1032	19.2	52					Nb = 0.2	1–3	12–22	
16 Cr/6 Ni/Mo L + Cu + Nb	–	SMAW	70	as welded	417	957	13.6	45	0.031	15.89	0.56	5.83	Cu = 1.6	–	–	100
				tempered 540/3h	633	988	18.3	43					Nb = 0.2			
17 Cr/5 Ni/Mo L + Cu + Nb	E 630	SMAW	70	as welded	742	943	5.3	40	0.042	16.63	0.42	4.48	Cu = 3.4	–	–	–
				quenched + tempered 1050/1h/air + 480/12h	1236	1382	14.2	14					Nb = 0.3			

well as the temperature at the beginning of martensite transformation (M_S point). The knowledge of this temperature is of importance when selecting preheat and interpass temperatures. As indicated by the mechanical properties of the different weld metal grades, low carbon martensitic weld metals already show in the as welded state substantially higher toughness values than the ferritic chromium alloyed weld metals without nickel addition (see Table 13, page 177). After annealing, toughness levels can be obtained which, considering their relatively high strength, are quite satisfactory and permit their application in the cryogenic range down to – 100° C the highest impact toughness values are obtained when welding the 12 Cr/ 4 Ni steel grades by the GTAW process. This is because of the high purity weld metal that is obtained from this process. In precipitation hardening low carbon martensitic weld metals, the yield strength and tensile strength values after solution annealing and precipitation hardening are even higher, especially if elements such as copper and niobium have been added (see also supplementary literature [93, 240, 241, 242, 280, 376, 377, 461]).

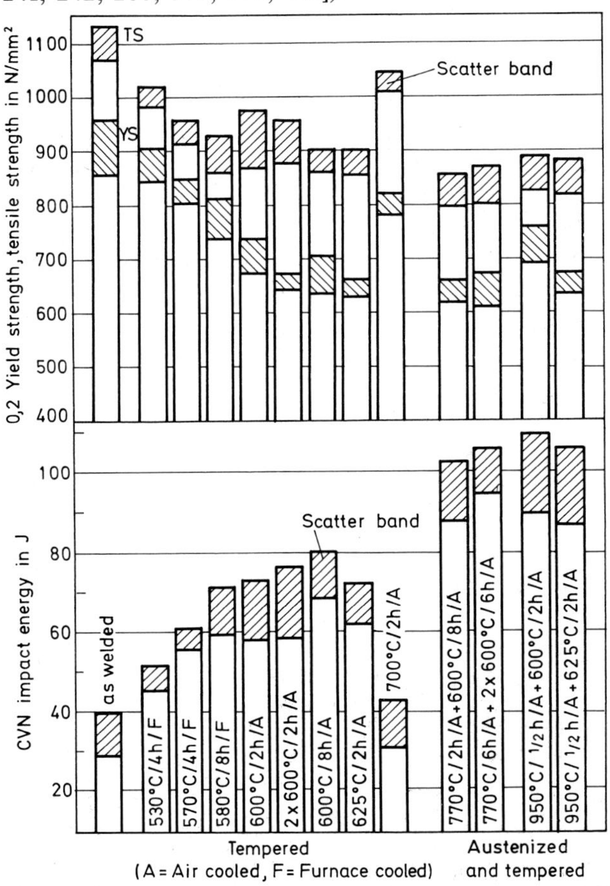

Fig. 104. The mechanical properties of low carbon martensitic 13/4 chromium-nickel SMAW weld metal in the as-welded-condition and also after different heat treatments according to Rabensteiner, Perteneder and Tösch [191]. Chemical composition in %: C = 0.031, Cr = 12.28, Mo = 0.50, Ni = 4.58, A = Air Cooling, F = Furnace cooling

Fig. 104 presents a survey of the mechanical properties of 13/4 chromium-nickel weld metal after different heat treatments according to Rabensteiner, Perteneder and Tösch [191]. After tempering from the as-welded-condition favourable mechanical properties can be achieved with an annealing temperature of 600° C and tempering times in the range of 2–6 hours. A double tempering process may also improve toughness values. With higher tempering temperatures, the impact energy values drop again, due to the formation of austenite which is transformed back into martensite during cooling. Quenching and tempering produces the highest impact toughness values. It should also be mentioned that it is possible to carry out an austenitizing treatment at 770° C which produces very good impact energy values in the quenched and tempered state.

7.3 Precipitation Phenomena in Low Carbon Soft Martensitic Stainless Steels and Weld Metals

Because of their low carbon contents, the soft martensitic Cr-Ni steels and weld metals show a lesser tendency towards carbide precipitation than do ferritic 12–17% Cr steels, because at higher temperatures carbon is completely dissolved in the austenitic matrix. This state is still maintained during transformation from austenite to martensite by way of a forced solution. Carbides only precipitate during tempering, partially in the form of coarse carbides at the grain boundaries, but mainly in the grain centers as fine $M_{23}C_6$ together with carbonitrides, producing in most cases a chromium depletion. For this reason, in low carbon soft martensitic steels the corrosion is not of an intergranular type but generally a corrosion of the surface of the grains. The above processes are discussed in detail together with the corrosion phenomena in section 4.1.3.2 which also includes a list of pertinent literature.

There is practically no embrittlement of the low carbon martensitic steels or their weld metals due to sigma phase precipitation because of the limited chromium content. Embrittlement could only possibly occur in steel grades with 17% chromium if held for long periods of time in the temperature range of about 550° C, however, due to the precipitation hardening effect at around 480° C of these grades, the practical application of such steels is limited to lower temperatures. Because of the relatively low chromium contents of low carbon martensitic materials, 475° C embrittlement will only take place after rather long periods of time (see Fig. 94c, page 142). Gümpel, Hook and Strom [461] investigated the embrittlement of 13 Cr/4 Ni steel during long term aging in the temperature range of 325–525° C at 20,000 hours. The greatest decrease in impact toughness was found at a temperature of 425° C. This reduction in toughness began after an aging time of about 100 hours and resulted in a drop in the impact energy values of about 85% after 20,000 hours. At 525° C the loss of impact toughness was negligible and at 325° C the impact toughness dropped about 25% after an aging time of 20,000 hours. It was found that fine carbide precipitations were formed during long term aging in the above mentioned temperature range. In 13/4 chromium/nickel steels no loss in toughness can normally be found at 300° C even after extremely long periods of time, but such a loss may happen in the 16/6 chromium-nickel steels (see also section 4.3).

8 Welding Metallurgy of Duplex Austenitic-Ferritic Stainless Steels[10]

Common austenitic chromium-nickel stainless steels show only a limited resistance to stress corrosion cracking in chloride containing media when compared with ferritic chromium stainless steels with more than 18% chromium. However on the other hand ferritic chromium stainless steels—including the socalled "Super Ferritics"—are relatively difficult to weld (see chapter 6). During welding the predominantly ferritic structure shows a tendency towards coarse grain formation, particularly in the HAZ. With the carbon content of the common ferritic chromium steels in the range of 0.10% and above, an additional amount of brittle martensite is formed in the structure. Both phenomena cause a drop in toughness and an increased cold cracking sensitivity in the welded joint. The older types of austenitic-ferritic duplex steels containing 0.1—0.2% carbon were also unsatisfactory in terms of todays demands on the toughness and corrosion resistance of weldments, mainly because of their high susceptibility to intergranular corrosion (IC).

This situation has led to the development of austenitic-ferritic Cr-Ni-Mo alloyed duplex steels with low carbon contents and quite often alloyed with nitrogen. They show good resistance towards intergranular, crevice and pitting corrosion as well as stress corrosion cracking and possess good weldability. In comparison with the conventional austenitic Cr-Ni-Mo steels, e.g. AISI 316 L low carbon duplex steels show better corrosion resistance and substantially higher yield and tensile strength values combined with good toughness, provided they are properly welded. Because of their two-phase structure, they are called duplex steels. Their structure consists roughly of equal parts of austenite and delta ferrite. Table 4 (page 118) represents a survey of the chemical compositions of the conventional duplex stainless steels. Table 16 shows the chemical composition requirements and the minimum values for the mechanical properties according to ASTM Standards (see also supplementary literature [477]). The carbon content is normally limited to 0.030% max. resp. 0.040% max. The chromium contents are generally in the range between 21 and 27% with nickel contents of between 4.5—7.5%, molybdenum contents of between 1.0—4.0% and often with nitrogen contents of between 0.1—0.2%. The steel grade type 22 Cr/5.5 Ni/3 Mo/N which is widely used today corresponds to the steel grade UNS S 31803 with the chemical composition as shown in Table 16.

---

10 The term duplex steel is generally used to define steels with a two-phase structure. With stainless steels it is common usage to employ this term for austenitic-ferritic steels with a high delta ferrite content in the range of 30—60%.

Table 16. *Chemical requirements and mechanical properties of the duplex stainless steels in accordance to ASTM Standards to Street [477].*

		Chemical composition requirements in %													
Specification	Designation*	C max.	Mn	P max.	S max.	Si	Ni	Cr	Mo	N	Others	0.2 YS min. N/mm^2	TS min. N/mm^2	El. min. %	RA min. %
ASTM A182	UNS S31200	0.030	2.00 max.	0.045	0.030	1.00 max.	5.50–6.50	24.0–26.0	1.20–2.00	0.14–0.20		450	690–900	25	50
ASTM A240												450	690	25	–
ASTM A789												450	690	25	–
ASTM A790												450	690	25	–
ASTM A182	UNS S31803	0.030	2.00 max.	0.030	0.020	1.00 max.	4.50–6.50	21.0–23.0	2.50–3.50	0.08–0.20		450	620	25	45
ASTM A240												450	620	25	–
ASTM A789												450	620	25	–
ASTM A790												450	620	25	–
ASTM A815												450	620	25	–
ASTM A240 ASTM A479	UNS S32550	0.04	1.5 max.	0.040	0.030	1.0 max.	4.50–6.50	24.0–27.0	2.0–4.0	0.10–0.25	Cu 1.5–2.5	550	760	15	–
ASTM A789 ASTM A790	UNS S32550	0.040	1.5 max.	0.040	0.030	1.0 max.	4.50–6.50	24.0–27.0	2.90–3.90	0.10–0.25	Cu 1.5–2.5	550	760	15	–
ASTM A240	UNS S32900	0.08	1.00 max.	0.040	0.030	0.75 max.	2.50–5.0	23.0–28.0	1.0–2.0	–		485	620	15	–
ASTM A744	CD-4MCu	0.04	1.00 max.	0.040	0.040	1.00 max.	4.75–6.00	24.5–26.5	1.75–2.25	–	Cu 2.75–3.25	485	690	16	–
ASTM A789 ASTM A790	UNS S31500	0.030	1.20–2.00	0.030	0.030	1.40–2.0	4.25–5.25	18.0–19.0	2.50–3.00	–		440	630	30	–
ASTM A789 ASTM A790	UNS S31250	0.030	1.00 max.	0.030	0.030	0.75 max.	5.50–7.50	24.0–26.0	2.50–3.50	0.10–0.30	Cu 0.20–0.80 W 0.10–0.50	440	630	30	–

* New designation established in accordance with ASTM E527 and SAE J1086, Practice for Numbering Metals and Alloys (UNS).

The literature on duplex stainless steels is rather comprehensive (see supplementary literature 19, 168, 170, 244, 247, 248, 380, 381, 382, 383, 384, 385, 386, 387, 388, 389, 463, 464, 465, 466, 467, 477, 478). Special attention should be drawn here to conference proceedings in which corrosion resistance, mechanical properties, weldability and special phenomena regarding the welding of duplex steels are discussed in great detail, i.e.: "International Symposium on Duplex Austenitic Ferritic Steels", St. Louis 1982, with 35 papers [469]; "Symposion on Advanced Corrosion Resistant Steels and Metal Alloys", Oslo 1986, with 29 papers [470]; and "International Conference on Duplex Stainless Steels", The Hague 1986, with 52 papers [471]. The properties of weld metals and the most suitable welding procedures are discussed by Perteneder, Rabensteiner and Tösch [468] and Tösch, Rabensteiner, and Reiterer [476]. Ferrite measurement and the problems of determining relatively high ferrite contents are dealt with by Kotecki [474] and Bonnefois, Blondeau and Catelin [475]. The problems arising from the welding of pipelines made from duplex stainless steels were discussed in six papers at the "Third International Conference of Welding and Performance of Pipelines", London 1986 [472]. A general survey on the properties of duplex stainless steels is given by Schläpfer and Weber [473].

The main decisive influence excerted on the most important properties of duplex stainless steel weld metal is that of the structural proportion of delta ferrite to austenite. In duplex stainless steel which because of solution annealing in the range between 1020° C and 1100° C is very close to a metallurgical equilibrium, the austenite-ferrite ratio is mainly determined by the ratio of ferrite forming elements Cr, Mo and Si to the austenite forming elements Ni, C and N, with carbon playing only a minor role due to its low content normally of 0.02–0.04%. In the course of steel making, this ratio can be set by a suitable modification to the chemical composition so that the steel will be in the range of 50% ferrite and 50% austenite after solution annealing. In order to understand the phenomena occurring during the welding of duplex steels, we shall first look at the equilibrium diagram in Fig. 6, page 8 (lefthand diagram) which shows a concentration profile of the ternary Fe-Cr-Ni constitution diagram at 70% iron. An alloy containing 25% chromium and 5% nickel which in metallurgical terms is very close to the widely used duplex stainless steel 22 Cr/3 Mo/5 Ni, solidifies completely to delta ferrite from the melt. During further cooling, δ-γ transformation starts at about 1200° C with the precipitation of predominantly nodular austenite at the ferrite grain boundaries. During further cooling to ambient temperature, there is only a partial transformation to austenite. The structure contains now some 60% primary delta ferrite and approx. 40% secondary precipitated austenite. If such an alloy is now subjected to a solution annealing treatment, this ratio can be shifted to slightly higher austenite contents. As mentioned earlier, in the practical application of duplex steels, in particular because of the addition of nitrogen, this ratio is around 50%. If the 25 Cr/5 Ni alloy of Fig. 6 is heated to a higher temperature level as is the case, for example, in the HAZ of a weldment, the austenite is transformed back into delta ferrite at temperatures above 1200° C.

Todays advanced duplex steels differ from the 25 Cr/5 Ni alloy discussed above in as much as they are normally alloyed for reasons of improved corrosion resistance with 3% molybdenum and the strong austenitizing element nitrogen. Nitrogen in particular has principally a positive effect on the behaviour of duplex steels during welding. The widely used duplex steel of the type 22 Cr/5.5 Ni/3 Mo which corresponds to ASTM grade S31803 (see Table 16, page 187), contains on average 0.14% nitrogen. For the calculation of the austenitizing effect of nitrogen as compared to nickel, the formula of Espy for nitrogen alloyed steels (see supplementary literature 170) may be employed:

$Ni_{equ} = \%Ni + 30 \times \%C + 0.87$ for $Mn + 0.33 \times \%Cu + (\%N - 0.045) \times 30$ when $N = 0.0$–$0.20\ \%$, $\times 22$ when $N = 0.21$–$0.25\ \%$, or $\times 20$ when $N = 0.25$–0.35%.

According to this formula, the austenitizing effect of nitrogen is equivalent at 0.10% N to the effect of 1.65% Ni, at 0.15% N to the effect of 3.15% Ni, at 0.20% N to the effect of 4.65% Ni and at 0.30% N to that of 5.1% Ni.

Due to the rapid heating and cooling rates experienced during electric arc welding, both the transformation of austenite into delta ferrite during heating as well as the retransformation of ferrite into austenite during cooling, are delayed in the area of the HAZ, which is heated to above 1200° C. Fig. 105 shows the effect of chemical composition and cooling rate on the share of austenite which is found in the weld metal after cooling. The beginning of δ–γ retransformation is illustrated by way of three alloys which differ in nitrogen and nickel content and which are held for 5 seconds at 1350° C for the purpose of ferritization. The cooling rates that normally occur during welding are located between the cooling curves a and b. The effect of cooling time $T_{12/8}$ between 1200° C and 800° C which is decisive for δ–γ

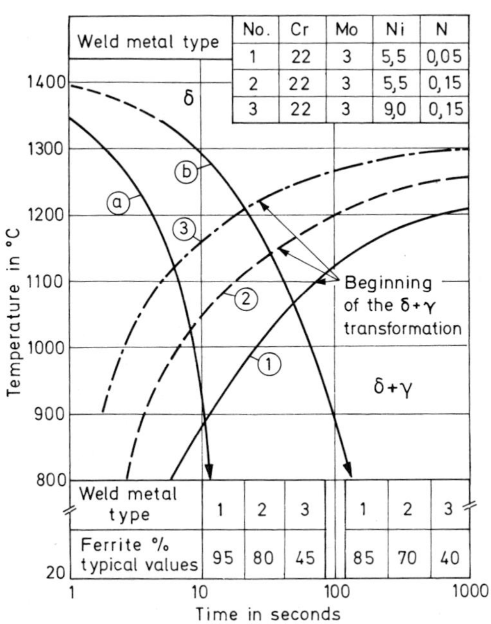

Fig. 105. The influence of the chemical composition of duplex SMAW weld metal on the beginning of the δ–γ retransformation during the cooling from 1350° C of a single weld bead of three weld metal grades 1, 2 and 3 at different cooling rates, *a* and *b*. The normal cooling rates that occur during welding are located between curve *a* and *b*. The rough ferrite contents after cooling to ambient temperature are stated at the bottom of the cooling curves (semischematical representation)

retransformation, can be taken from these curves. The ferrite content of the weld metal alloys found at ambient temperature are indicated at the bottom of the diagram. Alloy 1 of the type 22 Cr/3 Mo/5.5 Ni without the extra addition of nitrogen shows the highest ferrite values. The high cooling rates that occur during welding produce ferrite contents above 80%. The addition of 0.15% nitrogen shows positive effects. By raising the nickel equivalent by about three units, the beginning of $\delta-\gamma$ transformation is shifted to higher temperatures. This leads to higher austenite and lower ferrite contents after cooling to ambient temperature. The lowest ferrite contents are obtained with alloy 3 with 9% Ni and 0.15% N, i.e. with a nickel equivalent according to Espy [170] of about 12 using only the contents of Ni and N. The strong influence of the cooling rate on the delta ferrite content is greatly reduced by the effect of nitrogen and nickel (compare curve 3 in Fig. 105 to curve 1). A further advantage of alloying with nitrogen and nickel is only apparent in multi-pass welds which have not been incorporated into Fig. 105. As the $\delta-\gamma$ transformation rises to higher temperatures, a certain solution annealing effect can be achieved in the area of the weld which is reheated to around 1000° C by subsequent passes. In this way, a part of the forcibly supercooled ferrite may be transformed back to austenite. A further effect of nitrogen is produced by the fact that in the weld metal as well as in the high temperature zone of the HAZ which borders directly on to the weld metal, the time for coarse grain formation is reduced by the increase in temperature at the beginning of $\delta-\gamma$ transformation. So nitrogen stops further grain coarsening of the delta ferrite and leads to a finer micro structure.

Since both excessive ferrite contents and a coarse grain structure in multi-pass welds often excert a negative influence upon corrosion resistance and toughness, it is necessary to pay special attention to the weld metal cooling rate in the $\delta-\gamma$ transformation temperature range of between 1200 and 800° C. The cooling time $T_{12/8}$ should not be too short but neither too long. Since the cooling time $T_{12/8}$ is influenced by both heat input and material thickness, the latter must also be considered, it is not enough to consider just the heat input alone.

Advanced duplex stainless steels are mostly alloyed with nitrogen. The same also applies to weld metals but here the nickel content is also increased to within the range of 8–9%, in order to keep the delta ferrite content in the welded joint as constant as possible within a mean range of 30–40%. It must be mentioned here that in multi-pass welds the delta ferrite content is reduced by about 10% in comparison with single pass welds, this is because of the reheating effect produced by the subsequent weld passes, in which a part of the supercooled ferrite is again transformed into austenite.

Ferrite contents above 70% in the weld metal also increases the susceptibility towards hydrogen induced cold cracking, as shown by Fekken, Van Nassau and Verwey (471, paper 26). To avoid hydrogen induced cracking, it is necessary to rebake all coated electrodes and SAW fluxes prior to welding according to the instructions supplied by the filler metal manufacturer or at least to 250° C for two hours minimum. In the case of the GMAW and GTAW processes, it is necessary to avoid the use of shielding gases containing additions of hydrogen as may be sometimes be used, for example, in the automatic welding of stainless austenitic steel. According to Valtiera-Gallardo, Hawkins and Beech (471, paper 54), on the other hand, nitrogen may be added in small amounts to the shielding gas in order to increase the nitrogen content in the weld metal when welding low nitrogen containing steels without filler metals.

The common methods of ferrite measurement must be extended to cover the higher ferrite contents when welding duplex steels. A method which is frequently used is the metallographic method which permits measurement of the ferrite content in volume percent. But according to Kotecki ([471], paper 40]), Brantsma and Nijhof ([471], paper 45) and Bonnefois, Blondeau and Catelin ([471], paper 6), magnetic methods are also suitable. The extension of the ferrite number system to cover the higher ferrite contents found in duplex steels is reported on by Kotecki [474]. In the case of higher ferrite contents, however, greater scattering of the ferrite level must be expected which may be in the range of $\pm 10\%$.

Duplex steels possess a high corrosion resistance which is generally superior to austenitic Cr-Ni-Mo steels of the type 17 Cr/12 Ni/2.5 Mo/L (316 L). This applies in particular to crevice and pitting corrosion and stress corrosion cracking. A short survey of the corrosion behaviour of the duplex steels is given by Schläpfer and Weber [473]. The corrosion resistance of duplex steels is also discussed in detail in 16 papers of [471]. The alloying of duplex steels and weld metals with nitrogen is advantageous because of the improvement in the pitting resistance. The effect of nitrogen can be seen from the "pitting index" PRE_N[11] for nitrogen containing steels. A widely used formula for determining this index is:

$$PRE_N = \%\,Cr + 3.3 \times \%\,Mo + 16 \times \%\,N$$

PRE_N should be at least 30, in some cases even 35 or 40 (see [471], papers 3, 23, 34, 52). In the two-phase structure of duplex steel weldments, the austenite crystals of the weld metal and the HAZ are enriched with austenite forming elements such as nickel and nitrogen and the ferrite crystals with ferrite forming elements such as chromium and molybdenum. According to Liljas and Quarfort ([471], paper 2), in the ferrite crystals the contents of chromium are approx. 2–3% and those of molybdenum approx. 1–1.4% higher than they are in the austenite crystals, whereas in the austenite crystals the nickel contents are approx. 1–2.5% higher than they are in the ferrite crystals. The greatest differences are found however, with regard to nitrogen contents. In the HAZ the nitrogen contents in the ferrite crystals are in the range of 0.01–0.05% and in the austenite crystals between 0.21–0.29%. The average nitrogen content of the base metal could be established as being within the range of 0.13–0.18%. It is worth mentioning here, that the calculated pitting index PRE_N yields almost the same values for austenite as for ferrite because the lower amounts of Cr and Mo in austenite are equalized by the higher nitrogen content. According to Ishizawa and Inazumi ([471], paper 23), austenite shows roughly the same pitting resistance, despite the lower Cr and Mo values.

Excessive ferrite contents of above 60% in the welded joint may have a negative effect upon the corrosion resistance, because the excess ferrite often combines to give a stronger segregation of ferrite and austenite. In weld metals, this effect can be successfully counteracted by an increase in the nickel content to about 9%. Another

---

[11] PRE = Pitting resistance equivalent, N means nitrogen alloyed steel

unfavourable influence can be produced by coarse grain formation in the HAZ, a phenomenon which may occur if thin material is welded with a too high heat input (see [471], paper 21). It is often combined with higher dilution between the nickel rich weld metal and the base metal which is generally lower in nickel with the result that the effect of the higher nickel content in the all weld metal is partially lost (see also 471, paper 2). In general, both effects appear however, only with very slow cooling from the welding heat and especially when using the GTAW process in which the welded joint may be overheated. If cooling takes place too slowly, chromium nitride, and even chi phase may also precipitate in the temperature range around 800° C (see also Fig. 87, page 119). The toughness and also the corrosion resistance may be reduced because of the depletion of chromium due to the precipitation of chromium nitride at the ferrite grain boundaries as well as in the center of ferrite crystal grains. Such precipitations are normally only encountered in the HAZ of the base metal which is generally higher in ferrite content. This effect does not occur in a weld metal which is richer in nickel, because nickel slows down the precipitation rate. In order to avoid the precipitations in the HAZ of the base metal the cooling rate from the welding heat must be limited and the cooling time $T_{12/8}$ should not exceed 20 seconds particularly when welding thin materials.

According to Hertzman, Roberts and Lindenmo ([471], paper 30), the pitting resistance of duplex steel weldments increases with the decreasing precipitation of chromium nitride. According to the same authors, the precipitation of chromium nitride is also influenced by the structural configuration. With the denser arrangement of nodular austenite crystals and with increasing austenite contents, it is easier for nitrogen which is excessively dissolved in ferrite to be diffused into austenite with its much greater nitrogen solubility. In this way, the possibility of a chromium nitride precipitation in ferrite can thus be diminished. With a properly modified chemical composition (alloying with nitrogen and raising the nickel content) and the observance of certain rules during welding (see section 8.1), duplex steel weld metal has similar corrosion properties as the unwelded base material (see also 471, papers 2, 10, 17, 21, 24).

8.1 Practical Welding of Duplex Austenitic-Ferritic Stainless Steels

With the correct formulation of the chemical composition, the weldability of duplex stainless steels is quite satisfactory. The two-phase structure requires, however, the observance of certain welding procedures, aimed on the one hand at avoiding excessive ferrite contents particularly in the HAZ due to a too rapid cooling from the welding heat, and on the other hand excessive coarse grain formation and chromium nitride precipitation in the HAZ because of a too slow cooling rate. The weldability of duplex steels and the most suitable welding technology to use are discussed by Van Nassau [244], Petersen and Lang [368], Blumfield, Clark and Guha [387], Rabensteiner, Perteneder, Tösch and Schabereiter [267, 463, 476], Fiehn, Spähn, Wehner and Zürn [466] and ([471], papers 6, 21, 27, 38). The cracking resistance of duplex steel weldments is quite good, provided the ferrite content in the different heat affected zones of a weldment is limited to below 70% by suitable means. If this value is exceeded, hydrogen induced cold cracking may sometimes appear in the presence of higher welding stresses. In order for cold cracking to occur, however increased quantities of hydrogen would have to exist from the filler metals (see also [471], papers 21 and 26). Excessive dilution with the base metal which is lower in nickel must be avoided, to prevent the nickel content within the weld metal dropping too much. Otherwise, the resulting rise in the ferrite content would have a negative effect upon corrosion resistance, toughness and cracking resistance of the weld metal.

The following welding technology has been successfully employed for the welding of nitrogen alloyed austenitic-ferritic duplex steels:

— The filler metals should always be nitrogen alloyed types with suitably increased nickel contents (see Table 17, page 195).
— During welding, the cooling time in the HAZ and in the weld metal from temperatures between 1200 and 800° C ($T_{12/8}$) should not be too short. The cooling rate after welding should be adapted to the material thickness by selecting appropriate welding conditions. Thicker materials, especially above 20 mm, require a relatively high heat input, thinner materials, especially below 5 mm, a low one.
— Repair welds or welding on base material without the use of filler metals should not be carried out as this may produce very high delta ferrite contents in the fused areas.
— If the HAZ of a welded joint is small, fine grained and not too high in ferrite, the danger of a negative influence on the corrosion and toughness properties will also be very small.
— The dilution of the nickel-rich weld metal with the base metal which is normally lower in nickel should be kept to a minimum, if possible below 35%. The weld configuration and the welding procedure must be adapted accordingly.
— Coated electrodes and SAW fluxes must be rebaked prior to welding according to the manufacturer's instructions or at least to above 250° C, to keep the hydrogen supply during welding to a minimum.
— No hydrogen may be added to the shielding gas or to the root purge gas in order to avoid the formation of cold cracks.
— Preheating of duplex steels is normally not required. With thick materials, a preheating to 100–150° C may be advantageous.
— The interpass temperature should not exceed 150° C for material thicknesses below 12 mm and 180° C for thicknesses above 12 mm.
— A solution annealing treatment of the completed welded component is normally not necessary. If it is required, however, the temperature must be set according to the manufacturer's specification. With a solution annealing treatment it is possible to reverse any harmful structures in the HAZ that may have occurred during welding.
— In the course of production welding the ferrite content should be checked frequently.
— Stray arc strikes on the base material or on the completed welds must be absolutely avoided. This is because in the region of the arc strike high ferrite contents of 80% or more are produced due to the high cooling rates that occur.

These arc strikes may be the sites of severe local corrosion and lead to premature failure of the whole component. Grinding off the damaged material is normally not enough to eliminate completely the risk of corrosion.

— Dissimilar joints between unalloyed or low alloyed steels and duplex steels can —apart from the grades listed in Table 25, page 230—also be welded using duplex steel filler metals. The same applies to joint welding between austenitic and duplex steels. In general, the instructions given in chapter 12 should be followed. If in doubt, always ask the welding consumable manufacturer for their recommendations.

8.2 Mechanical Properties of Duplex Austenitic-Ferritic Stainless Steel Weld Metal

Typical mechanical properties of different nitrogen alloyed duplex weld metal grades, their delta ferrite contents and their chemical compositions are shown in Table 17, according to Perteneder, Tösch, Schabereiter and Rabensteiner [267]. Delta ferrite contents of about 30% produce favourable toughness values and good corrosion resistance. An increase in the yield strength to values of around 700 N/mm^2 is possible by the addition of manganese and by raising the nitrogen content to

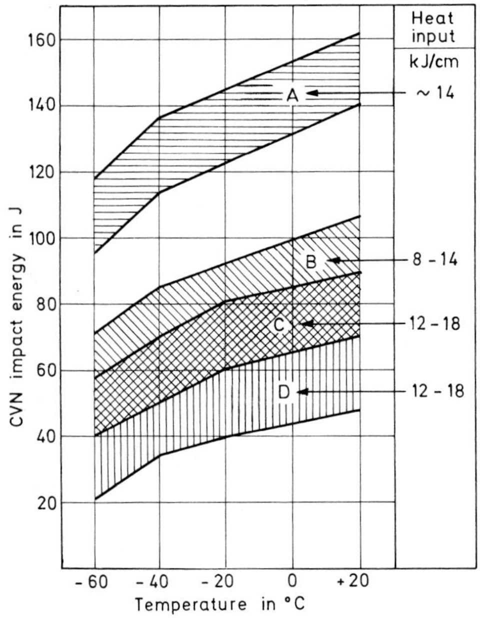

Fig. 106. Impact toughness versus test temperature for duplex stainless steel weld metal of type 22 Cr/8 Ni/3 Mo NL (UNS S31803). Welding with common electric arc processes. The delta ferrite content is in the range of 30–45%.
A = GTAW process (pure argon)
B = GMAW process (argon + 2.5% CO_2)
C = SAW and SMAW processes, basic flux and coating
D = SAW and SMAW processes, acid flux and rutile coating

Table 17. *Typical mechanical properties, delta ferrite contents and typical chemical compositions of nitrogen alloyed duplex stainless steel weld metals. Welding without preheating with an interpass temperature of 100–150° C.*
SMAW with a heat input of 9–11 kJ/cm
GMAW with a heat input of 20–22 kJ/cm and with Argon + 2.5% CO_2 as shielding gas

			Mechanical properties				Chemical composition in %					
No.	Delta ferrite content %	Weld. process	0.2 YS N/mm²	TS N/mm²	El. %	IE CVN J	C	Cr	Mo	Ni	N	0thers
1	17–23	SMAW	627	786	23.6	63	0.034	20.83	2.71	9.15	0.15	–
2	29–36	SMAW	622	792	23.1	59	0.036	22.21	2.82	8.81	0.13	–
3	43–55	SMAW	644	808	22.6	46	0.034	22.79	2.84	8.21	0.13	–
4	28–37	GMAW	608	809	26.8	98	0.037	22.52	3.03	8.51	0.14	–
5	27–36	SMAW	698	892	32	40	0.048	26.04	2.07	7.44	0.33	Mn–4.50

approx. 0.35% (see also [411 and 471] paper 14). Due to the beginning of 475° C embrittlement during long time exposure, the operating temperatures of the duplex steels and weldments are limited to 280° C (see also Fig. 94c, page 142 and [271], papers 8 and 27). The impact toughness versus temperature for duplex steel weld metal deposited by different welding processes is shown in Fig. 106. As indicated by the values shown in this diagram, the impact toughness is influenced by the type of welding process used and by the type of coating and welding flux used in the SMAW and SAW processes. This effect is mainly influenced by the possible oxygen pickup during welding. If the oxygen contents increase, the impact energy values decrease. The use of pure argon in the GTAW process does not cause any oxygen pickup in the weld metal. With this process, the highest impact toughness is obtained.

8.3 Precipitation Phenomena in Duplex Austenitic-Ferritic Stainless Steels and Weld Metals

Because of their low carbon contents in combination with alloying with nitrogen, duplex steels show only a moderate tendency towards the precipitation of carbides. Because of the increased chromium content, the chromium depletion that occurs is not enough to produce sensitivity of the grain boundaries which means that the Strauss test shows full resistance to intergranular corrosion even after sensitizing heat treatment of up to 30 hours. Fig. 87a (page 119) shows the precipitation behaviour of a duplex steel according to Herbsleb and Schwaab [245]. After only several minutes at 800° C the preferential precipitation of chi phase and chromium nitride begins. The result is an increase in the rate of corrosion as seen when tested in boiling strong nitric acid in accordance with the Huey test. The precipitation of sigma phase starts both in the steel (Fig. 87) and in the weld metal (Fig. 90g, page 126) after about 10–15 minutes. With heat treatments in the temperature range of between 650–950° C, the impact energy values quickly drop. The influence of temperature and time is thoroughly discussed by Müller-Stock [389], by Oppenheim and Chlibec [380], by Gümpel and Chlibec [478] and ([471], papers 3, 8 and 14). According to Herbsleb and Schwaab [245], 475° C embrittlement also starts after only a few minutes (see Fig. 94c), with the impact toughness dropping to low values. The same applies for the weld metal. The heat treatment of welded components within the temperature ranges of 300–500° C and 600–980° C should therefore be avoided. With regard to other phenomena, please refer to sections 4.1.3, 4.2 and 4.3 in which the precipitation processes of duplex materials are covered together with a discussion on the pertinent literature available.

9 Welding Metallurgy of Austenitic Stainless Steels

Stainless steels can generally be defined as steels with at least 11–12% chromium and carbon contents which are normally below 0.15%. With increasing demands on corrosion resistance, however, the chromium contents must be increased to above 16%. In ferritic stainless steels, the structure can be rather brittle which means poor weldability. Austenitic stainless steels show marked advantages here, particularly with regard to toughness properties and weldability. In stainless steels, the austenitic structure is generally produced by alloying with nickel. Nitrogen, a strong austenitizing element, has also gained increasing importance in recent years. One of the most important properties of austenitic stainless steels is their corrosion behaviour. The latter also determines their chemical composition. Beginning with the classical austenitic 18/8 chromium-nickel steel (AISI 302), Fig. 107 shows in a simplified but illustrative manner the effect that the basic elements and alloying additions have on the corrosion properties of the alloy, based on the representation of Sedriks [390]. For most types of corrosion an increase in the chromium content of the alloy above 18% will produce an improvement in corrosion resistance. A reduction in the carbon content to 0.030% max. or the stabilization of carbon by alloying with titanium or niobium will increase the alloys resistance against intergranular corrosion. Additions of molybdenum mainly improve the resistance to pitting and crevice corrosion. Increasing the nickel content in an austenitic matrix will increase the resistance to stress corrosion cracking, particularly in chloride containing media, where nickel based alloys with a content of more than 40% nickel give very good corrosion resistance values. On the other hand, a ferritic matrix is also insensitive to chloride induced stress corrosion cracking. This phenomenon is utilized, for example, in austenitic-ferritic duplex steels which show high resistance to stress corrosion cracking with delta ferrite contents above 25% (see Chapter 8). Finally, the controlled alloying with sulphur, lead or selenium must also be mentioned, a practice which improves machinability but greatly reduces weldability. For the improvement of strength and structural stability, nitrogen may also be added.

A well matched combination of alloying elements will produce a number of different corrosion resistant materials. Fig. 107 gives a rough idea of the corrosion behaviour of austenitic stainless steels. The level of the two major alloying elements chromium and molybdenum, and also that of niobium are limited by the precipitation of intermetallic phases (sigma phase and molybdenum and niobium rich phases). These phases exert a negative influence on corrosion resistance as well as on the toughness and high temperature properties (see Chapter 4). By increasing the nickel content to above 40%, it is possible to prevent the precipitation of sigma phase (see Fig. 12 and 13) and to shift the precipitation of other molybdenum and niobium containing phases to higher alloying contents which may produce decisive advantages under severe corrosive conditions. Higher nickel contents lead to the group of nickel based materials which are also indicated in Fig. 107. For further valuable information on the properties of stainless steels, please refer to the list of supplementary literature entries [233, 390–397]. A brief and simple description of the basic corrosion influences in stainless steels is supplied by Bäumel [479]. Latest findings regarding the resistance of austenitic stainless steels and weld metals towards pitting and crevice corrosion, and stress corrosion cracking are contained in the publications by Herbsleb and Westerfeld [402], Rockel and Renner [480], Risch [481], Herbsleb and Pfeiffer [482] and Mukai and Murata [483].

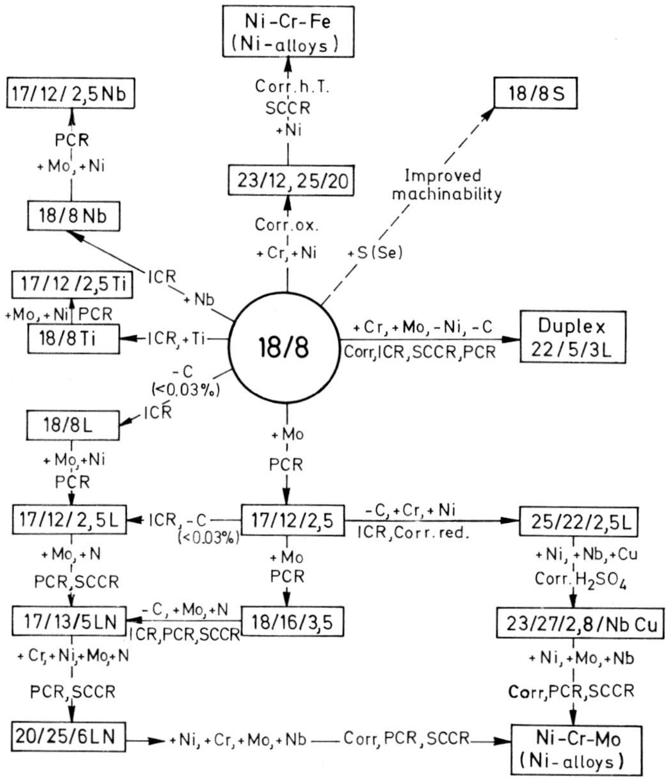

Fig. 107. Schematic representation of the effect of major alloying elements and the addition of minor elements on the corrosion resistance of 18/8 chromium-nickel steel (AISI 302). The major elements are stated in the order of Cr/Ni/Mo. Corr. = improved corrosion resistance (general), ox. = in oxidizing environments, red. = in reducing environments, h.t. = at high temperatures, ICR = improved intergranular corrosion resistance, PCR = improved pitting and crevice corrosion resistance, SCCR = improved stress corrosion cracking resistance, + = increase in content or the addition of an element, − = decrease in content of an element, L = low carbon content (normally 0.030% max.), N = addition of nitrogen (required for corrosion reasons)

9.1 Welding Metallurgy of Unstabilized Austenitic Stainless Steels

This group of steels contains the common austenitic stainless steels derived from the classical 18/8 chromium-nickel steel (AISI 302) which permit welding with similar filler metals and which produce an austenitic but ferrite containing weld metal. Steels with 2–3% molybdenum and grades with a low carbon content of 0.030% max. also belong to this group. Table 18 represents a survey of the chemical composition requirements of some major unstabilized steel grades with common standard analysis limits. These grades normally possess excellent weldability, provided they are welded with filler metals which yield an austenitic weld metal with delta ferrite contents in the range of about 5–15 FN. When welding these steels, however, it is necessary to follow certain procedures in order to achieve sufficient corrosion resistance, cracking resistance and toughness, these requirements shall be discussed below.

If the austenitic steels given in Table 18 are welded without filler metal, e.g. by the GTAW process, the result would be a weld metal of the same chemical composition as the base metal grade. It is common knowledge that austenitic steels are always subjected by the steel manufacturer to a solution annealing treatment, normally in the range of 1050–1100° C. In the course of this heat treatment, carbide $M_{23}C_6$, sigma phase and delta ferrite are completely dissolved and the annealing process produces a homogenous fully austenitic structure (see Fig. 6 and 9b). With a subsequent quenching treatment, this state is maintained down to ambient temperature which means that austenitic steels show prior to welding an austenitic structure without any delta ferrite. If such a steel grade is welded without a filler metal, a weld metal is produced with a structure which can be determined from its position in the Schaeffler or the DeLong diagram by calculating the chromium and nickel equivalents from the chemical composition of the steel grade. To obtain high hot cracking resistance, the weld metal structure produced should not be fully austenitic but rather austenitic with a delta ferrite content in the joint in the range of about 5–15 FN.

Table 18. *Chemical composition requirements for stainless austenitic Cr-Ni and Cr-Ni-Mo steels*

UNS-Designation* (unified number)	Chemical composition requirements in %			
	C_{max}	Cr	Mo	Ni
S30200	0.15	17.00–19.00	–	8.00–10.00
S30400	0.08	18.00–20.00	–	8.00–10.50
S30403	0.030	18.00–20.00	–	8.00–12.00
S30800	0.08	19.00–21.00	–	10.00–12.00
S31600	0.08	16.00–18.00	2.00–3.00	10.00–14.00
S31603	0.030	16.00–18.00	2.00–3.00	10.00–14.00

* New designation established in accordance with ASTM E527 und SAE J1086, Practice for Numbering Metals and Alloys (UNS = Unified Numbering System)

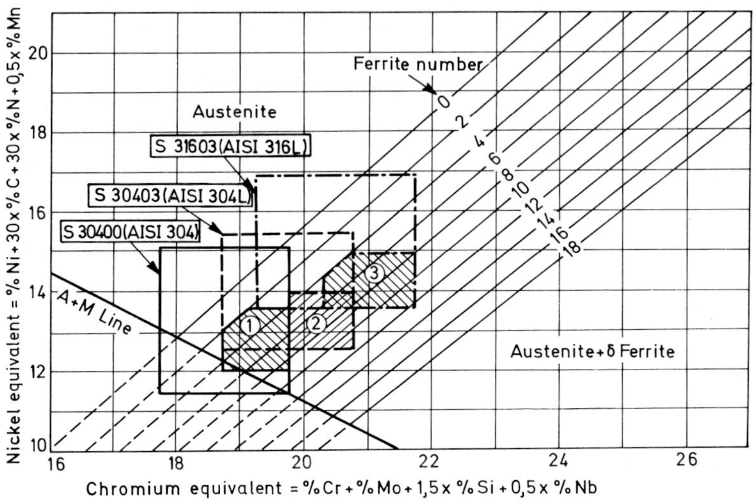

Fig. 108. The position of austenitic steels according to Table 18 (page 199) in the DeLong diagram with standard and restricted analysis limits. Chemical composition in %: Area No. 1—C = 0.05—0.07, Si = 0.5, Mn = 1.5, Cr = 18.0—19.0, Ni = 8.5—9.5, N = 0.04; Area No. 2—C = 0.02—0.03, Si = 0.5, Mn = 1.5, Cr = 19.0—20.0, Ni = 10.0—11.0, N = 0.04; Area No. 3—C = 0.02—0.03, Si = 0.5, Mn = 1.5, Cr = 17.5—18.5, Mo = 2.0—2.5, Ni = 11.0—12.0, N = 0.04

Fig. 108 shows the areas of three common steels using the DeLong diagram resulting from the possible chemical composition within the standard analysis limits. The areas are rather large and extend from a fully austenitic structure to delta ferrite contents of about 12 FN. The contents of silicon, manganese and nitrogen of 0.5%, 1.5% and 0.040% respectively were assumed to be constant. If a steel which is located in the fully austenitic range is welded without filler metal, the formation of hot cracks in the weld metal and the HAZ could possibly occur. To achieve crack-free weldments made from austenitic stainless steels without using filler metals requires a limitation of the chemical composition in such a way that after welding the weld solidifies to primary delta ferrite, thus producing an austenitic structure containing some delta ferrite after cooling to ambient temperature (see sections 5.3.1 and 5.3.2). Such restricted analysis ranges are shown as hatched areas 1—3 in Fig. 108, with the analysis limits which roughly coincide with these areas being stated in the text below Fig. 108.

If the narrowed analysis limits are compared to the values in Table 18, it is apparent that the chromium content must be kept in the upper range and the nickel content in the lower range to achieve a sufficient delta ferrite content in the steel being fused by welding. For processing and cost reasons, steel manufacturers often set lower values of chromium, nickel and also sometimes of the delta ferrite content within the standard analysis limits of the grade. Therefore, it is recommended that the steel manufacturer be contacted prior to welding large quantities of austenitic stainless steel if it is intended to weld the steel without the use of filler metal (e.g. autogenous production welding of tubes). A simple way of finding out whether or not an austenitic stainless steel becomes fully austenitic during welding without the use of a filler metal is to deposit a melt run on the base metal using the GTAW process and measuring the delta ferrite content in this weld. In general it is advisable that all welding shops which fabricate large quantities of austenitic stainless steel possess a ferrite measuring device together with the pertinent calibration standards (see also section 3.3 and 5.3).

When welding joints using filler metals few problems are normally encountered since the weld metal composition is set in such a way that it will primarily solidify to delta ferrite and thus contain a sufficient amount of delta ferrite at ambient temperature to guarantee high hot cracking resistance. There are a few exceptions, i.e. if a steel is located in the Schaeffler or the DeLong diagram far inside the austenitic range and there is high dilution with the base metal during welding, e.g. in GTA welding of root passes or in the application of high deposition welding processes using high arc energy such as GMAW and SAW. It is often possible, however, to remedy this problem by adapting the welding procedure to ensure the correct degree of dilution between the deposited weld metal and the base metal.

As previously mentioned, unstabilized austenitic stainless steels should usually be welded with matching filler metals. In some cases, where this is not possible, it is important to ensure that the weld metal is always more "noble" with regard to its chemical composition and corrosion resistance than the base metal. Then if due to the environmental conditions a simple galvanic cell may be formed between the differing compositions of the weld metal and base metal, galvanic corrosion may occur but will be transferred to the much larger anodic surface area of the parent plate material, where it can quite often be ignored. If on the other hand the parent plate material is, due to its chemical composition, more "noble" than the weld metal then the galvanic corrosion would be transferred to the weld metal which becomes the anodic region. Because the weld metal offers a by far smaller surface area than the rest of the welded component the corrosion here may be more severe. In this case it is recommended that the welding consumable manufacturer be contacted for advice before welding commences, as there may be combinations of base metal and weld metal that may give unsatisfactory corrosion resistance in service.

Weldments manufactured from austenitic stainless steel possess their highest corrosion resistance in the as welded or solution annealed condition. The solution annealing of finished components, however, can only be successfully achieved on rare occasions. With the normal cooling rates experienced during electric arc welding, however, we can assume that for weld metals and base metals in the as welded condition with carbon contents up to 0.06% no sensitivity towards intergranular corrosion will occur (see also diagram in Fig. 85b, page 108). If possible, the post weld heat treatment of welded components should be avoided (with the exception of a solution annealing treatment). If heat treatment cannot be avoided, however, special attention must be paid to the influence of carbide and phase precipitations on the corrosion resistance and toughness properties of the weld (see sections 4.1, 4.2, 4.3, 9.1.2 and 9.1.3). Valuable information regarding the influence that stress relieving in the temperature range of 550–750° C has on the corrosion resistance and toughness of unstabilized austenitic weld metal is given by Schabereiter [399] and Schabereiter and Rabensteiner [263, 264] and is discussed in detail in section 9.1.2 and 9.1.3. With reference to the limitation on the material service temperatures, the maximum values stated in the pertinent standards and material data sheets must be adhered too. However occasionally with suitable selection of materials, it is possible to exceed these values. When this is done, however, it is very important to always contact the material manufacturers and where applicable the appropriate inspection insurance authorities first for approval.

The question is often asked whether unstabilized low carbon (L) steels[12] can also be welded with niobium stabilized filler metals. In most cases this is possible, but according to Thier [397] there are cases where this is not recommended because low carbon steels may show a better resistance under certain corrosive conditions than the niobium stabilized materials. Stabilized steels, on the other hand, can normally be welded with low carbon filler metals, except in cases where a post weld heat treatment in the critical temperature range between 500 and 1000° C is to follow or where the anticipated service temperatures are between 350–400° C. In such cases it is advisable to contact the filler metal manufacturer to discuss what steps should be taken to safeguard against the possible detrimental influence that using such combinations may have on corrosion resistance and toughness properties.

For an improvement in the yield strength, nitrogen alloyed austenitic stainless steels with 0.12–0.20% nitrogen can also be employed (see for example grades with UNS designation 30451, 30452 and 31651). In general, these materials can be welded with the standard austenitic filler metals since the yield strength and ultimate tensile strength of the austenitic weld metal is sufficiently high in the as welded condition to match those of the nitrogen alloyed base material. Care must be taken, however, to avoid excessive dilution with the base metal because of the austenizing effect of nitrogen. These steels are often located relatively far inside the fully austenitic range of the Schaeffler or DeLong diagram which means that the delta ferrite content may drop too low in the diluted weld metal.

When welding austenitic stainless steels greater distortion of the welded component must be expected, this is due mainly to the fact that the austenitic stainless steels have a higher coefficient of thermal expansion and lower thermal conductivity than the ordinary ferritic steels. Because of the lower thermal conductivity of the austenitic stainless steels below 800° C, dissipation of the welding heat into the base metal takes place more slowly, therefore an accumulation of heat may occur in the weld which may lead to local overheating and severe distortion. Normally high deposition welding processes (GMAW, SAW) can also be employed for the welding of austenitic stainless steels, however, for the reasons stated above, an excessive degree of dilution must be avoided by selecting suitable welding parameters. When using these welding processes, the influence of the various shielding gases and fluxes upon the chemical composition of the weld metal must also be taken into consideration (according to Thier [398]). This can be done by plotting the effect of the additional alloying vectors for the shielding gas or welding flux being used into the Schaeffler or DeLong diagram. In principle, the GMAW process should only be carried out using shielding gases with low CO_2 contents, in order to keep the carburization of the weld metal by CO_2 to a minimum. If carburization is too strong,

[12] Low carbon stainless steels are generally designated as L (low carbon) steels. In general, these steel grades have a maximum carbon content of 0.03%, sometimes 0.02%. In weld metals, the maximum carbon content is normally 0.04% (see also US Standard ANSI/AWS A5.4–81 or DIN 8556)

the severe austenitizing effect of carbon may have a negative influence upon the hot cracking resistance of the weld and in the case of low carbon grades on the intergranular corrosion resistance. According to Kuwana, Kokawa and Tsujii [484] nitrogen additions to the shielding gas must be avoided because of the possible nitrogen pick up in the weld metal.

In order to give the welded components full corrosion resistance the welding operation must be followed by a suitable post weld surface treatment. The necessary steps which must be taken in post weld surface treatments are discussed in detail in section 10.3.

9.1.1 Practical Welding of Unstabilized Austenitic Stainless Steels

In general unstabilized austenitic stainless steels exhibit excellent weldability when using matching filler metals producing an austenitic weld metal containing small proportions of delta ferrite after cooling to ambient temperature. The following recommendations, if strictly adhered to, will ensure that the incidence of weld and parent metal defects will be largely avoided. Special emphasis should be paid to welding with a short arc in all welding processes to prevent nitrogen pick-up from the surrounding atmosphere. Because of the strong austenitizing effect of nitrogen, primary solidification to austenite may take place due to nitrogen pick-up, and this may lead to a fully austenitic weld metal with the associated danger of hot cracking. This effect is discussed in sections 5.3.1 through 5.3.5 and by Tösch and Schabereiter [323], see also Fig. 102, page 169.

The following welding procedures have produced excellent results when used for welding unstabilized austenitic stainless steels:

— In general use only matching filler metals for all welding operations producing weld metals with delta ferrite contents in the range of 5—15 FN (see also section 5.3.2 and Fig. 101, page 160).
— The arc should be kept as short as possible to avoid nitrogen pick-up from the surrounding atmosphere. This applies particularly to rutile coated electrodes and also to welding processes that use a shielding gas and which permit welding to be carried out with long arc lengths. Correct gas shielding should be maintained at all times when using gas shielded arc welding processes. With the GMAW process only shielding gases containing little or no CO_2 should be used, to keep carburization to a minimum.
— In general preheating is not necessary, however, for thick sections a preheat in the range of 100—150° C may be employed to aid fusion.
— An interpass temperature of 150° C should not be greatly exceeded.
— If the deposition of a backing run on the root side of the joint is not possible when gas shielded arc welding, then back purging should be provided during welding of the root pass using pure argon or an anti slag gas.
— If possible, dilution with the base metal should be kept below 30%. If it is greater than this amount, due to the type of welding process being used, a test weld should be deposited prior to the actual welding operation, using the same welding parameters and the same base metal, and the ferrite content of the diluted weld metal should be tested with a calibrated ferrite measuring device.

— If the execution of a test weld is not possible, the position of the base metal and the weld metal in the Schaeffler or the DeLong diagram should be determined from the chemical composition of both materials and the ferrite content of the diluted weld metal estimated taking into consideration the dilution level expected.
— A post weld heat treatment of welded components should be avoided. If this is not possible, it is essential to consider the intergranular corrosion and precipitation behaviour of the materials being used (see chapter 4 and sections 9.1.2 and 9.1.3).
— Due to the austenitic stainless steels having higher coefficients of thermal expansion and lower thermal conductivity than ordinary ferritic steels distorsion will be greater and therefore steps should be taken to safeguard against this (see section 10.1).
— Hot working of the component such as heating with an oxy/gas torch in order to straighten distorted sections should be avoided as this could lead to a local reduction in corrosion resistance.
— Before being put into service the post weld surface treatment of a completed welded component is mandatory, this is to ensure that optimum corrosion resistance is obtained. Instructions on such post weld surface treatments are given in section 10.3 and these should be followed carefully.
— Additions of nitrogen to the shielding gas in the GTAW and GMAW processes must be avoided, because nitrogen pick-up of the weld metal may produce a fully austenitic structure due to the strong austenitizing effect of nitrogen, which is much more susceptible to hot cracking than ferrite containing weld metal.

9.1.2 Mechanical Properties of Unstabilized Austenitic Stainless Steel Weld Metal

Table 19 shows some examples of the mechanical properties and the typical chemical compositions of unstabilized chromium-nickel-(molybdenum) weld metals with delta ferrite contents of approx. 8–12 FN. The values of 0.2% and 1.0% yield strength are considerably above the values of similar parent steel grades. For this reason, it is possible—with only a few exceptions—to weld nitrogen alloyed austenitic steels that have an improved yield strength with similar weld metals which are not nitrogen alloyed. However, if the weld metal too is also solution annealed, the yield strength values of the weld metal will drop to those of the lower level of the parent steel. The deviations in mechanical properties of the differently alloyed weld metal grades shown in Table 19 are rather small. The high elongation values of about 40% roughly coincide with those of austenitic stainless steels.

The various precipitation behaviours of the differently alloyed weld metals are normally only noticed after a heat treatment and become particularly obvious when carrying out impact tests. Fig. 109 shows three chromium-nickel alloyed weld metal grades (No. 1–3) which differ mainly with regard to their carbon and delta ferrite contents. After annealing the precipitation of the brittle sigma phase may produce a considerable drop in the impact energy values. For the weld metals shown in Fig. 109, the following rule applies: With increasing chromium, molybdenum and delta ferrite

Table 19. *All weld metal mechanical properties, chemical compositions and measured delta ferrite contents of unstabilized austenitic weld metal grades (SMAW process using 5 mm coated electrodes)*

No.	AWS* classi-fication	Mechanical properties 0.2 YS N/mm^2	1.0 YS N/mm^2	TS N/mm^2	El. %	IE J	Chemical composition in % C	Si	Mn	Cr	Mo	Ni	N	Delta ferrite FN
1	E 308	432	456	611	42	108	0.051	0.53	1.59	19.67	0.22	9.60	0.046	8.3
2	E 308L	420	432	569	44	101	0.030	0.57	1.02	19.94	0.18	10.47	0.043	9.1
3	E 308Mo	452	475	586	41	96	0.053	0.48	1.39	19.07	2.64	11.15	0.048	11.7
4	E 308MoL	459	478	589	40	87	0.032	0.60	1.05	18.60	2.58	11.70	0.042	11.1

* According to ANSI/AWS A5.4–81 standard

contents and a decreasing carbon content, the impact energy values decrease. In austenitic stainless steel weld metal, sigma phase will be precipitated predominantly from delta ferrite with maximum precipitation at an annealing in the range of 750–850° C. The precipitation of sigma phase from austenite starts later (see Fig. 90f, page 126). Delta ferrite disintegrates into sigma phase and also partly into austenite. With the precipitation of sigma phase, which is rich in chromium and deficient in nickel, the environment of the precipitated sigma phase crystals becomes deficient in chromium and richer in nickel and thus transforms to austenite. Therefore, it is quite easy to keep track of the precipitation of sigma phase by watching the reduction in the ferrite content of the weld metal (see lower diagram in Fig. 109). In the case of weld metal 1, sigma phase precipitation is largely retarded by the higher carbon content of 0.054% and practically no drop in impact energy values can be noted even after ten hours annealing between 600 and 900° C. A slight drop in the toughness values is caused by the precipitation of the carbide $M_{23}C_6$ from 550° C onwards.

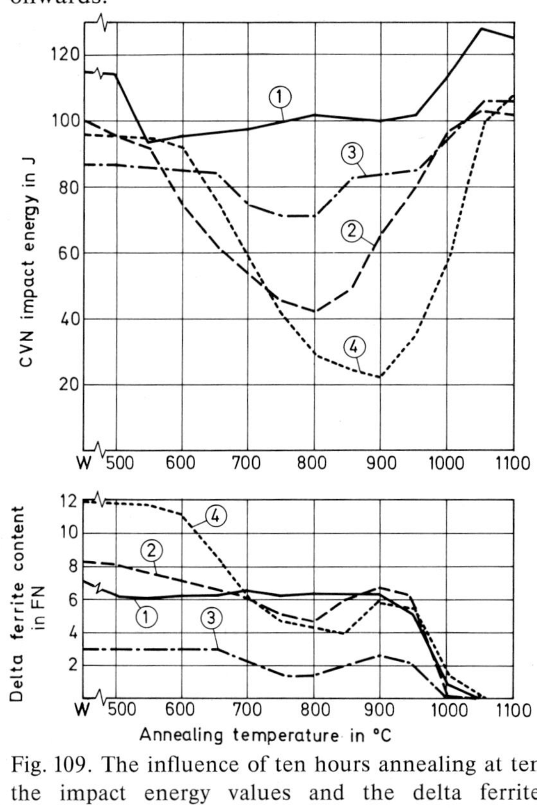

Fig. 109. The influence of ten hours annealing at temperatures between 500 and 1100° C on the impact energy values and the delta ferrite contents of unstabilized austenitic chromium-nickel and chromium-nickel-molybdenum weld metal; W = as welded condition. Chemical composition in %: No. 1—C = 0.054, Si = 0.65, Mn = 1.26, Cr = 18.91, Mo = 0.15, Ni = 9.45, N = 0.046, delta ferrite FN = 7.2; No. 2—C = 0.030, Si = 0.57, Mn = 1.02, Cr = 19.94, Mo = 0.18, Ni = 10.47, N = 0.043, delta ferrite FN = 9.1; No. 3—C = 0.028, Si = 0.65, Mn = 1.71, Cr = 19.92, Mo = 0.22, Ni = 12.67, N = 0.056, delta ferrite FN = 3.0; No. 4—C = 0.032, Si = 0.76, Mn = 0.75, Cr = 19.22, Mo = 2.49, Ni = 12.32, N = 0.049, delta ferrite FN = 11.6

The re-dissolution of the precipitated carbide $M_{23}C_6$, sigma phase and delta ferrite begins at about 950° C. At the normal solution annealing temperature of 1050° C, all phases disappear. Now, the structure consists of homogenous austenite which results in an increase in the impact energy values. In the weld metal grades containing 0.03 carbon (No. 2) the impact energy values no longer indicate the presence of $M_{23}C_6$ precipitation at 550–600° C, however, in the temperature range between 650–950° C sigma phase precipitation which because of the reduced carbon content is delayed less than that of weld metall No. 1, is felt stronger. With a delta ferrite content of 9.1 FN (weld metal No. 2), the drop in the impact energy values is much greater than that of weld metal No. 3 with 3 FN, this is because more sigma phase can precipitate from the higher delta ferrite content. The precipitation of sigma phase causes at the same time a drop in the delta ferrite content. The influence of molybdenum on the re-dissolution temperature of phases can be seen by comparing the behaviour of weld metal No. 2 (without Mo) with No. 4 containing about 2.5% Mo. With the molybdenum addition, the precipitation of sigma and chi phases is increased and shifted to higher temperatures, as indicated by the course of the impact energy values. A higher solution annealing temperature is required for the re-dissolution of these phases than is required for the practically Mo-free weld metals Nos. 1, 2 and 3.

9.1.3 Precipitation Phenomena in Unstabilized Austenitic Stainless Steels and Weld Metals

The precipitation kinetics of the chromium-iron carbide $M_{23}C_6$ are of particular importance for the resistance of weld metal to intergranular corrosion (IC). Unstabilized weld metal performs here in the same way as similar austenitic stainless steel grades although it contains, compared to the solution annealed parent steel, certain amounts of delta ferrite in which the carbide $M_{23}C_6$ can precipitate faster than in the purely austenitic structure.

Consequently, delta ferrite containing weld metal ought to be more sensitive to IC than the fully austenitic stainless steel grade. Because of the segregation processes that occur during solidification of the weld metal and in particular during the δ–γ transformation, the delta ferrite contains considerably higher chromium contents than the austenitic part of the structure of the weld metal which consists on average of about 90% austenite and 10% delta ferrite by volume. Table 6 (page 133) shows some examples of the structural analysis of delta ferrite and austenite together with the average analysis of the weld metal. Nos. 1 and 2 represent unstabilized weld metals with delta ferrite contents of 10 and 3 FN respectively. Delta ferrite contains 23.6% chromium compared with about 20% chromium for the average weld metal analysis and 19.9% chromium for the austenitic part of the structure. Due to the fact that increased chromium contents will shift the area of intergranular corrosion resistance to longer times, this compensates for the increased tendency towards $M_{23}C_6$ precipitation in the delta ferrite, which means that the ferrite containing austenitic weld metal does not differ much from a similar ferrite free austenitic steel with regard to its intergranular corrosion behaviour.

The influence that different alloying elements have on IC attack is shown in Fig. 85 (page 108) by way of TTC diagrams for different steel grades. The influence of carbon is quite strong (Fig. 85b). By lowering the carbon content to below 0.04%, the start of IC attack can be delayed in the 18/8 chromium-nickel steels by more than one hour. This means that the socalled "L" materials are IC resistant after welding, i.e. the cooling rates experienced during the welding of austenitic stainless steels (curves 1 and 2) are located within such short periods of time that no increased intergranular corrosion sensitivity resulting from the welding process being used is anticipated for the common unstabilized low carbon steels and weld metals. According to Edström and Ljundberg [217], molybdenum together with low carbon contents below 0.03% will shift the beginning of IC attack to slightly longer periods of time (see also Fig. 86, page 111).

The intergranular corrosion behaviour of low carbon unstabilized weld metal is shown in Fig. 89 (page 123) for grades Nos. 1 and 2 containing 18 and 20% chromium respectively. This roughly coincides with the behaviour of similar austenitic steel grades. Schabereiter [399] investigated the IC resistance behaviour of various austenitic weld metal grades after tempering in the temperature range between 550 and 750° C for different periods of time. It was found that IC attack is strongest in the range between 550–650° C. A similar behaviour is apparent from the IC attack diagrams in Fig. 89 (page 123). The danger of intergranular corrosion induced by the common short time heat treatment processes is only expected at temperatures above 500° C. In the case of long time high temperature resistance, however, IC may occur at lower temperatures e.g. if the welded component is subjected to long term high temperature service under corrosive conditions. Depending on the chemical composition, IC resistance of unstabilized austenitic weld metal can only be guaranteed up to temperatures in the range of 300–350° C.

The precipitation of sigma phase is the most important of all precipitation phenomena apart from $M_{23}C_6$ precipitation, particularly with regard to the mechanical properties. Due to the higher chromium content of delta ferrite, the weld metal, containing delta ferrite, is often more precipitation prone than a steel of similar composition which, because of a solution annealing treatment, does not contain any delta ferrite. The influence of sigma phase precipitation on toughness properties is discussed in section 9.1.2. Due to the chromium depletion of the matrix by the precipitation of the chromium rich sigma phase, a more severe reduction in the corrosion resistance could be expected. This occurs only rarely, however, in weld metal when testing in accordance with the Strauss test because of the fact that the delta ferrite—from which sigma phase precipitates—is enriched with chromium which partially compensates for the chromium depletion of the matrix. Furthermore, the precipitation range of sigma phase is located at such high temperatures that a re-diffusion of chromium into the depleted zones is possible.

With long-term exposure to elevated temperatures, the detrimental effect of sigma phase precipitation and the precipitation of molybdenum containing phases can be felt in the form of embrittlement. This is also the explanation, why, if the operating temperature and the time at temperature gives rise to the danger of the precipitation of brittle phases, in creep resistant steel grades the chromium content in molybdenum free steels is limited to about 18% and in steels containing 2% Molybdenum to about 16%. The processes that take place during phase precipitation, their chemical compositions and the precipitation kinetics after long term post weld heat treatment have been investigated for a 316 weld metal by Gill et al. [485].

Apart from $M_{23}C_6$ and sigma phase, also the Mo containing chi phase was identified. Cieslak, Ritter and Savage [439] found, that under unfavourable conditions chi phase may even be precipitated in the as welded condition in 16 Cr/8 Ni/2 Mo cast steel weldents (see also sections 4.2.3, 4.2.4 and Fig. 92, page 136).

9.2 Welding Metallurgy of Stabilized Austenitic Stainless Steels

This group of steels contains grades which are alloyed with titanium or niobium in order to improve their IC resistance properties and which permit welding with matching filler metals giving a ferrite containing austenitic weld deposit. Because of the excessive and uncontrollable burn off of titanium across the arc, particularly when welding with coated electrodes, stabilized filler metals are normally alloyed with niobium. This element burns off only a little during welding. Therefore sufficient amounts of it can be successfully transferred across the arc during welding from the filler metal into the weld metal. Titanium stabilized steels can be welded without restrictions using niobium stabilized filler metals. If it is necessary to weld titanium stabilized austenitic stainless steel without the addition of a filler metal then this is normally carried out using either the GTAW or the plasma arc welding process, where the titanium burn off is rather small due to the inert shielding gas protection. These processes require full protection from the surrounding atmosphere to prevent any excessive titanium burn off.

Table 20 represents a survey of the chemical compositional requirements of some of the major stabilized austenitic stainless steels. Like the unstabilized steels the stabilized steels are generally distinguished by their good weldability, provided they are welded with filler metals which produce sufficient quantities of delta ferrite in which case it is possible to obtain high hot cracking resistance in the weld metal. From Fig. 101 (page 160) it appears that the minimum ferrite content required in the stabilized weld metal grades must be somewhat higher than that of the unstabilized grades to produce the same hot cracking resistance. If, however, stabilized weld metal solidifies to primary austenite crystals which means that it is free from delta ferrite after cooling to ambient temperature, it performs less favourable with regard

Table 20. *Chemical composition requirements for stabilized austenitic chromium-nickel steels*

UNS Designation* (unified number)	Chemical composition requirements in %			
	C_{max}	Cr	Ni	Others
S32100	0.08	17.00–19.00	9.00–12.00	Ti = 5 × C min.
S34700	0.08	17.00–19.00	9.00–13.00	Nb = 10 × C min.
S34800	0.08	17.00–19.00	9.00–13.00	Nb = 10 × C min. Co = 0.20 max. Ta = 0.10 max.
S30940	0.08	22.00–24.00	12.00–15.00	Nb = 10 × C min.

* New designation established in accordance with ASTM E527 and SAE J1086, Practice for Numbering Metals and Alloys (UNS = Unified Numbering System)

to hot cracking susceptibility than unstabilized fully austenitic weld metal. This is the result of the formation of low melting niobium containing eutectics (see section 5.3.4). For this reason, the recommendations for the avoidance of fully austenitic weld metal that have already been given in section 9.1.1 for unstabilized weld metal must be closely followed (e.g. arc as short as possible, control of delta ferrite content etc.).

Stabilized steels and weld metals show their highest corrosion resistance in the as welded condition or in the solution annealed state. In welded components, however, the latter can only be obtained in rare cases. Therefore, other heat treatments apart from solution annealing should be avoided whenever possible. If it is essential that a post weld heat treatment must be carried out, e.g. stress relieving treatments, it is possible to achieve sufficient IC resistance by a suitable analytical modification of the weld metal. In this way, the beginning of the critical range at which IC attack takes place is not only shifted to relatively long periods of time but also limited with regard to the temperature range used (see section 9.2.3). If heat treatment processes are carried out in the critical temperature range of the $M_{23}C_6$ precipitation from 550—700° C, a higher stabilization ratio must be employed for the weld metal in order to avoid knife line corrosion in the HAZ of a multi layer weld metal (see section 4.1.4.1). In order to maintain high hot cracking resistance the niobium content must be kept as low as possible resulting in the carbon content also being kept as low as possible. Stabilized steels and weld metals are somewhat more susceptible to sigma phase precipitation, this is because carbon is bonded as NbC in such a strong way that its sigma phase precipitation delaying effect is weakened (see section 9.2.3).

The welding of stabilized steels should generally be carried out with niobium containing filler metals. Unstabilized low carbon niobium free filler metals can also be used, if attention is paid to the lower service temperatures of the latter. With stabilized austenitic steels, the standard service temperature upper limit is generally around 400° C which is about 50—100° C higher than the upper temperature limit for unstabilized low carbon steels. In order to give the welded components full corrosion resistance the welding operation must be followed by a suitable post weld surface treatment. The necessary steps which must be taken in this respect are discussed in detail in section 10.3.

9.2.1 Practical Welding of Stabilized Austenitic Stainless Steels

If sufficient delta ferrite contents are maintained in the weld metal (see Fig. 101, page 160), stabilized steels show practically the same good weldability as unstabilized ones. In principle, the same rules apply here as those stated in section 9.1.1 for unstabilized steels, with a few additions being necessary, however, which are as follows:

- Titanium or niobium stabilized steels should be welded with similar niobium stabilized filler metals which produce weld metals with delta ferrite contents in the range of 7—15 FN.
- The use of unstabilized low carbon filler metal is normally possible, there are exceptions, however, because these types are not recommended for high temperature components due to their lower service temperature range.

— Whenever possible, heat treatment after welding should be avoided. If it cannot be avoided, it is important to contact the welding consumables manufacturer who will recommend the use of special filler metals.
— A post weld surface treatment of the completed welded component must be carried out carefully, following the instructions stated in section 10.3 to guarantee full corrosion resistance.

9.2.2 Mechanical Properties of Stabilized Austenitic Stainless Steel Weld Metal

Table 21 shows the mechanical properties and the chemical compositions of stabilized chromium-nickel(-molybdenum) weld metals with delta ferrite contents of about 8—11 FN. When compared with unstabilized weld metal (Table 19, page 205) the values for yield strength and tensile strength are slightly higher, those for elongation and impact energy are a little lower. Electrodes with basic coatings show higher impact energy values than do rutile coated electrodes. The corrosion and toughness behaviour of stabilized chromium-nickel weld metal after heat treatment is thoroughly discussed by Schabereiter [253] and by Schabereiter and Rabensteiner

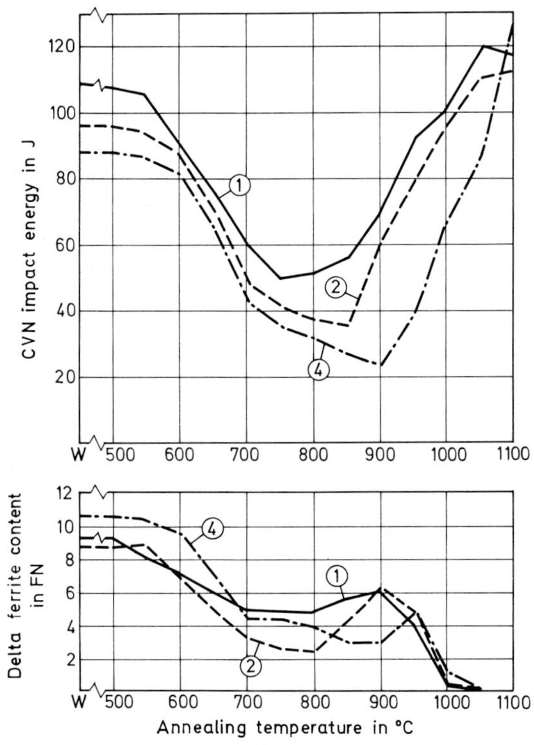

Fig. 110. The influence of ten hours annealing at temperatures between 500 and 1100° C on the impact energy values and delta ferrite contents of stabilized austenitic chromium-nickel (-molybdenum) weld metal, weld metal grades 1, 2 and 4, see Table 21 (page 212) for chemical composition, W = as welded condition

Table 21. *All weld metal mechanical properties, chemical compositions and measured delta ferrite contents of niobium stabilized austenitic weld metal grades. SMAW process using 5 mm coated electrodes, No. 1, 2 and 4 basic coated electrodes, No. 3 and 5 rutile coated electrodes*

No.	AWS classification*	Mechanical properties 0.2 YS N/mm^2	1.0 YS N/mm^2	TS N/mm^2	El. %	IE J	Chemical composition in % C	Si	Mn	Cr	Mo	Ni	Nb	N	Delta ferrite FN
1	E 347	484	514	650	37.4	109	0.043	0.58	1.76	20.13	0.22	10.36	0.55	0.051	9.1
2	E 347 L	471	503	637	36.0	95	0.026	0.73	1.55	20.27	0.26	10.32	0.60	0.050	8.6
3	E 347 L	500	516	640	36.2	68	0.018	1.08	0.97	19.80	0.28	10.26	0.39	0.063	11.3
4	E 318	497	519	646	38.0	88	0.040	0.39	1.50	18.56	2.81	11.18	0.48	0.049	10.5
5	E 318 L	535	554	650	34.0	66	0.022	0.84	0.72	18.62	2.94	11.41	0.39	0.061	9.3

* According to ANSI/AWS A 5.4—81 standard.

[263, 264]. Low carbon weld metals which are also stabilized, e.g. grades 2, 3 and 5 in Table 21, show the best values for IC resistance, with the level of the chromium content also being of decisive importance. The behaviour of stabilized weld metal is also indicated by the TTC diagram in Fig. 89 (page 123).

Fig. 110 shows the influence that the annealing temperature has on the impact toughness and delta ferrite content after a heat treatment period of ten hours. This diagram contains the chromium nickel weld metal grades 1 and 2 and the chromium-nickel-molybdenum weld metal grade 4 from Table 21 (page 212). Stabilized weld metal shows a slightly stronger tendency towards embrittlement by sigma phase precipitation than does an unstabilized one. The carbon induced slow-down of sigma phase precipitation is rather insignificant because carbon is largely bonded by niobium as niobium carbide. In the case of molybdenum alloyed weld metal No. 4, the embrittlement range is shifted to higher temperatures. Therefore a higher solution annealing temperature is required in this case than for the practically Mo free weld metals Nos. 1 and 2.

9.2.3 Precipitation Phenomena in Stabilized Austenitic Stainless Steels and Weld Metals

Contrary to the steel, the weld metal is normally not solution annealed. In the course of solution annealing apart from the dissolution of the carbide $M_{23}C_6$, intermetallic phases and delta ferrite, carbon bonds mainly with niobium to form the stable niobium carbide NbC. According to Fig. 39 (page 32), some 70–80% of the carbon present in the matrix is bonded with niobium during solution annealing treatment at 1050° C/1 hour. In the event of a surplus of niobium, the share of retained carbon will be increased still further. With weld metal which is not solution annealed, the process of carbon retention must take place during the very short cooling period following the welding process which lasts only a few seconds. Simultaneously in multi run welds niobium carbide becomes re-dissolved in the HAZ of the previously deposited weld bead at temperatures above 1300° C, i.e. the effect on the stable compound of carbon with niobium as NbC is reversed again in a very small zone of the HAZ which is indicated in Fig. 39 (page 32) by the re-dissolution range of weld metal No. 5. In other words, with stabilized weld metal the retention of carbon takes place with more irregularity and with less controllability than that experienced during the well defined solution annealing of the parent steel. In weld metal, this fact can be compensated for by increasing the niobium content and decreasing the carbon content. The intergranular corrosion resistance diagram for the different weld metal grades shown in Fig. 89 (page 123) indicates that with the stabilized grades Nos. 3–5, the beginning of IC attack in the critical temperature range of around 550° C is shifted by a suitable modification of the chemical composition to relatively longer periods of time together with a narrowing of the temperature range within which IC attack occurs.

In this way there are three possibilities that are available to ensure that a post weld heat treatment keeps outside the IC attack temperature ranges, these are:

— to select the heat treatment temperature below the IC attack range
— to select this temperature above the IC attack range

— to ensure that the annealing time is kept below the time that IC attack commences.

When using the annealing temperature above the temperature range of IC attack, the possibility of sigma phase precipitation must be taken into account. In practical applications, unstabilized weld metal which shows a lesser tendency towards sigma phase precipitation is often annealed at temperatures above the IC attack temperature range. With stabilized weld metal, the selected annealing time is often kept below the range where IC attack commences, i.e. the heat treatment ends prior to the beginning of the IC attack range. With regard to corrosion behaviour, however, it is best in all cases to completely avoid all heat treatment processes with the exception of solution annealing.

The precipitation of sigma phase from delta ferrite which originally takes place in the weld metal (see Fig. 90 f., page 126) will start during normal heat treatment processes only at temperatures above 600° C. The resulting damage to the weld metal can be taken from Fig. 110 (page 211). The drop in impact energy values also limits the possibility of post weld heat treatments being applied at temperatures above the range of IC attack. For this reason, a compromise is often required between IC attack behaviour and sigma phase embrittlement, as both ranges border or even overlap each other. If components made from stabilized austenitic stainless steels and their weld metals are not subjected to post weld heat treatments no IC attack or sigma phase precipitation at all will appear, the exception being those that are subjected to high service temperatures. In exceptional cases, there is the possibility of subjecting the finished welded component as a whole to a solution annealing treatment at 1050–1100° C followed by quenching or rapid cooling, in which case the detrimental effects caused by carbide and phase precipitations can again be eliminated. However this procedure has serious practical limitations e.g. distortion of the component.

9.3 Welding Metallurgy of Fully Austenitic Stainless Steels

This group of stainless steels contains all austenitic grades which have a stable austenitic structure and which must normally be welded with fully austenitic filler metals that do not produce any ferrite in the weld deposit. A stable austenitic structure is often required due to increased demands on corrosion resistance. It also offers advantages with regard to avoiding the precipitation of brittle or corrosion promoting phases. In addition, fully austenitic materials are paramagnetic (nonmagnetic) and often possess excellent toughness at subzero temperatures. Within the field of stainless steels, this group contains all grades for which improved resistance to crevice and pitting corrosion, stress corrosion cracking and severe corrosion environments in oxidizing and reducing media is required.

A stable austenitic structure can be obtained by increasing the nickel content or by alloying with nitrogen. The disadvantage of this type of structure is that it can practically only be achieved by primary solidification of the weld metal to austenite crystals which increases the hot cracking sensitivity as compared to ferrite containing austenitic weld metal (see also Chapter 5 and in particular section 5.3.4).

It is also possible for solidification to take place by way of the eutectic triangle ($\delta + \gamma + L$) provided that there is a sound knowledge of the phase boundaries of the eutectic groove and the subsequent—transformation (see Figs. 6 and 7, page 8) which may bring about advantages with regard to structure and hot cracking resistance according to Ritter, Henry and Savage [443] and Ogawa, Aoki, Sakamoto and Zaizen [404, part II]. Fully austenitic steels also show a significant tendency towards the formation of liquation cracks in the HAZ, especially if they are stabilized with niobium or titanium. Table 22 shows some examples of the chemical compositional ranges for fully austenitic steels which are distinguished by their excellent corrosion resistance. The following tendencies can be noted here:

— Increase in the molybdenum content for improved resistance to pitting and crevice corrosion.
— Increase in the chromium content for improved resistance to acids and acidic media.
— Increase in the nickel content for improved resistance to stress corrosion cracking.
— Alloying with copper for improved resistance to sulphuric acid.
— Alloying with nitrogen for the prevention of undesirable precipitations, for stabilization of the austenitic structure and for improvement of the yield strength and tensile strength. For application in the cryogenic range, however, lower nitrogen contents are normally required.

With regard to corrosion resistance please refer to the pertinent literature which also deals with the welding of fully austenitic steels. The properties, weldability, corrosion resistance and heat treatment of nitrogen alloyed fully austenitic steels are discussed by Randak, Wessling, Bock, Steinmaurer and Faust [400], Wessling and Bock [215], Kügler [218], Becker, Hilbrans and Köhlert [401], Herbsleb and Westerfeld [402], Gerlach, Kautz and Schäffler [403], Espy [170], Ogawa, Aoki, Sakamoto and Zaizen [404] and Lundin and Chou [456]. The weld metal properties of highly corrosion resistant filler metals, their corrosion resistance and hot cracking resistance as well as the most suitable welding technology to be utilized are covered by Schabereiter, Rabensteiner and Folkhard [301], Diebold and Weingerl [208], Backman and Lundquist [405], Gottschalk [406], Rabensteiner and Schabereiter [266, 407], and Rabensteiner, Tösch and Schabereiter [348]. The properties and the possible applications of steel No. 2 in Table 22, page 216 (17 Cr/13 Ni/4.5 Mo/NL) are discussed by Thier [38], Bäumel, Horn and Siebers [408] as well as Bäumel [409]. The welding metallurgy of fully austenitic CrNiMo steels is discussed in detail by Rabensteiner and Tösch [486], including the hot cracking resistance of fully austenitic weld metal (see also Fig. 99, page 155 and section 5.3.4).

The high alloyed steel No. 6 in Table 22 containing about 6% molybdenum is distinguished by its high resistance towards pitting and crevice corrosion as well as stress corrosion cracking. It has opened up new fields of application in marine engineering and is used in advanced processes in the paper and pulp industry. In the proceedings of the symposium on marine engineering [410 and 470], several papers are dedicated to this particular steel grade and its most suitable application. Kohl, Rabensteiner and Hochörtler [411] and Kohl, Hochörtler, Kriszt and Koren [248]

Table 22. *Chemical composition requirements for fully austenitic chromium-nickel and chromium-nickel-molybdenum steels*

No.	Grade Short designation	Chemical composition in %				
		C_{max}	Cr	Mo	Ni	Others
1	18 Cr/16 Ni/3.5 Mo/L	0.030	17.0–19.0	3.0–4.0	14.0–17.0	–
2	17 Cr/13 Ni/4.5 Mo/NL	0.030	16.5–18.5	4.0–5.0	12.5–14.5	N = 0.12–0.22
3	24 Cr/20 Ni/Nb/L	0.030	23.0–25.0	–	19.0–21.0	Nb ≥ 10 × % C
4	25 Cr/22 Ni/2 Mo/NL	0.030	24.0–26.0	2.0–2.5	21.0–24.0	N = 0.10–0.16
5	23 Cr/27 Ni/3 Mo/3 Cu/NbL	0.030	22.0–24.0	2.5–3.0	26.0–28.0	Cu = 2.5–3.5, Nb = 0.6–0.8
6	20 Cr/25 Ni/6 Mo/2 Cu/NL	0.030	19.0–21.0	5.5–7.0	24.0–26.0	Cu = 1.0–2.0, N = 0.13–0.17

deal with the development of high alloyed, molybdenum containing austenitic steels together with their corrosion resistance and their weldability.

An interesting comparison between the properties of ferritic ELI steels with extremely low carbon and nitrogen contents, ferritic-austenitic duplex steels and high molybdenum alloyed fully austenitic steels is supplied by Vicentini, Sinigaglia et al. [412]. Garner [413–415] deals with the corrosion resistance, in particular the resistance to pitting and crevice corrosion and discusses questions raised in combination with the processing and welding of high molybdenum containing fully austenitic steels. The properties of fully austenitic weld metal with about 6% molybdenum are reported on by Rabensteiner and Schabereiter [266] and Rabensteiner and Tösch [486].

Higher nitrogen contents permit an improvement in the tensile properties of austenitic stainless steels. Such steels with nitrogen contents up to 0.40% and 0.2% yield strength values above 500 N/mm^2 are discussed by Kohl [36], Wedl and Kohl [211], Lennartz and Kiesheyer [416] and Ritter, Henry and Savage [443]. The properties of weld metals employed for the welding of such steels are given by Rabensteiner and Schabereiter [407] and Rabensteiner and Tösch [486].

Attention must be drawn here to a special feature of the high molybdenum alloyed austenitic materials, i. e. the relatively strong tendency towards the segregation of the alloying element molybdenum. This may result in a situation where there is a marked difference in the molybdenum content within the same single crystal resulting in the formation of a layered crystal structure (see also section 2.7). This penomenon is discussed by Bäumel [133], Weingerl, Straube and Blöch [134], Garner [414], Cieslak, Ritter and Savage [439], and Ritter, Henry and Savage [443]. The strong tendency towards molybdenum segregation must already be considered during the melting, hot forming and heat treatment of high molybdenum alloyed fully austenitic stainless steels. In welds where corrosive stresses are not too extreme, e.g. sea water, there are normally no problems associated with the occurrence of segregations in the weld metal, provided that the final weld runs of the welded component that are exposed to the corrosive medium are not welded with too high a heat input in order to keep grain size and thus segregations to a minimum. If, however, the corrosive conditions are located at the upper resistance limit of the material, welding with a filler metal grade that possesses higher corrosion resistance may be advantageous, e.g. welding of steel No. 2 in Table 22 (page 216), containing about 4.5% molybdenum with a filler metal with about 6% molybdenum (weld metal grade No. 6 in Table 23, page 220) and welding of steel No. 6 in Table 22 (page 216), containing about 6% molybdenum with a nickel-based filler metal containing 8–9% molybdenum (e. g. grade E NiCrMo-3 to AWS A5.11–83). To obtain full corrosion resistance from a welded component it is necessary to carry out a proper post weld surface treatment which is discussed in detail in section 10.3.

Similar to delta ferrite containing weld metal, fully austenitic weld metal too shows its highest corrosion resistance in the as welded condition and also after solution annealing and quenching. Other heat treatment procedures after the welding operation should be avoided whenever possible. If they cannot be avoided, it is absolutely essential to consider the intergranular corrosion behaviour and precipitation behaviour of the materials being used (see section 9.3.3).

9.3.1 Practical Welding of Fully Austenitic Stainless Steels with Fully Austenitic Stainless Steel Filler Metals

Because of their homogenous structure, fully austenitic steels and weld metals possess excellent corrosion resistance and toughness properties. The major problem during welding is caused by a relatively strong tendency towards hot cracking, both in the form of solidification and liquation cracking. With proper modification of the chemical compositions of both weld metal and steel grade combined with a sufficient degree of cleanness, it is possible, to obtain satisfactory hot cracking resistance in fully austenitic weld metals and steels. The necessary measures are discussed in detail in section 5.3.4. Nevertheless, for constitutional reasons, fully austenitic materials show a stronger tendency towards hot cracking than the austenitic weld metal grades containing small amounts of delta ferrite after cooling to ambient temperature. For this reason it is necessary at both the design and the manufacture of fully austenitic component to carry out a number of steps in order to minimize the risks arising from the materials hot cracking sensitivity, a situation which is also discussed by Gooch and Honeycomb [303], Killing [359] and Thiesen and Poweleit [360].

Below, we give a number of practical suggestions regarding special procedures that must be followed when welding fully austenitic stainless steels using fully austenitic stainless steel filler metals in order to obtain as far as possible crack free components:

- Both the material thickness and the rigidity of the component excert a major influence upon the formation of hot cracks. Therefore, measures should be taken at the design stage to reduce as much as possible material thickness, rigidity and accumulation of welds. In order to achieve optimum results it is recommended that the design department discusses any anticipated problem areas with the welding engineering/metallurgy department early on at the design stage. It is vitally important to avoid all notch effects both internally and externally as well as sharp changes in material thickness and sections as these represent areas of potential high stress and therefore points of increased hot cracking sensitivity.
- The heat input during welding should be limited, to keep the grain size of the primary austenite crystals small. As the size of the primary crystals increase, the tendency towards the formation of segregations, liquid phases and consequently hot cracks also rises.
- To reduce the hot cracking sensitivity, it is advisable to maintain a weld width to depth ratio 1.5 : 1 to 2 : 1.
- Hot cracking sensitivity can normally be reduced by selecting welding procedures which reduce the total heat input into the workpeace. This can be achieved, for example, by limiting the interpass temperature or by staggering the welding sequence to allow the component to cool whilst other seams are welded elsewhere.
- Preheating has normally only little influence upon the formation of solidification cracks during the welding of fully austenitic steels using fully austenitic filler metals. It can be advantageous, however, in the case of steels and weld

metals which are susceptible to liquation cracking, in particular when welding thick sections.

— Apart from these measures the manipulation of the electrode or filler rod during welding can be of great importance. Here stringer beads should be preferably used, as even slight weaving would produce larger weld pools and so increase the susceptibility to hot cracking.
— Every precaution should be taken to avoid weld defects as hot cracks will often initiate from such defects. Subsequent repair of such defects will not necessarily eliminiate micro cracks and may induce new hot cracks.
— It is important to grind all stop start positions and also to lightly grind the surface of the weld prior to welding the next pass in order to minimize defects.
— Weld joint edge preparations must be prepared carefully and correctly to ensure the safe and easy deposition of the weld runs. The plate edge preparations must be smooth and clean. Correct tack welding is of particular importance (see section 10.1).
— Back grinding and welding of backing runs should be carried out whenever possible. If this is not possible, full penetration welding of the root run is facilitated by utilizing the GTAW process in combination with back purging using argon or an anti slag gas. This process offers definite advantages for all welds where the root side of the weld is inaccessible.
— The welding of fully austenitic stainless steels should be carried out only by experienced and well trained welders that possess sufficient knowledge in the field of stainless steel welding.
— Special attention must be paid to ensure that the welding procedures are correctly controlled before, during and after the welding cycle. The supervision and control of welding operations should only be carried out by well trained personnel that are familiar with the correct procedures that must be followed when welding fully austenitic stainless steels using fully austenitic stainless steel filler metals.
— When welding high molybdenum alloyed fully austenitic stainless steels using matching filler metals special care must be taken to ensure that all weld capping runs that are exposed to the corrosive environment are welded using a low heat input. This will help to keep the segregations, in particular those of molybdenum, to a minimum.
— A post weld surface treatment of the complete welded component must be carried out carefully, following the instructions stated in section 10.3 to guarantee full corrosion resistance.

9.3.2 Mechanical Properties of Fully Austenitic Stainless Steel Weld Metal

Table 23 shows the mechanical properties and the chemical compositions of various fully austenitic stainless steel all weld metal samples. In order to improve the materials hot cracking resistance, manganese contents are increased and are normally, if permitted by corrosion resistance, in the range of between 4 and 5%. With molybdenum contents above 4%, fully austenitic weld metal must be additionally alloyed with nitrogen, this is to prevent the precipitation of undesirable phases

Table 23. *All weld metal mechanical properties and chemical compositions of fully austenitic stainless steel weld metal grades. SMAW process using 4 mm basic coated electrodes*

No.	Grade short designation	Mechanical properties					Chemical composition in %							
		0.2 YS N/mm^2	1.0 YS N/mm^2	TS N/mm^2	El %	IE J	C	Si	Mn	Cr	Mo	Ni	N	Others
1	18 Cr/16 Ni/2 Mo/L	426	448	564	40.0	98	0.038	0.25	4.50	18.45	2.20	15.52	0.047	–
2	18 Cr/16 Ni/3.5 Mo/L	431	456	589	37.7	91	0.030	0.25	3.48	17.50	3.53	15.81	0.048	–
3	18 Cr/16 Ni/4.5 Mo/NL	469	499	659	31.6	116	0.033	0.45	2.58	18.05	4.56	16.65	0.165	–
4	25 Cr/20 Ni/2 Mo/L	452	484	637	37.0	75	0.026	0.49	4.98	25.12	1.97	19.92	0.151	–
5	20 Cr/34 Ni/3 Mo/3 Cu/NbL	404	439	621	37.4	95	0.033	0.29	2.59	20.32	2.51	33.72	0.048	Cu = 3.01 Nb = 0.36
6	20 Cr/25 Ni/6 Mo/2 Cu/NL	448	481	675	39.1	87	0.031	0.38	4.02	19.37	6.44	24.66	0.150	Cu = 2.48
7	25 Cr/23 Ni/4 Mo/NL	560	583	790	42.0	94	0.030	0.50	5.63	24.37	4.27	23.18	0.391	–

during welding. The mechanical properties fall in a similar range as those given for the unstabilized weld metals (see Table 19, page 205), with the exception of weld metal No. 7 in which the values of the yield strength and tensile strength have been increased by about 100–150 N/mm^2 due to the high nitrogen content of 0.391%. The properties of weld metal No. 7 are already relatively close to the tensile properties of the ferritic-austenitic duplex weld metal (see Table 17, page 195). The development of fully austenitic high tensile weld metal grades and the influence of different alloying elements is discussed by Rabensteiner and Schabereiter [407] and Rabensteiner and Tösch [486].

The influence on the impact energy values after 10 hours annealing in the temperature range of 500–1100° C is shown in Fig. 111. This diagram contains weld metal grades with molybdenum contents of 2.2–6.4%. Weld metal No. 1 containing 2.2% molybdenum does not show a drop in its impact energy values after annealing for 10 hours. According to Schabereiter [265] such a drop only occurs after annealing for 400 hours at an annealing temperature of 750° C. Weld metal No. 2 containing about 3.5% molybdenum shows a drop in its impact energy values to about half of its original value after annealing for 10 hours at 800° C. In the case of weld metal No. 3 containing about 4.5% molybdenum, the drop in impact energy is only slightly more pronounced, this is because of additional alloying with 0.165% nitrogen. However the reduction in impact toughness is extended to higher annealing temperatures. This weld metal requires rapid cooling from the solution annealing temperature of 1100° C to obtain high impact energy values, as indicated by a comparison between curve 3 (air cooling) and 3a (furnace cooling) after annealing at temperatures above 1000° C. Weld metal No. 6 containing about 6.4% molybdenum shows a marked

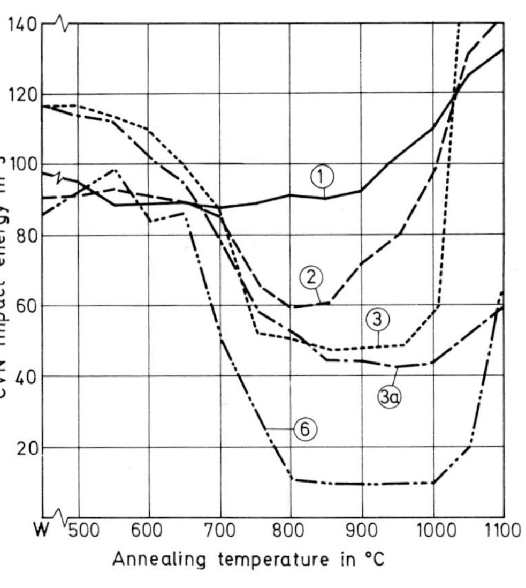

Fig. 111. Influence of a ten hour annealing treatment at temperatures between 500 and 1100° C on the impact energy values of fully austenitic stainless Cr-Ni-Mo weld metal. For the chemical composition of weld metal grades 1, 2, 3 and 6 see Table 23, page 220; Air cooling after annealing–grades 1, 2, 3 and 6, Furnace cooling after annealing–grade 3a

drop in impact energy toughness which extends up to 1100° C. Here, the solution annealing temperature must be raised to 1150° C to obtain a precipitation free fully austenitic structure.

9.3.3 Precipitation Phenomena in Fully Austenitic Stainless Steels and Weld Metals

The precipitation behaviour of the molybdenum alloyed austenitic stainless steels is relatively complex as described earlier in section 4.2.4. The weld metal types included in Table 23, page 220, are in the as welded condition fully resistant to intergranular corrosion when tested in accordance with the Strauss test, this due to their fully austenitic structure and their low carbon content. However such corrosion resistance is of little relevance when compared with more stringent demands, such as resistance against pitting and crevice corrosion and stress corrosion cracking. In the event of a post weld heat treatment, the IC attack range is relatively strongly influenced by the chromium and nitrogen contents, as indicated by Figs. 85d and 85e (page 108). By increasing the contents of both elements, the beginning of the IC attack range is shifted to longer periods of time. Of great importance to the fully austenitic molybdenum containing stainless steel weld metal alloys is their resistance against pitting and crevice corrosion. A precondition for good pitting corrosion resistance is the formation of a fully austenitic precipitation free structure. The presence of only a small amount of sigma phase or chi phase precipitation is sufficient to give a marked drop in the pitting resistance. The same applies to the corrosion resistance when tested in a strong boiling nitric acid solution (Huey test).

Fig. 92 (page 136) shows the precipitation behaviour of molybdenum containing austenitic stainless steels, with weld metal behaving in a similar manner as the steel grades. With a molybdenum content above 4% it is necessary to additionally alloy with nitrogen to safely avoid phase precipitations during welding (see Fig. 92d through 92f, page 136). With the addition of nitrogen, IC attack due to $M_{23}C_6$ precipitation is also shifted to considerably longer periods of time (see Fig. 85e, page 108). Regarding the precipitations in molybdenum containing austenitic steels, please refer to section 4.2.4 and Fig. 92 (page 136) and to the great wealth of literature on the corrosion and precipitation behaviour of chromium-nickel-molybdenum steels and weld metals, in particular the following entries in our list of supplementary literature [33, 38, 41, 42, 44, 45, 46, 213, 214, 215, 216, 218, 219, 265, 266, 400, 401, 402, 407, 408, 409, 411, 413, 414, 415, 443, 447, 448, 470, 480, 482, 485, 486]. A 475° C embrittlement does not occur in fully austenitic weld metal because this only occurs in the presence of a ferritic structure.

10 General Instructions for the Welding and Post-Weld Surface Treatments of Fabrications and Welded Components Made from Austenitic Stainless Steel

10.1 Welding Procedure

When welding austenitic stainless steels the tendency towards greater distortion should be taken into consideration. Depending on the component design and dimensions, distortion can be counteracted and the need for straightening after welding reduced by following these guidelines:

— Use clamping fixtures where possible, particularly when welding thin material.
— Avoid corner welds, instead where possible bend corners and then butt weld the joints.
— Use tack welds, with the maximum distance between tacks being 100 mm. The length of tack welds should be 30–40 mm.
— The welding sequence should start from the center of the joint working outwards or be carried out using the back step technique.
— Use double V butt joints even when welding thin materials.

Joint preparations are much the same for austenitic stainless steels as they are for the standard mild steels, with due care being taken to ensure that the joint design is suitable for the type of welding process that is going to be used. It is important to make sure that the joint faces and the surrounding weld areas are absolutely clean. Any contamination in the form of scale, grease, oil, paint etc. may have detrimental effects upon the quality of the weld. Weld stop-start positions and excess weld metal should be carefully ground off in order to facilitate the deposition of subsequent defect free welds. When using resin bonded wheels for back grinding etc. ensure that the wheel itself does not smear the material surface, in most cases it is preferable to use ceramic bonded low sulphur wheels. The use of ordinary steel wire brushes should not be allowed, instead use only stainless steel wire brushes for all surface cleaning.

Particular attention should be paid to avoid the possibility of "stray arcing" on the workpeace surface which is often caused by faulty insulation on electrode holders or cables and also by poor connections at the return current cable clamp. The consequences of such stray arc strikes can be serious, as the localized over heating can often lead to the formation of micro cracks on the material surface which in turn may result in crevice corrosion once the component is put into service.

In workshop conditions such "stray arcing" is often overlooked as being trivial, but for high quality weldments demanding maximum service life every effort should be taken to eliminate this occurrence.

10.2 Welding of Austenitic Stainless Steel Castings

Austenitic stainless steel castings can be welded in the same way as rolled or forged austenitic Cr–Ni steels. In general, we are confronted here with the joining together of austenitic steel castings, the joining of austenitic steel castings to rolled or forged components or the repair welding of austenitic steel castings. The electric arc welding processes that are mainly used for this application are SMAW, GTAW and GMAW. Austenitic stainless steels castings are always supplied in the solution annealed and quenched condition.

Even after a solution annealing treatment, generally these castings still show pronounced grain boundaries in their structure particularly in the case of heavy sections. During welding there may be an increased tendency towards hot cracking due to the possible presence of low melting grain boundary substances and as a result of thermal stresses set up during welding. However by taking metallurgical measures aimed at obtaining grain refinement and a higher degree of cleanliness, these disadvantages can be largely eliminated. In addition it has been found that crack sensitivity may be partially reduced by carrying out a cold working operation on the joint faces prior to welding, such cold working can be successfully performed by using a pneumatic chipping hammer. This cold working causes the grain boundaries to break and in the welding operation that follows the recrystallization temperature will be exceeded, then new grain formation takes place along the slip lines and in the gaps between the primary grain boundaries. In this way, a fine grained rim is produced at the edge of the weld which shields the weld from the rest of the casting and thus reduces the danger of crack formation.

In general, only matching filler metals are used which in the case of electric arc welding consist mainly of coated electrodes. Steps should be taken to reduce the welding stresses where possible by employing suitable welding sequences. In many cases it is possible to obtain good results by buttering the joint faces with a layer of matching weld metal prior to joining. In the case of heavy wall thickness, we recommend a slight preheat in the range of 100–150° C and in particular a peening of each individual weld run to help reduce the shrinkage stresses. Air cooling after welding is in most cases recommended. Quenching after welding is normally only necessary when welding cast steel grades with carbon contents above 0.06% or components of intricate shape. If faulty steel castings are repaired by welding a full heat treatment, i.e. solution annealing and quenching may be recommended by the steel manufacturers or in many cases by the manufacturers codes to which the component is made.

10.3 Post Weld Surface Treatments

In order to achieve maximum corrosion resistance it is absolutely essential that the complete surface is metallically bright. This not only means that all slag, spatter and scale is removed but the temper colours must also be removed. Such cleaning can be done mechanically by grinding, polishing, brushing or sand blasting or chemically by pickling. If weld metal also has to be removed, this can be achieved by rough grinding followed by finish grinding. Rough grinding is carried out using corundum type wheels of suitable coarseness, whereas finish grinding is carried out with felt wheels impregnated with a fine corundum grit, typically 80, 120 or 180. Corrosion resistance is directly proportional to the smoothness of the surface finish. Any brushing or polishing is best carried out using rotary fibre brushes and polishing wax. For fine finish polishing, lapping wheels and polishing wax or paste are used. If the finished components are to be used immediately after grinding or polishing in a corrosive environment, then passivation of the polished surfaces with a 5–20% nitric acid solution is absolutely necessary.

A commonly used process is pickling, with dip pickling being possible, however, only in rare cases. Pickling is normally restricted to a localized "wipe pickling" operation, where only the weld and its immediate surroundings are treated. This is carried out using special pickling solutions or pickling pastes which are applied by brush or with a cloth or cotton pad. The effect of the treatment can be increased if the pad is dipped additionally in silica sand or diatomaceous earth, i.e. if the chemical action is supported by the effect of mechanical abrasion. Pickling agents are highly caustic and the use of rubber gloves, rubber aprons, breathing apparatus and goggles is mandatory. The pad soaked in the pickling solution should be manipulated using a special holder or tongues made from a chemical resistant steel to make sure that the operator does not come into contact with the pickling agent. After pickling, the component must be carefully flushed with water and subjected to a final passivation treatment using a 5–20% nitric acid solution, thus creating the optimum conditions for good corrosion resistance. After passivation too, the component must be carefully cleaned with water and then dried to avoid the formation of blotches. After pickling and often also after passivation, weld seams or areas repaired by welding normally become visible due to the unavoidable difference in colour between the base metal and the weld metal. Sand blasting is used when grinding or pickling are not practicable. This may only be carried out using steel grit (shot peening) made of stainless steel or silica sand. Both methods produce a metallically bright but somewhat rough surface which is more difficult to keep clean. A sand blasting or shot peening treatment should also be followed by a passivation treatment.

In all kinds of cold working processes, e.g. in sand blasting as well as grinding operations, cold deformation takes place at the components surface and is normally accompanied by considerable surface stresses. According to Risch [417] after grinding these stresses may extend to a depth of 0.2 mm below the material surface and may cause stress corrosion cracking when subjected to an aggressive corrosive environment. As a counter measure, Risch recommends a stress relieving or a strong pickling treatment after grinding. This post grinding treatment is only necessary, however, if the component is to be exposed to a corrosive medium which may cause stress corrosion cracking. For more details, please refer to the original work [417]. Other valuable information regarding the preventive measures to be taken to safeguard against chloride induced stress corrosion cracking in austenitic steels and weld metals is given by Risch [481].

11 Welding Metallurgy of Heat Resisting Steels

Heat resisting steels are normally alloyed with chromium, silicon, aluminium and often also with nickel, with chromium being the most important element with regard to the scaling resistance. There exist ferritic, ferritic-austenitic und austenitic heat resisting steels. The latter contain higher amounts of nickel, normally in the range of approx. 4—35%. A comprehensive survey of the effect of alloying elements and scaling behaviour in different corrosive media is given by Rapatz [8], Houdremont [9], Fontana and Green [233] and Sedrics [390]. Table 24 shows some examples of the chemical composition requirements and scaling temperature limits of heat resisting steels. Compared to the corrosion resisting steels (see Tables 4, 12, 14, 16, 18, 20 and 22) not only are the silicon and aluminium contents increased but also the carbon content is often increased. If the ferritizing effect of 1% silicon and 1% aluminium is considered, these alloys will sometimes show high delta ferrite contents which in turn will lead to a strong tendency towards coarse grain formation at elevated operating temperatures. The ferritic heat resisting steels No. 1—3 thus show a strong tendency towards embrittlement which often appears only after cooling to ambient temperature. The presence of austenite counteracts coarse grain formation and such steels show less embrittlement than the ferritic grades No. 1—3. During operation, the fully austenitic heat resisting grades Nos. 5—9 also become coarse grained, however, they retain their toughness when cooled to ambient temperature because of their austenitic structure. The ferritic heat resisting grades Nos. 1—3 show good resistance in oxidizing, reducing and in particular sulphur containing gases. During welding, they show a tendency towards hardness increase and coarse grain formation and behave in a similar manner to ferritic stainless steels (see Chapter 6). In principle, the same welding procedures apply for these steels as apply to the welding of ferritic chromium stainless steels (see section 6.1), i.e. preheating to 200—300° C and subsequent annealing at 650—700° C for an improvement in toughness values. Normally, these steels are welded with austenitic filler metals of the type of 25% chromium and 20% nickel (AWS E 310) with a chemical composition of the weld metal similar to steels 6 and 7 in Table 24. However, in order to avoid any damage due to the formation of the low melting nickel sulphide (see also Fig. 44, page 39), the atmosphere under service conditions should not contain sulphur. Matching filler metals are rarely used since weld metals with a similar alloying formulation are normally brittle. In the presence of a sulphur containing atmosphere, a ferritic-austenitic 25/4 chromium-nickel filler metal with a chemical composition similar to steel No. 4 is used which produces a ferritic-austenitic weld metal structure and represents a compromise solution.

Table 24. *Chemical composition requirements and scaling temperature limits for heat resisting steels. Steels 1—3 ferritic; steel 4 ferritic-austenitic, steels 5—9 austenitic*

No.	Grade Short designation	Chemical composition in %							Scaling temperature limit in air, °C
		C	Si	Mn_{max}	Al	Cr	Ni	Others	
1	10 Cr/1 Al	0.12	0.7—1.4	1.0	0.7—1.2	12.0—14.0	—	—	850
2	18 Cr/1 Al	0.12	0.7—1.4	1.0	0.7—1.2	17.0—19.0	—	—	1000
3	25 Cr/1.5 Al	0.12	0.7—1.4	1.0	1.2—1.7	23.0—26.0	—	—	1150
4	25 Cr/4 Ni/1 Si	0.10—0.20	0.8—1.5	2.0	—	24.0—27.0	3.5— 5.5	—	1100
5	20 Cr/12 Ni/2 Si	0.20	1.5—2.5	2.0	—	19.0—21.0	11.0—13.0	—	1000
6	25 Cr/20 Ni	0.15	≤0.75	2.0	—	24.0—26.0	19.0—22.0	—	1050
7	25 Cr/20 Ni/2 Si	0.20	1.5—2.0	2.0	—	24.0—26.0	19.0—22.0	—	1150
8	16 Cr/34 Ni/1.5 Si	0.15	1.0—2.0	2.0	—	15.0—17.0	33.0—37.0	—	1100
9	20 Cr/32 Ni/Al/Ti	0.12	≤1.0	2.0	0.2—0.6	19.0—23.0	30.0—34.0	Ti = 0.2—0.6	1100

Compared to the purely ferritic steel grades, the ferritic-austenitic steel which is equivalent to grade No. 4 in Table 24, shows a somewhat better weldability because of its lower tendency towards coarse grain formation. It is generally also resistant to sulphur containing gases but shows a tendency due to its high delta ferrite content towards 475° C embrittlement in the temperature range between 300 and 500° C and towards sigma phase embrittlement in the range between 600 and 900° C. Such steels are normally welded with maching filler metals but with a welding technology which is equivalent to that of the ferritic chromium stainless steels, i.e. preheating to 200–300° C.

The welding of the heat resisting austenitic steels (Nos. 5–9 in Table 24) does not cause any particular difficulties providing that the instructions given for the welding of fully austenitic stainless steels (section 9.3.1) are followed. Normally, these steels are welded with matching filler metals (e.g. E 310). When welding components which are exposed to severe temperature changes, the observance of the instructions stated in section 9.3.1 is of great importance.

The chemical compositions of the heat resisting stainless steel weld metals generally coincide with the values given for parent steel grades as shown in Table 24. This table can therefore also be used for these weld metals. Normally, only filler metals which correspond to the steel grades Nos. 4–9 are used. Due to their increased silicon contents weld metal grades which correspond to the steel grades Nos. 4–7 show a tendency towards sigma phase embrittlement in the temperature range of 600–900° C, a fact which must be considered when heat treatments or operating temperatures fall in this range. This applies also to the welding of the ferritic steels Nos. 1–3 when using such filler metal grades. Weld metal grades which correspond to the alloys Nos. 8 and 9 can, however, due to their high nickel contents be used for all operating temperatures up to the scaling temperature limit.

12 Welding Metallurgy of Austenitic-Ferritic Dissimilar Joints

When welding stainless steel components, particularly in plant engineering applications, it may be necessary to join unalloyed or low alloyed ferritic steels to austenitic chromium-nickel-(molybdenum) stainless steels. Such dissimilar joints are also called "black-and-white joints". With regard to corrosion resistance, there are normally no particular requirements demanded from this type of joint, because the low alloyed steel is normally not corrosion resistant. Because of the different structure between the types of steels to be joined, there may be certain difficulties experienced during welding. Low alloyed steels, for example, normally require controlled preheat and interpass temperatures during welding, slow cooling after welding and often a final tempering or stress relieving heat treatment. Austenitic steels on the other hand should be welded "cold", the welds should cool rapidly and post weld heat treatments should be avoided wherever possible.

The major requirement of a dissimilar joint is the selection of a weld metal which is as tough as possible and to obtain tough, crack-free heat effected zones. Because of the basic differences between the base metals structures and the great differences in the alloying compositions, this requires a number of suitable measures to be taken. No brittle zones, produced for example by martensite formation, should be allowed to occur at the transition between high alloyed austenitic stainless steel and the low alloyed ferritic steel, this is despite the fact that the Schaeffler diagram shows a wide martensitic structure area between the areas of α ferrite and austenite.

In order to counteract the contradicting requirements of the base metals it is necessary to carefully select the type of welding process, choice of welding filler metals, welding procedures to be used, and any subsequent post-weld heat treatment requirements, to suit as near as possible the requirement of the steels to be joined.

Due to the great variety of material combinations available, it is not possible here to establish any generally applicable rules for dissimilar joint welding. However we can provide guidelines which may be helpful in avoiding the possible formation of defects. These guidelines should be used with due regard to the peculiarities of each of the steel grades to be joined. It is also impossible to list even tentatively the vast body of literature published on this topic in recent years. Please refer to a survey supplied by Lundin [418] published in 1982, in which 160 works are discussed. A good survey covering the state of the art in dissimilar joint welding between low alloyed and austenitic stainless steels was published in 1982 by Faber and Gooch [419]. It contains many practical welding hints together with references to another 25 publications.

Table 25. *All weld metal chemical compositions, delta ferrite contents and coefficients of thermal expansion (rough average) of filler metals used in the welding of dissimilar austenitic-ferritic steel joints*

No.	Grade short designation	AWS Standard*	Chemical composition in %							Delta ferrite content Volume %	Mean linear coefficient of thermal expansion 20–500° C, 10^{-6}m/m.K
			C	Si	Mn	Cr	Mo	Ni	Nb		
1	19 Cr/9 Ni/3 Mo	~ E 308 Mo	0.05	0.8	1.0	19.0	3.4	9.0	–	18	18.0
2	23 Cr/12 Ni/L	E 309 L	0.02	0.9	1.2	23.0	–	12.0	–	18	17.5
3	23 Cr/12 Ni/3 Mo/L	E 309 L Mo	0.02	0.9	1.0	23.0	2.7	12.0	–	25	17.4
4	22 Cr/18 Ni/5 Mn		0.13	0.9	4.5	22.5	–	18.0	–	0	18.0
5	16 Cr/70 Ni/6 Mn/2 Cb	E NiCrFe 3	0.05	0.5	6.0	16.0–20.0	–	70.0	2.2	0	14.4

* No. 1–3 according to AWS A5.4–81, No. 5 according to AWS A5.11–83
~ = only roughly equivalent

12.1 Selection of Filler Metals and the Importance of Dilution

A detailed discussion on the criteria employed in the selection of filler metals with consideration to the post weld heat treatment of welded components is supplied by Widowitz and Folkhard [420]. In general, only high alloyed filler metals are used for the joining of unalloyed or low alloyed steels to austenitic stainless steels. Table 25 shows some examples of the all weld metal chemical compositions and delta ferrite contents for the common filler metals used for austenitic-ferritic dissimilar joints. The alloys No. 1 to 3 represent austenitic filler materials with increased delta ferrite contents, with No. 2 and 3 being overalloyed with chromium and nickel to compensate for alloy loss due to the effect of dilution with the low alloyed ferritic steel grade.

Alloy No. 4 represents a fully austenitic filler metal with 4–5% manganese which has performed very well in the welding of low alloyed ferritic steels containing higher carbon contents. Filler metal No. 5 is a nickel-based alloy, used in particular for components which require a heat treatment after welding and components exposed to high service temperatures. Compared to filler metals Nos. 1–4, this alloy offers the advantage that carbon diffusion from the low alloyed ferritic steel to the high alloyed weld metal is slowed down by the high nickel content. Also this weld metal does not tend to precipitate brittle phases during service at high operating temperatures or during the post weld heat treatment of welded assemblies. It is also important to mention that the coefficient of thermal expansion of the nickel-based weld metal is similar to that of the low alloyed ferritic steel. This is of particular importance for ferritic-austenitic dissimilar joints exposed to alternating thermal stresses.

The major task of any dissimilar welding filler metal is to obtain a tough crack-free welded joint between the ferritic steel and the austenitic stainless steel. A major aid in understanding the processes which take place during the welding of austenitic-ferritic dissimilar joints is the Schaeffler diagram, developed 1947 by Schaeffler [163–165] for this purpose. Its practical application can be taken from Fig. 112. Let us assume, for example, that a mild steel containing 0.2% carbon (point A_1) is to be joined to an austenitic stainless steel of the type 18 Cr/10 Ni (point B). The filler metal to be used is an overalloyed 23 Cr/12 Ni coated electrode (AWS E 309 L, alloy No. 2 in Table 25). The all weld metal analysis corresponds to point C in Fig. 112. It is necessary to use an over-alloyed filler metal, because during welding the weld metal becomes partly diluted with the unalloyed steel. In the example it is assumed that equal quantities of mild steel and austenitic stainless steel will be diluted, i.e. the over-alloyed weld metal is diluted with alloy D_1 which consists of equal parts of mild steel and austenitic stainless steel of the type 18 Cr/10 Ni. If the points D_1 and C in Fig. 112 are now connected by a straight line, we arrive, depending upon the percentage of dilution at different weld metal grades. Working on the assumption that dilution from the SMAW process will be in the range of 20–30% we arrive at a weld metal structure that is austenite and containing approx. 10% delta ferrite (hatched area), this structure would possess high hot cracking resistance.

If the dilution was equal to 40%, the weld metal would contain only 4% delta ferrite and would be at the lower limit of sufficient hot cracking resistance. With dilution

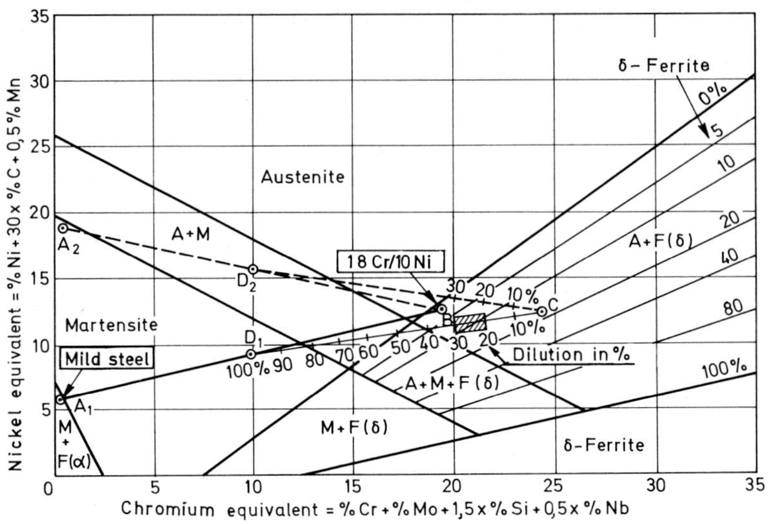

Fig. 112. Use of the Schaeffler diagram for the determination of the type of structure and delta ferrite content of weld metal when joining unalloyed or low alloyed steels to austenitic stainless steels.

Examples – A_1 mild steel with 0.2% carbon
A_2 carbon steel with 0.6% carbon
B austenitic steel of type 18 Cr/10 Ni
C over-alloyed filler metal of type 23 Cr/12 Ni (alloy No. 2 in Table 25)

equal to 50%, the zero ferrite line is reached. At the same time, apart from austenite martensite is formed in the structure. Due to the primary austenite solidification which results in lack of ferrite at ambient temperature, this weld metal becomes sensitive to hot cracking. Due to the presence of martensite in the structure, it is also susceptible to cold cracking. In other words, there is a high probability that the welded joint would crack at 50% dilution. These examples illustrate the great importance of dilution on the cracking resistance of austenitic-ferritic dissimilar joints. In the case of austenitic molybdenum alloyed chromium nickel steels, higher dilution and fully austenitic stainless steels, the Cr-Ni-Mo alloyed weld metal No. 3 with a higher ferrite content should be used. This filler metal grade is also suitable for root passes where greater fluctuations in the percentage of dilution may occur.

Where it is necessary to join ferritic steels with a high nickel equivalent (according to Schaeffler) such as quenched and tempered 9% nickel steels, or high carbon manganese steels, to austenitic stainless steels the position of the dilution line in the Schaeffler diagram will change. For example in Fig. 112 a carbon steel (point A_2) containing approx. 0.6% carbon and having thus a nickel equivalent of approx. 18 is plotted to be joined to an austenitic 18 Cr/10 Ni stainless steel (point B), using the 23 Cr/12 Ni filler metal denoted by point C. The dilution line connecting the weld metal alloy C to point D_2 (which represents equal proportions of material A_2 and B) reaches the zero ferrite line with dilution equal to only approx. 30%, therefore the weld metal becomes fully austenitic at this point and sensitive to hot cracking. In such cases the high ferrite containing weld metal or also the high manganese containing fully austenitic weld metal as shown in Table 25 (Nos. 3 and 4 respectively) have given good results in the welding of such material combinations. Due to its high manganese content combined with a higher carbon content, weld metal 4 shows sufficient hot cracking resistance even though it has a fully austenitic structure.

Nickel-based filler metals, e.g. No. 5 in Table 25, generally show good hot cracking resistance due largely to their high degree of cleanliness. This can normally be maintained with a dilution level up to a maximum of approx. 35%. With higher levels of dilution, nickel-base weld metal may become sensitive to hot cracking due to the fusion with the less clean ferritic steel and especially due to the pick-up of sulphur and phosphorus. If the welding of an austenitic-ferritic dissimilar joint is started using nickel-based filler metals then it must be finished with them. The use of austenitic chromium-nickel steel filler metals, e.g. for the fill and cap runs of a joint when the root run has been welded using nickel-based electrodes, will lead with great probability to a fully austenitic structure due to the nickel pick-up in the subsequent austenitic weld metal passes and thus to the formation of hot cracks. The reverse order, however, in many cases is possible, e.g. deposition of nickel based weld metal on previously deposited austenitic Cr-Ni steel weld metal.

12.2 Practical Welding of Austenitic-Ferritic Dissimilar Joints

Austenitic-ferritic dissimilar joints are best SMA welded using coated electrodes similar to those given in Table 25 (page 230) or by the GTAW and GMAW processes. The root pass is the most critical part of the weld because of the possibility of high dilution levels of about 50% and more which may easily occur, especially when using the GTAW process. As described above, there may be a high probability of crack formation in the weld metal. Another critical situation occurs with single V butt joints in which high dilution levels must be expected in the root pass, particularly in the event of poor joint preparation. It is recommended therefore where possible to use double V butt joints instead of single V joints even for relatively thin wall sections and to ensure that all joint preparations are accurate with good fit up. Furthermore, all joint root passes which are to be welded with austenitic filler metals should be welded using over-alloyed filler metals with relatively high delta ferrite contents (e.g. No. 3 in Table 25, page 230 or also 29 Cr/9 Ni [E 312] type filler metals). Thorough back grinding of the root run from the reverse side and the subsequent deposition of a backing run using small diameter electrodes in the prepared groove will be a positive step towards the production of crack-free austenitic-ferritic dissimilar joints. The deposition of subsequent passes should always take place in such a way that about the same amount of ferritic and austenitic material is fused, i.e. the welding operation takes place as close as possible to point D_1 (or D_2) in Fig. 112. If this is not the case e.g. if mainly ferritic material is fused then point D_1 will be shifted closer towards point A, resulting also in the shifting of the dilution line C—D_1 and with 30% dilution possibly producing a weld metal that contains zero ferrite and which possibly extends into the martensite range. For more details see the supplementary literature entry [419 and 420].

In the event of a dissimilar joint between an austenitic stainless steel and a ferritic steel that has limited weldability and which requires preheating to above 150° C, it is sometimes advisable to first deposit a buttering layer* on the joint face of the ferritic material using an over-alloyed austenitic electrode of type No. 3 in Table 25 (page 230). It must be underlined that only one layer of buttering is required on the ferritic joint face. Buttering is carried out not only using the correct filler metal but also the correct preheating temperature. Then after this the dissimilar joint is finally welded. By utilizing such a buffer layer, the high welding stresses that are produced during joint welding are thus shifted from the critical HAZ region of the ferritic steel to the joint region between the austenitic buttering layer and the austenitic weld metal. The buttering layer must be deposited in such a way that the first weld run laid directly on the preheated ferritic steel joint face is deposited using only small diameter electrodes (e.g. 2.5 mm) in combination with a low heat input. The subsequent passes can then be deposited using larger diameter electrodes with greater overlapping, i.e. aiming for equal proportions of the ferritic steel and the previously deposited buffer run to be diluted by the following weld run. After buttering the joint face the joint welding should be carried out using standard austenitic electrodes (e.g. E 308 or 316 types), as the joint itself is now produced between the austenitic buffer layer and the austenitic parent steel. Therefore preheating is no longer necessary.

As a rule, only rebaked electrodes should be used. If the low alloyed steel possesses only limited weldability, it is important that only austenitic electrodes that give a deposited weld metal with a low hydrogen content are used. Otherwise, hydrogen may diffuse into the HAZ of the low alloyed steel as long as it is austenitic during cooling. Then, when the austenitic structure of the HAZ transforms partly or completely into martensite during further cooling, hydrogen induced cold cracking may occur if the HAZ has picked up too much hydrogen from the austenitic weld metal. An important step to take in order to avoid the possible occurrence of hydrogen induced cold cracking is the correct selection and control of the preheating and interpass temperatures, which are selected taking into consideration the chemical composition and the carbon equivalent value of the ferritic steel in question (for more details see the supplementary literature entry [88]).

The following welding procedures have produced good results when used for the welding of austenitic-ferritic dissimilar joints:

- Weld with a minimum heat input.
- Interpass temperature of 150° C max. is advisable.
- If the low alloyed steel requires preheating and an interpass temperature above 150° C, first deposit a buttering layer on the joint face using over-alloyed rebaked electrodes. After buttering, the welding of the joint can be completed with normal austenitic stainless steel filler metals at lower interpass temperatures.
- Moderate waving of electrodes to a maximum of twice the electrode diameter may be employed.
- Back grinding of the root pass and the deposition of a backing run is advisable.
- Provision of double V butt joints even for thin sections of materials is advisable.
- If heat treatments above 600° C are necessary after the welding operation, pay close attention to the information given in section 12.3.

* This term is usually used for a weld overlay on joint faces prior to joining. "Buffer layer" means the same.

Apart from the supplementary literature entries [418–420], already mentioned, the following authors supply practical information on the welding of austenitic-ferritic dissimilar joints: Flemming [421], Hennemann and Schütte [422], Thorneycroft [423], Wirtz [424], Gooch [425], Klueh and King [426] and Shifrin [427].

12.3 Criteria for the Heat Treatment of Austenitic-Ferritic Dissimilar Joints

The post weld heat treatment of austenitic-ferritic dissimilar joints or weldments should be avoided whenever possible. However heat treatments such as annealing or stress relieving may at times be unavoidable or even mandatory. They should always be adapted to suit the requirements of the low alloyed steel section of the joint. Often, the steels in question are low alloyed creep resistant steel types as used in boiler and pressure vessel construction which demand a post weld heat treatment to suit each particular steel grade. In austenitic-ferritic dissimilar joints, such heat treatments may lead to the occurrence of the following phenomena:

— Carbon enrichment in the weld metal due to the diffusion of carbon from the low alloyed steel into the austenitic weld metal.
— As a consequence thereof–carbon depletion in the HAZ of the low alloyed steel.
— Coarse grain formation in the HAZ of the low alloyed steel due to recrystallization processes.
— Embrittlement of the austenitic weld metal due to the precipitation of brittle phases, e.g. sigma phase (this does not occur with nickel-based weld metal).

All these processes are time and temperature dependent. If we consider the normal annealing times in the range of 2–10 hours, eventual damage will normally only occur depending upon the material combination, at temperatures above 600° C.

Chromium has a strong carbide forming effect and possesses a higher affinity to carbon than does iron. For this reason, carbon begins to diffuse from the lower alloyed ferritic steel containing normally 0.15–0.20% carbon to the chromium rich austenitic weld metal the moment that the optimum temperature conditions are reached to enable carbon diffusion to take place. Adjacent to the fusion boundary of the weld a rim of carbon rich chromium carbide may be formed in the austenitic weld metal which reach, e.g. with annealing temperatures of 750° C, carbon contents of about 1%. In nickel-based weld metals, carbon diffusion is impeded because of the fact that nickel possesses practically no solubility for carbon (see section 1.3.1). Although carbon diffusion is not entirely prevented, the zone of carbon enrichment is narrowed. Parallel to the carbon enrichment in the weld metal, there may also be a carbon depletion in the HAZ of the ferritic steel. At the same time as carbon diffusion occurs but largely independent of it, coarse grain formation due to recrystallization takes place in the HAZ of the ferritic steel, this is induced or speeded up by shrinkage processes or cold working which in turn are produced by the differences in the coefficient of thermal expansion between ferrite and austenite (approx. 13 and 18.10^{-6} m/m.K respectively, see also Table 25, page 230).

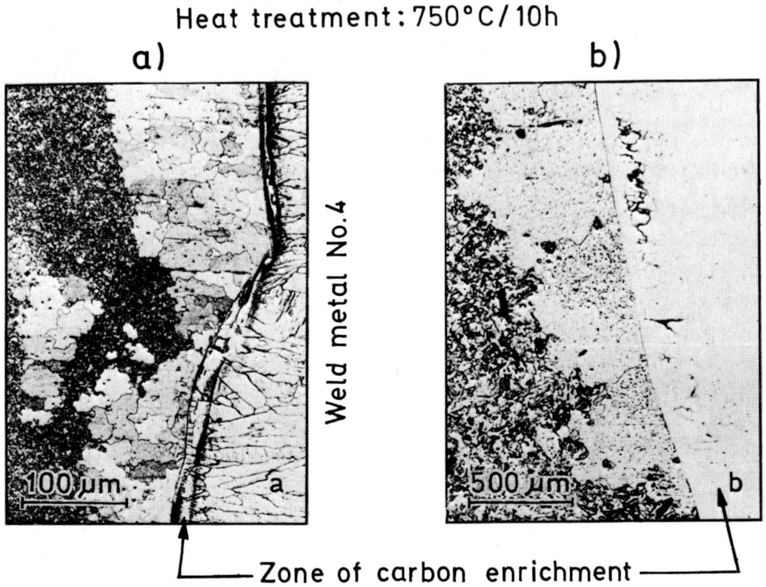

Fig. 113. Micro structure of the weld transition from the low alloyed steel of type 2.25 Cr/1 Mo to: *a* austenitic weld metal grade 22 Cr/18 Ni/5 Mn (No. 4 in Table 25, page 230), *b* nickel-based weld metal (No. 5 in Table 25). Etchant: ethanol/aqua regia. Note the different magnification factors of figures a) and b)

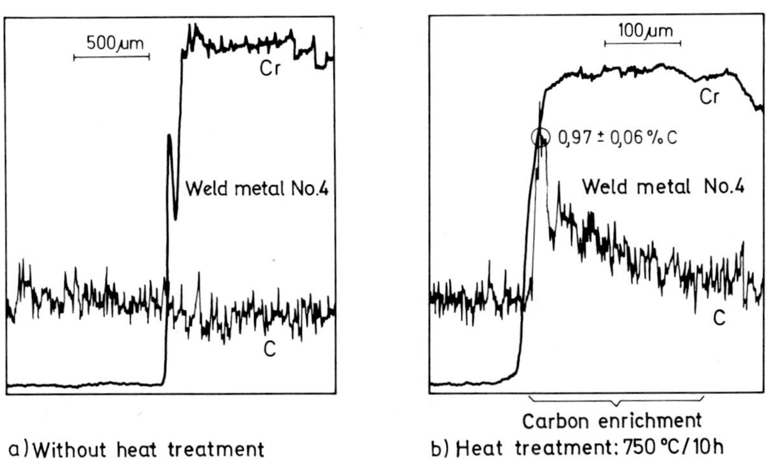

Fig. 114. Carbon and chromium concentration profile of weld transition from the low alloyed steel type 2.25 Cr/1 Mo to austenitic weld metal grade 22 Cr/18 Ni/5 Mn (No. 4 in Table 25, page 230). *a* without heat treatment, *b* after annealing at 750° C for 10 hours

Carbon excerts an indirect influence inasmuch as coarse grain formation is slowed down by the nucleating effect of carbides. Thus, a HAZ which is depleted, for example, of carbon and carbides may easily form a relatively coarse grain structure due to recrystallization which means, that the HAZ experiences a drop in ductility.

Fig. 113 a and b show the micro structures of the transition zones between a ferritic creep resistant 2.25 Cr/1 Mo low alloyed steel and a high alloyed weld metal after ten hours annealing at 750° C (at × 100 and × 500 magnification). The micrographs are taken from a joint work with the Welding Subcommitee of Eisenhütte Österreich [428]. The welding was carried out with austenitic and nickel-based electrodes, the weld metal compositions correspond to Nos. 4 and 5 in Table 25, page 230. The micrographs in Fig. 113 show clearly the coarse grain formation in the ferritic steel and the carbon enriched carbide rim in the austenitic weld metal. When using nickel-based electrodes, the coarse grain zone and the carbide rim are less pronounced. Fig. 114 shows the electron probe microanalysis measurements of the distribution of the carbon and chromium contents at the transition zone of the austenitic weld metal a) without heat treatment and b) after ten hours annealing at 750° C. We can clearly distinguish the strong increase in carbon to 0.97% in the austenitic weld metal right next to the fusion line which has been produced by carbon diffusion from the ferritic steel during heat treatment. From the above investigations [428] it appears that coarse grain formation is caused primarily by recrystallization and is only indirectly connected with the processes of carbon depletion in the HAZ of the ferritic steel.

The embrittlement which is produced by sigma phase precipitation in austenitic weld metal due to annealing has been discussed in detail in sections 4.2, 9.1–9.3. It mainly effects the impact toughness of austenitic weld metal as indicated in Figs. 109–111. Nickel-based weld metal such as alloy 5 in Table 25 (page 230) is, due to its high nickel content outside the range of sigma phase precipitation (see also Figs. 12 and 13, page 13) and shows no signs of embrittlement during heat treatments.

Nickel-based weld metal which corresponds to alloy No. 5 in Table 25, behaves better than austenitic weld metal in cases where a subsequent post weld heat treatment is to be applied to an austenitic-ferritic dissimilar joint. For this reason, all austenitic-ferritic weldments which must be subjected to a subsequent heat treatment at temperatures above 600° C should, whenever possible, be welded using nickel-based filler metals. The filler metal grade No. 5 listed in Table 25 represents a well tested alloy for the welding of austenitic-ferritic dissimilar joints which require subsequent heat treatment. When heat treating whole components or vessels, it is important to consider that the corrosion resistance of the austenitic section of the assembly may be considerably impaired by annealing above 500° C because of carbide and phase precipitations (see Fig. 84, page 105 and Fig. 90, page 126). It is therefore advisable in many cases, to first join an austenitic transition piece to the ferritic side of the weldment and to carry out the full heat treatment in accordance with the requirements of the ferritic steel grade. The actual joining of the parts of the component which must withstand severe corrosive stresses to the austenitic transition piece are then carried out after this heat treatment. As the welding now only takes place between the austenitic transition piece and the actual austenitic part of the component, a post weld heat treatment is normally not required.

12.4 Mechanical Properties of Austenitic-Ferritic Dissimilar Joints

When transverse tensile testing austenitic-ferritic dissimilar joints, the specimen generally fractures outside the weld area, normally in the austenitic stainless steel side of the joint. In joints between a stainless steel and an unalloyed mild steel the fracture may also occur in the mild steel. This means that when testing transversely to the weld bead the tensile properties of the weldment are mainly determined by the properties of the steel grades to be joined. In the event of long exposures to elevated temperatures, the fracture may also occur in the coarse grained HAZ of the ferritic steel. In the case of high tensile creep resisting steels which are not covered, however, within the context of this book, the influence of the coarse grain zone on the creep rupture behaviour at elevated temperatures may be quite significant, but as far as the common service temperatures of stainless steels are concerned, the influence of the coarse grain zone on the creep rupture behaviour is of minor significance.

A property which is of particular interest when welding austenitic-ferritic joints is the toughness behaviour of the welds, because here the different zones of the weld may show different effects. This applies in particular to the transition between the high alloyed weld metal and the low alloyed ferritic steel, in which after a heat treatment a carbon enriched zone may occur in the weld metal and a coarse grain zone may occur in the HAZ of the ferritic steel (see Fig. 113, page 236). In the weld metal itself, a heat treatment may also produce embrittlement due to sigma phase precipitation. Fig. 115 shows the impact energy values of austenitic-ferritic joints between a ferritic creep resisting 2.25 Cr/1 Mo steel and an austenitic 18 Cr/10 Ni steel (AISI 304). The welding operation was carried out by the SMAW process with covered electrodes corresponding to alloys Nos. 2, 4 and 5 in Table 25, page 230. The scope of testing included the impact toughness of the weld metal and the transition zone between the high alloyed weld metal and the ferritic steel type 2.25 Cr/1 Mo. The type of weld and the position of the notch can be taken from Fig. 115. A single bevel V butt joint was chosen to permit the testing of impact energy at the transition zone. Testing was carried out in the as welded condition “S” and also after annealing at 650, 700 and 750° C for 2 and 10 hours. The evaluation of test results from specimen with the notch position in the transition zone between the particular weld metal and the ferritic steel can be quite difficult as the course of the fracture will not always follow the weld transition zone. Deviations of the course of fracture into the weld metal or the base metal may influence the test results and sometimes prevent clear results. For this reason, the course of fracture of the notched bar impact specimen should be included in the evaluation of impact energy values as a supplement to the test results.

Weld metal grade No. 2 type 23 Cr/12 Ni shows satisfactory impact energy values in the as welded state whereas we can note after annealing particularly with annealing times of 10 hours a considerable drop in impact energy values both in the weld metal and the transition zone. The fully austenitic weld metal No. 4 of the type 22 Cr/18 Ni/5 Mn shows no drop in impact energy values in the weld metal after post weld heat treatment, there is however a definite drop in the transition zone impact energy values which are still higher, however, than those of weld metal No. 2

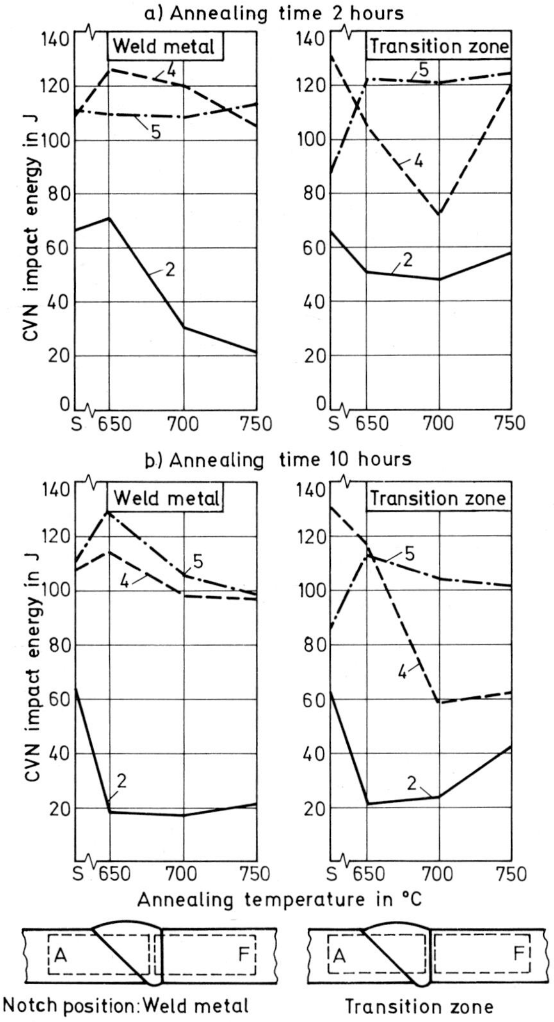

Fig. 115. Correlation of impact energy and heat treatment for an austenitic-ferritic dissimilar joint, tested in the high alloyed weld metal and in the transition zone between the weld metal and the low alloyed ferritic steel of the type 2.25 Cr/1 Mo with the following chemical composition:

C = 0.13%, Cr = 2.41%, Mo = 1.08%
S = as welded condition
2 = weld metal grade 23 Cr/12 Ni (AWS E309L)
4 = weld metal grade 22 Cr/18 Ni/5 Mn
5 = weld metal grade 71 Ni/19 Cr/2 Nb (AWS E NiCrFe 3)

For the chemical composition of the weld metals please refer to No. 2, 4 and 5 in Table 25, page 230 containing delta ferrite. The nickel-based weld metal No. 5 shows high impact energy values after annealing, both in the weld metal and the transition zone. These individual results confirm the practical experience that the use of nickel-based filler metals is recommended, if austenitic-ferritic dissimilar joints are to be subjected to post weld heat treatments at temperatures above 600° C.

Appendix

Abbreviations and Short Designations

1 Abbreviations (in alphabetic order)

A	Austenite
AISI	American Iron and Steel Institute
Ann	Annealed
ANSI	American National Standard Institute
ASTM	American Society for Testing Materials
AWS	American Welding Society
C	Carbide (in Diagrams) or Carbon content in %
Corr	Corrosion resistance (general)
CVN	Charpy-V-Notch (specimen for testing of impact energy)
d	Direction of crystallisation
El	Elongation in tensile testing (measured in %)
F	Ferrite
FN	Ferrite Number (for indicating delta ferrite content)
FP	Fusion Point
GMAW	Gas Metal Arc Welding (formerly MIG welding)
GTAW	Gas Tungsten Arc Welding (formerly TIG welding)
HAZ	Heat Affected Zone
H_{DM}	Hydrogen content of Deposited Metal
HT	Heat Treatment
IC	Intergranular Corrosion
ICR	Intergranular Corrosion Resistance
IE	Impact Energy (also called impact strength or impact toughness) measured in Joule
IPT	Interpass Temperature
J	Joule
L	Liquid phase (in diagrams) or Low carbon content in short designations
M	Martensite
M_s	Start of Martensite transformation
M_f	End of Martensite transformation
N/mm^2	Newton per square millimeter (=MPa/Mega Pascal)
ox	in oxydizing environments
%	always weight (mass) percent when used in diagrams or chemical composition
P	Pearlite
PC	Pitting Corrosion
PCE	Pitting Corrosion Equivalent
PCR	Pitting Corrosion Resistance
PHT	Preheating Temperature
PWHT	Post Weld Heat Treatment
RA	Reduction of Area (in tensile testing)
red	in reducing environment
RT	Room (ambient) Temperature
S	Solid phase
SAE	Society of Automotive Engineers
SAW	Submerged Arc Welding
SCC	Stress Corrosion Cracking
SCCR	Stress Corrosion Cracking Resistance
SMAW	Shieded Metal Arc Welding (formerly manual arc welding with coated electrodes)
$T_{12/8}$	Time for cooling from 1200° C–800° C, e.g. after welding
Temp	Tempered
TS	ultimate Tensile Strength (measured in N/mm^2 = MPa/Mega Pascal)

TTC Time Temperature Corrosion (diagram)

TTP Time Temperature Precipitation (diagram)

TTT Time Temperature Transformation (diagram)

UNS Unified Numbering System, new designation system established in accordance with ASTM E 527 and SAE J 1087: Practice for numbering metals and alloys

W as Welded condition

0.2 YS Yield strength at 0.2% elongation (also called 0.2 proof stress); Conventional yield strength measured in N/mm^2 (= MPa)

2 Short designation for chemical composition (approximate mean values)

19 Cr/9 Ni N = 19% Cr, 9% Ni, nitrogen alloyed (e.g. AISI 304 N)

18 Cr/10 Ni Nb = 18% Cr, 10% Ni, niobium alloyed (e.g. AISI 347)

17 Cr/12 Ni/2.5 Mo L = 17% Cr, 12% Ni, 2.5% Mo, low carbon content (e.g. AISI 316 L with max. 0.030% carbon)

22 Cr/5 Ni/3 Mo NL = 22% Cr, 5% Ni, 3% Mo, nitrogen alloyed, max. 0.030% C (e.g. UNS 31803)

References

[1] Strauß, B., Maurer, E.: Die hochlegierten Chrom-Nickel-Stähle als nichtrostende Stähle. Kruppsche Monatsh. **1**, 120–146 (1920).

[2] Hansen, M., Anderko, K.: Constitution of Binary Alloys. New York-Toronto-London: McGraw-Hill Book Comp. 1958*.

[3] Elliott, R. P.: Constitution of Binary Alloys. First Supplement. New York-St. Louis-San Francisco-London-Sydney: McGraw-Hill Book Comp. 1965*.

[4] Shunk, F. A.: Constitution of Binary Alloys. Second Supplement. New York-St. Louis-San Francisco-London-Sydney-Toronto-Mexico-Panama: McGraw-Hill Book Comp. 1969*.

[5] Kubaschewski, O.: Iron-Binary Phase Diagrams. Berlin-Heidelberg-New York: Springer; Düsseldorf: Verlag Stahleisen m. b. H. 1982.

[6] Rhines, F. N.: Phase Diagrams in Metallurgy, Their Development and Application. New York-Toronto-London: McGraw-Hill Book Comp. 1956.

[7] Schumann, H.: Metallographie, 10. Aufl. Leipzig: VEB Verlag für Grundstoffindustrie 1980.

[8] Rapatz, F., unter Mitwirkung von Krainer, H., Frehser, J.: Die Edelstähle, 5. Aufl. Berlin-Göttingen-Heidelberg: Springer 1962.

[9] Houdremont, E., unter Mitarbeit von Wiester, H. J.: Handbuch der Sonderstahlkunde, 3. Aufl. Berlin-Göttingen-Heidelberg: Springer; Düsseldorf: Verlag Stahleisen m. b. H. 1956.

[10] Wever, F., Jellinghaus, W.: Das Dreistoffsystem Eisen-Chrom-Nickel. Mitt. K.-Wilh.-Inst. Eisenforsch. **13**, 98–108 (1931).

[11] Jenkins, C. H. M., Bucknall, E. H., Austin, C. R., Mellor, G. A.: Some Alloys for Use at High Temperatures. Part IV: The Constitution of the Alloys of Nickel, Chromium and Iron. J. Iron Steel Inst. **136**, 187–222 (1937).

[12] Schürmann, E., Brauckmann, J.: Untersuchungen über die Schmelzgleichgewichte in der Eisenecke des Dreistoffsystems Eisen-Chrom-Nickel. Arch. Eisenhüttenwes. **48**, 3–7 (1977).

[13] Perteneder, E.: Vorgänge der Primärkristallisation und der Phasenumwandlung von austenitischen Chrom-Nickel-Schweißgutlegierungen. Jubiläumsschrift „50 Jahre Böhler Schweißtechnik". Kapfenberg: VEW 1977.

[14] Perteneder, E., Jeglitsch, F.: Primär- und Sekundärgefüge im austenitischen Schweißgut und ihr Einfluß auf die Heißrißbildung. Prakt. Metallogr. **19**, 573–591 (1982).

[15] Perteneder, E.: Untersuchungen zur Primärkristallisation und Phasenumwandlung an austenitischem Schweißgut vom Typ X 3 CrNi 20 10. Dr.-Ing.-Dissertation Montanuniversität Leoben 1981.

[16] Schafmeister, P., Ergang, R.: Das Zustandsschaubild Eisen-Chrom-Nickel unter besonderer Berücksichtigung des nach Dauerglühung auftretenden spröden Gefügebestandteils. Arch. Eisenhüttenwes. **12**, 459–464 (1939).

---

* The diagrams taken from this book are reproduced with permission of McGraw-Hill Book Company.

[17] Barin, I., Knacke, O., Kubaschewski, O.: Thermochemical Properties of Inorganic Substances, Supplement, S. 195. Berlin-Heidelberg-New York: Springer; Düsseldorf: Verlag Stahleisen m. b. H. 1977.

[18] Brewer, L., Chang, Shih-Ger: Constitution of Ternary Alloys. Metals Handbook, Vol. 8, 8th ed. (Lyman, Taylor, eds.), S. 421, 423, 424, 425. American Society for Metals 1973.

[19] Hoffmeister, H., Mundt, R.: Untersuchungen zum Einfluß der Schweißparameter und der Legierungszusammensetzung auf den Deltaferritgehalt des Schweißgutes hochlegierter Chrom-Nickel-Stähle. Schweißen u. Schneiden **30**, 214–218 (1978).

[20] Lippold, J. C., Savage, W. F.: Solidification of Austenitic Stainless Steel Weldment. Part I: A Proposed Mechanism. Weld. J. **58**, Res. Suppl., 362s–374s (1979).

[21] Bungardt, K., Kunze, E., Horn, E.: Untersuchungen über den Aufbau des Systems Eisen-Chrom-Kohlenstoff. Arch. Eisenhüttenwes. **29**, 193–203 (1958).

[22] Kunze, E.: Über einen besonderen Legierungsbereich im Vierstoffsystem Eisen-Kohlenstoff-Chrom-Nickel. Thyssen Edelstahl Techn. Ber. **2**, 70–74 (1976).

[23] Bungardt, K., Kunze, E., Horn, E.: Einfluß verschiedener Legierungselemente auf die Größe des Gammaraumes im System Eisen-Chrom-Kohlenstoff. Arch. Eisenhüttenwes. **39**, 863–867 (1968).

[24] Nicholson, M. E., Samans, C. H., Shortleeve, F. J.: Composition Limits of Sigma Formation in Nickel-Chromium Steels at 1200 °F (650 °C). Trans. ASME **64**, 601–620 (1957).

[25] Westendorp, R.: Die Anwendung von austenitischen Stählen für hohe Temperaturen. Z. Schweißtechn. (Basel) **52**, 863–867 (1962).

[26] Gerlach, H.: Gefüge- und Eigenschaftsänderungen hitzebeständiger austenitischer Stähle bei hohen Temperaturen. Bänder, Bleche, Rohre **12**, 516–522 (1971).

[27] Staudinger, H.: Anschauliche Erklärung des Zustandsschaubildes Eisen-Kohlenstoff. Werkstattstechn. **59**, 229–234 (1969).

[28] Eichelmann, G. H., Hull, F. C.: The Effect of Composition of Spontaneous Transformation of Austenite to Martensite in 18-8-Type Stainless Steel. Trans. A.S.M. **45**, 77–104 (1953).

[29] Tama, H., Vyklicky, M., Löbl, K.: Kohlenstoffaktivität und -löslichkeit in austenitischen Chrom-Nickel-Stählen mit rund 18% Chrom. Arch. Eisenhüttenwes. **41**, 983–988 (1970).

[30] Gerlach, H.: Wesentliche Gesichtspunkte bei der Wärmebehandlung hochlegierter Chrom- und Chrom-Nickel-Stähle. Drahtwelt **56**, 25–32 (1970).

[31] Hoffmeister, H., Mundt, R.: Untersuchungen zum Einfluß des Kohlenstoffs und des Stickstoffs auf die Delta-Gamma-Umwandlung ferritisch-austenitischer Chrom-Nickel-Stähle. Arch. Eisenhüttenwes. **52**, 159–164 (1981).

[32] Ergang, R., Günther, G., Weik, H.: Der Stand der Kenntnisse über die Sigmaphase in Eisen-Chrom-Nickel-Legierungen. Stahl u. Eisen **79**, 46–51 (1959).

[33] Wiegand, H., Doruk, M.: Einfluß von Kohlenstoff und Molybdän auf die Ausscheidungsvorgänge, besonders auf die Bildung intermetallischer Phasen in austenitischen Chrom-Nickel-Stählen. Arch. Eisenhüttenwes. **33**, 559–566 (1962).

[34] Tösch, J., Schabereiter, H.: Einfluß der Schweißbedingungen auf den Ferritgehalt und die Heißrißanfälligkeit austenitischer Schweißgutlegierungen. Jubiläumsheft „50 Jahre Böhler Schweißtechnik". Kapfenberg: VEW 1977.

[35] Humbert, C. J., Elliott, J. F.: The Solubility of Nitrogen in Liquid Fe-Cr-Ni Alloys. Trans. Metallurg. Soc. AIME **218**, 1076–1088 (1960).

[36] Kohl, H.: Austenitische korrosionsbeständige Stähle mit hoher Festigkeit im lösungsgeglühten Zustand. Berg- u. hüttenm. Mh. **113**, 377–388 (1968).

[37] Bungardt, K., Laddach, H., Lennartz, G.: Über die Löslichkeit von Stickstoff in austenitischen Chrom-Nickel-Stählen. DEW – Technische Berichte **12**, 134–154 (1972); **13**, 75–84 (1973).

[38] Thier, H.: Der Einfluß von Stickstoff auf das Ausscheidungsverhalten des austenitischen Chrom-Nickel-Stahles X 5 CrNiMo 17 13. Dr.-Ing.-Dissertation, Techn. Hochschule Aachen 1967.

[39] Gerlach, H., Schmidtmann, E.: Einfluß von Kohlenstoff, Stickstoff und Bor auf das Ausscheidungsverhalten eines austenitischen Stahles mit rund 16% Chrom, 2% Molybdän, 16% Nickel und Niob. Arch. Eisenhüttenwes. **39**, 139–149 (1968).

[40] Baerlecken, E., Fischer, W. A., Lorenz, K.: Untersuchungen über das Umwandlungsverhalten, die Kerbschlagzähigkeit und die Neigung zur interkristallinen Korrosion von Eisen-Chrom-Legierungen mit Chromgehalten bis 30%. Stahl u. Eisen **81**, 768–778 (1961).

[41] Thier, H., Bäumel, A., Schmidtmann, E.: Einfluß von Stickstoff auf das Ausscheidungsverhalten des Stahles X 5 CrNiMo 17 13. Arch. Eisenhüttenwes. **40**, 333–339 (1969).

[42] Gerlach, H., Kautz, H. R., Schäffler, F.: Untersuchungen über die Wirkung von Stickstoff und Niob in austenitischen korrosionsbeständigen Stählen. Bänder, Bleche, Rohre **10**, 418–425 (1969).

[43] Bechtoldt, C. J., Vacher, H. C.: Phase-Diagram Study of Alloys in the Iron-Chromium-Molybdenum-Nickel System. J. Res. Nat. Bur. Stand. **58**, 7–19 (1957).

[44] Kügler, A., Bäumel, A., Schmidtmann, E.: Einfluß der Gefügeausbildung auf das Korrosionsverhalten von hochlegierten molybdänhaltigen austenitischen Chrom-Nickel-Stählen. Arch. Eisenhüttenwes. **48**, 353–358 (1977).

[45] Weiss, B., Stickler, R.: Phase Instabilities During High Temperature Exposure of 316 Austenitic Stainless Steel. Metallurg. Trans. **3**, 851–866 (1972).

[46] Kügler, A.: Wirkung von Molybdän in austenitischen Chrom-Nickel-Stählen. Berg- u. hüttenm. Mh. **122**, 507–513 (1977).

[47] Bungardt, K., Lennartz, G.: Einfluß von Molybdän auf Gefüge, Härte und Kerbschlagzähigkeit niobhaltiger austenitischer Chrom-Nickel-Stähle nach Wärmebehandlung bei 650–850 °C. Arch. Eisenhüttenwes. **33**, 251–259 (1962).

[48] Bungardt, K., Kunze, E., Horn, E.: Einfluß verschiedener Legierungselemente auf die Größe des Gammaraumes im System Eisen-Chrom-Kohlenstoff. Teil I: System Eisen-Chrom-Kohlenstoff-Molybdän. Arch. Eisenhüttenwes. **38**, 309–320 (1967).

[49] Schafmeister, P., Ergang, R.: Die spröde Sigmaphase im Dreistoffsystem Eisen-Chrom-Mangan. Arch. Eisenhüttenwes. **12**, 507–510 (1939).

[50] Krainer, H., Mirt, O.: Das Zustandsschaubild stickstoffhaltiger Chrom- und Chrom-Mangan-Stähle. Arch. Eisenhüttenwes. **15**, 467–472 (1942).

[51] Wentrup, H., Reif, O.: Über die Löslichkeit von Stickstoff in Eisenschmelzen mit Chrom-, Mangan- und Nickelzusätzen. Arch. Eisenhüttenwes. **20**, 359–362 (1949).

[52] Ruttmann, W., Gerlach, H., Kautz, H. R.: Einfluß des Stabilisierungsverhältnisses auf das Zeitstand- und Ausscheidungsverhalten hochwarmfester austenitischer Stähle mit rund 16% Chrom, 13–16% Nickel, 2% Molybdän und Niobzusatz. Arch. Eisenhüttenwes. **38**, 304–307 (1967).

[53] Irvine, K. J., Murray, J. D., Pickering, F. B.: The Effect of Heat-Treatment and Microstructure on the High-Temperature Ductility of 18% Cr-12% Ni-1% Nb-Steels. J. Iron Steel Inst. **196**, 166–179 (1960).

[54] Kautz, H. R., Klärner, H. F., Schmidtmann, E.: Einfluß von Wärmebehandlung und Warmformgebung auf das Ausscheidungsverhalten und die Zeitstandfestigkeit des Stahles X 8 CrNiMoVNb 16 13. Arch. Eisenhüttenwes. **36**, 571–582 (1965).

[55] Hull, F. C.: Effects of Composition on Embrittlement of Austenitic Stainless Steels. Weld. J. **52**, Res. Suppl., 104s–113s (1973).

[56] Dahl, W., Düren, C., Müsch, H.: Einfluß verschiedener Legierungselemente auf die Heißrißanfälligkeit eines austenitischen Stahles beim Schweißen. Stahl u. Eisen **93**, 813–822 (1973).

[57] Dahl, W., Düren, C., Müsch, H.: Ursachen der Heißrißbildung in Schweißverbindungen eines niobstabilisierten Stahles mit 16% Chrom und 16% Nickel. Stahl u. Eisen **93**, 805–812 (1973).
[58] Tamura, H., Watanabe, T.: Mechanism of Liquation Cracking in Weld Heat-Affected Zone of Austenitic Stainless Steels. Trans. Japan Weld. Soc. **4**, 30–42 (1973).
[59] Schüller, H. J.: Metallographische Untersuchungen zur Ermittlung des Temperaturbereiches beginnender Aufschmelzungen von austenitischen Titan- und Niob-stabilisierten Chrom-Nickel-Stählen. Arch. Eisenhüttenwes. **34**, 61–68 (1963).
[60] Hattersley, B., Hume-Rothery, W.: Constitution of Certain Austenitic Steels. J. Iron Steel Inst. **204**, 683–701 (1966).
[61] Scheil, E.: Mitt. Forsch. Inst. Ver. Stahlw. Dortmund **1**, 1–12 (1928/30).
[62] Schüller, H. J.: Über die Lage des Temperaturbereiches der Sigmaphase in ferritischen Chrom-Stählen. Arch. Eisenhüttenwes. **36**, 513–516 (1965).
[63] Horn, E. M., Kügler, A.: Entwicklung, Eigenschaften, Verarbeitung und Einsatz des hochsiliziumhaltigen austenitischen Stahles X 2 CrNiSi 18 15, Teil I. Z. Werkst.-Techn. **8**, 362–370 (1977).
[64] Masumoto, I., Imai, Y.: Sci. Rep. Res. Inst. Tohuku Univ. Nagoya, Ser. A., 16–32 (1964).
[65] Ogawa, T., Tsunetomi, E.: Hot Cracking Susceptibility of Austenitic Stainless Steels. Weld. J. **61**, Res. Suppl., 82s–93s (1982).
[66] Sadowsky, S.: Schweißen dicker vollaustenitischer Stahlbleche u. Schmiedestücke. Schweißen und Schneiden **16**, 217–228 (1964).
[67] Brown, R. S., Koch, J. B.: Development of a Solidification Crack Resistant Weld Filler Metal for a Stabilized Fully Austenitic Alloy. Weld. J. **57**, Res. Suppl., 38s–42s (1978).
[68] Matsuda, F., Nakagawa, H., Katayama, S., Arata, Y.: Solidification Crack Susceptibility in Weld Metals of Fully Austenitic Stainless Steels, Report IV. Trans. Japan Weld. Res. Inst. **11**, 79–94 (1982).
[69] Brooks, J. A., Lambert, F. J.: The Effect of Phosphorus, Sulfur and Ferrite Content on Weld Cracking of Type 309 Stainless Steel. Weld. J. **57**, Res. Suppl., 139s–143s (1978).
[70] Bernstein, A., Carlen, J. C., Rick, L.: Influence of Phosphorus and Sulfur on the Properties of Weld Metal in Certain Austenitic Stainless Steels. Weld. J. **44**, Res. Suppl., 504s–508s (1965).
[71] Hull, C. F.: Effects of Alloying Additions on Hot Cracking of Austenitic Chromium-Nickel Stainless Steels. Proc. American Soc. Test. Mat. **60**, 667–690 (1960).
[72] Robitsch, J.: Interstitiell eintretende Elemente und ihre Verbindungen sowie intermetallische Ausscheidungen in Edelstählen. Habilitationsschrift Montanuniversität Leoben 1971.
[73] Weingerl, H.: Einfluß von Bor und Silizium auf die Eigenschaften korrosionsbeständiger Stähle. Berg- u. hüttenm. Mh. **117**, 317–321 (1972).
[74] Swoboda, K., Malissa, H.: Bestimmung der Verteilung einiger Elemente in hochborhaltigen Eisen-Chrom-Nickel-Legierungen. Mikrochim. Acta (Wien) **1965**, 532–547.
[75] Malissa, H., Grasserbauer, M., Hoke, E., Draxler, H., Weingerl, H.: Beitrag zur Charakterisierung von hochborhaltigen Chrom-Nickel-Molybdän-Kupfer-Stählen mit Hilfe der stereometrischen Analyse. Mikrochim. Acta (Wien) **1977**, 73–91.
[76] Kügler, A.: Bleche aus Bor-legierten Stählen für die Kerntechnik. Berg- u. hüttenm. Mh. **126**, 411–413 (1971).
[77] Donati, J. R., Guttmann, D., Zacharie, G.: Influence de la teneur en bore sur la tendence à la fissuration à chaud dans les zones affectées par soudage d'aciers inoxydables 18/10. Rev. Métallurg. **71**, 917–930 (1974).
[78] Brooks, J. A.: Effect of Alloy Modification on HAZ-Cracking of A-286 Stainless Steel. Weld. J. **53**, Res. Suppl., 517s–523s (1974).

[79] Williams, T. M., Talks, M. G.: Effect of Boron on the Precipitated Phases in Aged Type 316 Austenitic Stainless Steel. J. Iron Steel Inst. **210**, 870–879 (1972).

[80] Knüppel, H.: Desoxydation und Vakuumbehandlung von Stahlschmelzen. Band I, S. 194: Thermodynamische und kinetische Grundlagen. Düsseldorf: Verlag Stahleisen m. b. H. 1970.

[81] Rabensteiner, G.: Die metallurgischen Vorgänge beim Lichtbogenschweißen mit unlegierten Mantelelektroden. Dr.-Ing.-Dissertation, Montanuniversität Leoben 1972.

[82] Widgery, D. J.: Deoxydation Practice and Toughness of Mild Steel Weld Metal. Ph.D. Thesis, University of Cambridge 1974.

[83] Kojima, Y., Sakao, H., Sano, K.: Chromoxyd im Gleichgewicht mit Eisen-Chrom-Legierungen bei 1600 °C. Arch. Eisenhüttenwes. **39**, 187–190 (1968).

[84] Plessing, R.: Sauerstofffrischen legierter Stähle im basischen Lichtbogenofen. Radex Rdsch. **1955**, 604–624.

[85] Smialowski, M.: Hydrogen in Steel. Oxford-London-New York-Paris: Pergamon Press 1969.

[86] Alefeld, G., Völkl, J.: Hydrogen in Metals, Bd. 1 und 2. (Topics in Applied Physics, Vol. 29.) Berlin-Heidelberg-New York: Springer 1969.

[87] Ortner, H. M.: Gase in Metallen. Conference proceedings. Darmstadt 1982, organized by GDBM and Deutsche Ges. f. Metallk. (H. M. Ortner, editor).

[88] Coe, F. R.: Welding Steels Without Hydrogen Cracking. Abington-Hall Cambridge: The Welding Institute 1973.

[89] Geller, W., Sun, T. H.: Einfluß von Legierungszusätzen auf die Wasserstoffdiffusion im Eisen und Beitrag zum System Eisen-Wasserstoff. Arch. Eisenhüttenwes. **21**, 423–430 (1950).

[90] Luckemeyer-Hasse, L., Schenck, H.: Löslichkeit von Wasserstoff in einigen Metallen und Legierungen. Arch. Eisenhüttenwes. **6**, 209–214 (1932).

[91] Schenck, H., Wünsch, H.: Über die Gleichgewichtslöslichkeit des Wasserstoffs im flüssigen reinen Nickel und Eisen und die Beeinflussung im Eisen durch Sauerstoff. Arch. Eisenhüttenwes. **32**, 779–790 (1961).

[92] Schwarz, W., Zitter, H.: Löslichkeit und Diffusion von Wasserstoff in Eisenlegierungen. Arch. Eisenhüttenwes. **36**, 343–349 (1965).

[93] Brezina, P.: Martensitische Chrom-Nickel-Stähle mit tiefem Kohlenstoffgehalt. Escher Wyss Mitteilungen **1980**, 218–236. Zürich: Escher Wyss.

[94] Tösch, J., Perteneder, E., Rabensteiner, G.: Schweißzusatzwerkstoffe zum Schweißen weichmartensitischer Stähle. Schweißtechnik (Wien) **35**, 2–6 (1981).

[95] Narita, N., Altstetter, C. J., Birnbaum, H. K.: Hydrogen-Related Phase Transformations in Austenitic Stainless Steels. Metallurg. Trans. A. of AIME **13**, 1355–1365 (1982).

[96] Wittke, K.: Modellierung der Primärkristallisation beim Schmelzschweißen. Schweißtechnik (Berlin) **18**, 295–299 (1968).

[97] Savage, W. F., Lundin, C. D., Aronson, A. H.: Weld Metal Solidification Mechanics. Weld. J. **44**, Res. Suppl., 175s–181s (1965).

[98] Eichhorn, F., Engel, A.: Primärkristallisation beim Schmelzschweißen. Schweißen u. Schneiden **25**, 495–499 (1973).

[99] Leitner, F., Plöckinger, E.: Die Edelstahlerzeugung. 2. Aufl. von Plöckinger, E., Straube, H., S. 728–731. Wien-New York: Springer 1965.

[100] Kurz, W., Lux, B.: Über Ursache, Auswirkung und Beeinflussung der Seigerungen bei der Erstarrung. Schweiz. Archiv **35**, 49–65 (1969).

[101] Liquid Metals and Solidification. Conference proceedings. Cleveland, Ohio: American Soc. for Metals 1958.

[102] Winegard, W. C.: Fundaments of the Solidification of Metals. Metallurg. Rev. **6**, 57–99 (1961).

[103] The Solidification of Metals. Conference proceedings. London: Iron and Steel Inst. 1968.

[104] Erstarrung metallischer Schmelzen. Conference proceedings. Oberursel: Deutsche Ges. für Metallkunde 1982.

[105] Kurz, W., Sahm, P. R.: Gerichtet erstarrte eutektische Werkstoffe. Berlin-Heidelberg-New York: Springer 1975.

[106] Lücke, K.: Wesen und Eigenschaften der Korngrenzen polykristalliner Metallkörper. Z. Metallkde. **44**, 370–378, 418–426 (1953).

[107] Gueussier, A.: Altérations intergranulaires des aciers alliés. Mém. Sci. Rev. Metallurg. **72**, 373–396 (1975).

[108] Briant, C. L., Banerji, S. K.: Intergranular Failure in Steel: The Role of Grainboundary Composition. Intern. Metals Rev. **1978**, Rev. 232, 164–199.

[109] Chalmers, B.: Principles of Solidification. New York-London-Sydney: John Wiley & Sons 1967.

[110] Flemings, M. C.: Solidification Processing. New York: McGraw-Hill Book Comp. 1974.

[111] Tiller, W. A.: Solute Segregation During Ingot Solidification. J. Iron Steel Inst. **192**, 338–350 (1959).

[112] Eckstein, H. J.: Wärmebehandlung von Stahl. Metallkundliche Grundlagen, S. 92, 130. Leipzig: VEB Deutscher Verlag für Grundstoffindustrie 1969.

[113] Wittke, K.: Gesetzmäßigkeiten der Primärkristallisation beim Schweißen. Schweißtechnik (Berlin) **16**, 158–164 (1966).

[114] Matsuda, F., Hashimoto, T., Senda, T.: Fundamental Investigations on Solidification Structure in Weld Metal. Trans. Nat. Res. Inst. Metals **11**, 43–58 (1969).

[115] Arata, Y., Matsuda, F., Nakata, K.: Effect of Solidification Rate on Solidification Structure in Weld Metal. Trans. Japan Weld. Res. Inst. **5**, 47–52 (1976).

[116] Savage, W. F.: Solidification, Segregation and Weld Imperfections. Weld. World **18**, 89–114 (1980).

[117] Kulmburg, A.: Überlegungen zur Erstellung von Erstarrungsschaubildern. Mikrochim. Acta (Wien), Suppl. **6**, 25–48 (1975).

[118] Oberhoffer, P.: Zur Kenntnis der Primärätzung. Stahl u. Eisen **45**, 223–224 (1925).

[119] Beraha, E.: New Metallographic Reagent for Stainless Steel and Heat-Resisting Alloys. J. Iron Steel Inst. **204**, 248–251 (1966).

[120] Lichtenegger, P., Blöch, R.: Beitrag zur Farbätzung hochlegierter Stähle. Prakt. Metallogr. **12**, 567–573 (1975).

[121] Perteneder, E., Lichtenegger, P.: Beitrag zur Gefügeausbildung in austenitischen Schweißungen. Sonderband 4 der Prakt. Metallographie **1975**, 104–111.

[122] Beraha, E., Shpigler, B.: Color Metallography. Cleveland, Ohio: American Soc. f. Metals 1977.

[123] Beraha, E.: Farbätzung für rostfreien Stahl, korrosions- und hitzebeständige Legierungen. Prakt. Metallographie **8**, 443–454 (1968).

[124] Weck, E., Leistner, E.: Metallographic Instructions for Colour Etching by Immersion. Part I: Colour Etchants for Non-Alloy Steels and Manganese Steel, Cast Iron and Copper Materials. Part II: Colour Etchants and their Variants for Low-Alloy and High-Alloy Steels, Nickel-Based Alloys and Cobalt-Based Alloys. Fachbuchreihe Schweißtechn. Band 77. Düsseldorf: Deutscher Verlag für Schweißtechnik (1983).

[125] Flemings, M. C.: New Solidification Processes and Products. Solidification and Casting of Metals. London: Metals Soc., Book 192, 479–485 (1977).

[126] Baikie, B. L., Yapp, D.: Orientated Structures and Properties in Type 316 Stainless-Steel Weld Metal. Solidification and Casting of Metals. London: Metals Soc., Book 192, 438–442 (1977).

[127] Suutala, N.: Effect of Solidification Conditions on the Solidification Mode in Austenitic Stainless Steels. Metallurg. Trans. A. **14 A**, 191–197 (1983).

[128] Zitter, H.: Prüfung geschweißter und ungeschweißter Chrom-Nickel-Stähle auf interkristalline Korrosion. Arch. Eisenhüttenwes. **28**, 401–416 (1957).

[129] Nippes, E. F.: The Weld Heat-Affected Zone. Weld. J. **38**, Res. Suppl., 1s–18s (1959).

[130] Suutala, N., Sivonen, S., Moisio, T.: Einfluß der Erstarrungsgeschwindigkeit auf die Gefügeausbildung des Schweißgutes aus austenitischem Chrom-Nickel-Molybdän-Stahl. Schweißen u. Schneiden **30**, 170–173 (1978).

[131] David, S. A., Goodwin, G. M., Braski, D. N.: Solidification Behavior of Austenitic Stainless Steel Filler Metals. Weld. J. **58**, 330s–336s (1979).

[132] Velkov, K., Schelew, A.: Bestimmung der Zustandsänderung und Phasenumwandlung unter Einwirkung eines simulierten Schweißzyklus. Schweißtechn. (Berlin) **32**, 542–543 (1982).

[133] Bäumel, A.: Einfluß von Seigerungen auf das Korrosionsverhalten von austenitischem nichtrostendem Schweißgut. Werkst. u. Korrosion **27**, 687–693 (1976).

[134] Weingerl, H., Straube, H., Blöch, R.: Über die Auswirkung von Seigerungen auf das Korrosionsverhalten austenitischer Chrom-Nickel-Molybdän-Stähle. Werkst. u. Korrosion **27**, 69–77 (1976).

[135] Blanc, G., Tricot, R.: Solidification, ségrégation et homogénisation des aciers inoxydables austénitiques contenant de la ferrite delta. Mém. Sci. Rev. Métallurg. **68**, 735–753 (1971).

[136] Kulmburg, A., Blöch, R., Swoboda, K.: Beitrag zur Frage des Einflusses der Kristallseigerungen legierter Stähle auf deren Umwandlungsverhalten. Berg- u. hüttenm. Mh. **111**, 420–428 (1966).

[137] Hoffmeister, H.: Kristallseigerungen und Deltaferritbildung in austenitischem Schweißgut. DVS-Bericht Nr. 3, 2–8 (1973); siehe auch Schweißen u. Schneiden **25**, 164–166 (1973).

[138] Masumoto, I., Tamaki, K., Kutsuna, M.: Hot Cracking of Austenitic Steel Weld Metal. J. Japan Weld. Soc. **41**, 1306–1314 (1972).

[139] Mitsche, R., Niessner, M.: Angewandte Metallographie. Leipzig: Johann Ambrosius Barth 1939.

[140] Schürmann, E., Voss, H. J.: Seigerungsverhalten der Legierungselemente in Eisen-Chrom-Nickel-Schmelzen und bei Zusatz von Molybdän und Vanadin. Arch. Eisenhüttenwes. **48**, 129–132 (1977).

[141] Siegel, U., Günzel, M.: Mikroseigerungen in austenitischen Chrom-Nickel-Stählen. Neue Hütte **18**, 422–429, 599–602 (1973).

[142] Fridberg, J., Törndahl, L. E., Hillert, M.: Diffusion in Iron. Jernkontor. Ann. **153**, 263–276 (1969).

[143] Cieslak, M. J., Ritter, A. M., Savage, W. F.: Solidification Cracking and Analytical Electron Microscopy of Stainless Steel Weld Metals. Weld. J. **61**, Res. Suppl., 1s–8s (1982).

[144] Héritier, J., Lévy, J.: Fusion et solidification des alliages Fe-Cr-Ni. Mém. Sci. Rev. Métallurg. **73**, 523–535 (1976).

[145] Takalo, T., Suutala, N., Moisio, T.: Austenitic Solidification Mode in Austenitic Stainless Steel Welds. Metallurg. Trans. **10 A**, 1173–1181 (1979).

[146] Savage, W. F., Nippes, E. F., Miller, T. W.: Microsegregation in 70Cu–30Ni Weld Metal. Weld. J. **55**, Res. Suppl., 165s–173s (1976).

[147] Frederiksson, H.: The Solidification Sequence in an 18/8 Stainless Steel, Investigated by Directional Solidification. Metallurg. Trans. **3**, 2989–2997 (1972).

[148] Perteneder, E., Blöch, R., Wedl, F.: Beitrag zur Gefügeausbildung des Deltaferrits im Schweißgut hochlegierter Chrom-Nickel-Stähle. Sonderband 14 „Fortschritte in der Metallographie“ der Prakt. Metallographie **1983**, 345–352.

[149] Mundt, R., Hoffmeister, H.: δ-γ-Phase Equilibria in Iron-Chromium-Nickel Alloys at Temperatures Between 1200 and 1350 °C. Arch. Eisenhüttenwes. **54**, 253–256 (1983).

[150] Kujanpää, W., Suutala, N., Takalo, T., Moisio, T.: Correlation Between Solidification Cracking and Microstructure in Austenitic and Austenitic-Ferritic Stainless Steel Welds. Weld. World **9**, 55–75 (1979).

[151] Takalo, T., Suutala, N., Moisio, T.: Influence of Ferrite Content on its Morphology in Some Austenitic Weld Metals. Metallurg. Trans. **7 A**, 1591–1592 (1976).

[152] Suutala, N., Takalo, T., Moisio, T.: The Relationship Between Solidification and Microstructure in Austenitic and Austenitic-Ferritic Stainless Steel Welds. Metallurg. Trans. **10 A**, 512–514 (1979).

[153] Suutala, N., Takalo, T., Moisio, T.: Single-Phase Ferritic Solidification Mode in Austenitic-Ferritic Stainless Steel Welds. Metallurg. Trans. **10 A**, 1183–1190 (1979).

[154] Lyman, C. E.: Analytical Electron Microscopy of Stainless Steel Weld Metal. Weld. J. **58**, Res. Suppl., 189s–194s (1979).

[155] Suutala, N., Takalo, T., Moisio, T.: Technical Note: Comment on the Transformation $\delta \rightarrow \gamma$ by a Massive Mechanism in Austenitic Stainless Steel. Weld. J. **60**, Res. Suppl., 92s–93s (1981).

[156] David, S. A.: Ferrite Morphology and Variations in Ferrite Content in Austenitic Stainless Steel Weldments. Weld. J. **60**, Res. Suppl., 63s–71s (1981).

[157] Lippold, J. C., Savage, F.: Solidification of Austenitic Stainless Steel Weldment. Part 2: The Effect of Alloy Composition on Ferrite Morphology. Weld. J. **79**, Res. Suppl., 48s–58s (1980).

[158] Leone, G. L., Kerr, H. W.: The Ferrite to Austenite Transformation in Stainless Steels. Weld. J. **61**, Res. Suppl., 13s–21s (1982).

[159] Weber, H.: Einfluß der Abkühlgeschwindigkeit auf die Deltaferrit-Ausscheidung im austenitischen Schweißgut. Z. Werkstofftechn. **10**, 221–229 (1979).

[160] Eckstein, H. J., Viehrig, H. W.: Beitrag zum Umwandlungsverhalten ferritisch-austenitischer Chrom-Nickel-Stähle unter besonderer Berücksichtigung des Schweißens – Auflösungsverhalten. Neue Hütte **27**, 409–414 (1982).

[161] Mundt, R., Hoffmeister, H.: The Continuous δ-γ-Transformation During Cooling of Ferritic-Austenitic Iron-Chromium-Nickel Alloys. Arch. Eisenhüttenwes. **54**, 333–336 (1983).

[162] Scherer, R., Riedrich, G., Hoch, G.: Einfluß eines Gehaltes an Ferrit in austenitischen Chrom-Nickel-Stählen auf den Kornzerfall. Arch. Eisenhüttenwes. **13**, 53–57 (1939).

[163] Schaeffler, A. L.: Selection of Austenitic Electrodes for Welding Dissimilar Metals. Weld. J. **26**, Res. Suppl., 603s–620s (1947).

[164] Schaeffler, A. L.: Welding Dissimilar Metals with Stainless Electrodes. Iron Age **162**, 73–79 (1948).

[165] Schaeffler, A. L.: Constitution Diagram for Stainless Steel Weld Metal. Metal Progr. **56**, 680, 680 b (1949).

[166] Welding Handbook, Vol. 4: Metals and Their Weldability. 6. Aufl., S. 65.9. Miami, Florida: American Welding Soc. 1972.

[167] Ornig, H.: Das Schaeffler-Diagramm. Aufbau – Anwendung – Genauigkeit. Z. Schweißtechn. (Basel) **55**, 307–317 (1968).

[168] Mundt, R., Hoffmeister, H.: Einfluß der chemischen Zusammensetzung und der Schweißparameter auf die δ-Ferrit-Gehalte des Schweißguts hochlegierter ferritisch-austenitischer Chrom-Nickel-Stähle. Stahl u. Eisen **103**, 611–614 (1983).

[169] Hull, F. C.: Delta Ferrite and Martensite Formation in Stainless Steels. Weld. J. **52**, Res. Suppl., 193s–203s (1973).

[170] Espy, R. H.: Weldability of Nitrogen-Strengthened Stainless Steels. Weld. J. **61**, Res. Suppl., 149s–156s (1982).

[171] Krainer, E.: Zur Problematik der Ferritbestimmung in austenitischen Schweißverbindungen. Z. Schweißtechn. (Basel) **52**, 1–11 (1965).

[172] Bungardt, K., Dietrich, H., Arntz, H. E.: Beitrag zur magnetischen Ferritbestimmung in austenitischen Werkstoffen, besonders in austenitischem Schweißgut. DEW-Techn. Berichte **10**, 298–325 (1970).

[173] Rettenbacher, H., Fuchs, K.: Die Ferritgehaltmessung und ihre Bedeutung für die Beurteilung von austenitischem Schweißgut. Jubiläumsschrift „50 Jahre Böhler Schweißtechnik". Kapfenberg: VEW 1977.

[174] Recommended Method for the Metallographic Determination of Delta-Ferrite in Chromium-Nickel Austenitic Weld Metals by Means of Normal Optical Microscopy and Visual Comparison with an Atlas. Weld. World **13**, 219–223 (1975).

[175] Lüning, R., Tauchert, H. G.: Metallographischer Nachweis von Deltaferrit und Sigmaphase in austenitischem Stahl. Schweißen u. Schneiden **30**, 301–304 (1978).

[176] Blöch, R., Huszar, R.: Untersuchungen über den Zerfall des Deltaferrits im Schweißgut X 3 CrNiMoN 25 20. Sonderband 12, Prakt. Metallogr., S. 159–167.

[177] Ratz, G. A., Gunia, R. B.: How Accurate Are Methods for Measuring Ferrite? Metal Progr. **76**, 76–80 (1969).

[178] Final Report on the Cooperative Testing Programme 1971–1972 on the Measurement of Ferrite in Austenitic Stainless Weld Metal. Weld. World **12**, 17–26 (1974).

[179] Ornig, H., Kohl, H., Rabensteiner, G., Rettenbacher, H., Weberberger, H.: Bestimmung des Ferritgehaltes in austenitischem Schweißgut in Ferritnummern (FN). Schweißtechn. (Wien) **35**, 110–112 (1981).

[180] DeLong, W. T.: Ferrite in Austenitic Stainless Steel Weld Metal. Weld. J. **53**, Res. Suppl., 273s–286s (1974).

[181] Recommended Standard Method for the Determination of the Ferrite Number in Austenitic Weld Metal Deposited by Cr-Ni-Steel Electrodes. Weld. World **20**, 7–14 (1982).

[182] Robinson, L. L., Rabensteiner, G., Neff, F.: Procedures Used to Prepare Secondary Standards for Delta-Ferrite in Austenitic Stainless Steel Weld Metal. Weld. World **20**, 15–22 (1982).

[183] Ornig, H., Kohl, H., Rabensteiner, G., Rettenbacher, H., Weberberger, H.: Messung des Ferritgehaltes in austenitischem Schweißgut in Ferritnummern (FN). Berg- u. hüttenm. Mh. **125**, 221–228 (1980).

[183a] Merinov, P., Rettenbacher, H., Rabensteiner, G., Neff, F.: Ferritgehalt im austenitischen CrNi-Schweißgut unter Verwendung verschiedener Meßmethoden. Schweißtechn. (Wien) **37**, 166–168 (1983).

[184] Long, C. L., DeLong, W. T.: The Ferrite Content of Austenitic Stainless Steel Weld Metal. Weld. J. **52**, Res. Suppl., 281s–297s (1973).

[185] Suutala, N.: Effect of Manganese and Nitrogen on the Solidification Mode in Austenitic Stainless Steel Welds. Metallurg. Trans. **13 A**, 2121–2129 (1982).

[186] Rocha, H. J., Lennartz, G.: Die Aktivierungspotentiale von Eisen-Chrom-Legierungen und ihre Beziehung zu der chemischen Beständigkeit in Schwefelsäure. Arch. Eisenhüttenwes. **26**, 117–123 (1955).

[187] Kulmburg, A., Sölkner, W., Korntheuer, F., Schmid, H. E.: Das Umwandlungsverhalten komplexlegierter Stähle mit 12 bis 18% Chrom. Berg- u. hüttenm. Mh. **124**, 400–406 (1979).

[188] Irvine, K. J., Crowe, D. J., Pickering, F. B.: The Physical Metallurgy of 12% Chromium Steels. J. Iron Steel Inst. **193**, 386–405 (1960).

[189] Irvine, K. J., Llewellyn, D. T., Pickering, F. B.: Controlled-Transformation Stainless Steels. J. Iron Steel Inst. **192**, 218–238 (1959).

[190] Castro, R., Tricot, R.: Etude des transformations isothermes dans les acies inoxydables semi-ferritiques à 17% de Chrome. Mém. Sci. Rev. Métallurg. Part 1: **59**, 571–586 (1962); Part 2: **59**, 587–596 (1962); Part 3: **63**, 657–684 (1966).

[191] Rabensteiner, G., Perteneder, E., Tösch, J.: Optimierung von weichmartensitischen Chrom-Nickel-Schweißgutlegierungen für den Einsatz im Kernreaktorbau. DVS-Bericht **75**, 261–265 (1982). Düsseldorf: Deutscher Verlag für Schweißtechnik.

[192] Koepke, W., Skuin, K., Herfert, S.: Untersuchungen zum Umwandlungsverhalten 13%iger Chromstähle mit geringem Kohlenstoffgehalt. Neue Hütte **15**, 165–170 (1970).

[193] Woltron, K.: Schweißsprödigkeit ferritischer Chromstähle. Berg- u. hüttenm. Mh. **116**, 429–437 (1971).

[194] Kulmburg, A., Korntheuer, F., Koren, M., Gründler, O., Hutterer, K.: Umwandlungs- und Ausscheidungsverhalten von weichmartensitischen und hochfesten korrosionsbeständigen Stählen. Berg- u. hüttenm. Mh. **126**, 104–108 (1981).

[195] Leymonie, C., Ottmann, M.-C., Risacher, R., Thauvin, G.: Contribution à l'étude des transformations structurales des aciers à 13% Cr–4% Ni. Rev. Métallurg. **76**, 815–826 (1979).

[196] Leymonie, C., Lecocq, M. C., Ottmann, M.-C.: Relations entre la structure et les propriétés mécaniques de deux aciers à 16% Cr–4% Ni. Traitement thermique **165**, 57–62 (1982).

[197] Eckstein, H. J.: Umwandlungsverhalten und Eigenschaften nichtrostender ferritisch-austenitischer Stähle, insbesondere der Marke X 5 CrNiTi 26.6. Berg- u. hüttenm. Mh. **122**, 358–368 (1977).

[198] Jolly, P., Hochmann, J.: Evolution structurale d'un acier inoxydable austeno-ferritique par maintien entre 600 et 1150 °C. Mém. Sci. Rev. Métallurg. **70**, 117–124 (1973).

[199] Niederau, H. J.: Entwicklungsstand nichtrostender weichmartensitischer Chrom-Nickel-Stähle unter besonderer Berücksichtigung des Stahles X 5 CrNi 13 4. Stahl u. Eisen **98**, 385–392 (1978).

[200] Wiester, H. J., Pier, G.: Untersuchungen über die interkristalline Korrosion austenitischer Chrom-Nickel-Stähle nach langdauernder Beanspruchung zwischen 450 und 800 °C. Arch. Eisenhüttenwes. **30**, 293–297 (1959).

[201] Herbsleb, G., Schüller, H. J., Schwaab, P.: Ausscheidungs- und Korrosionsverhalten unstabilisierter und stabilisierter 18/10-Chrom-Nickel-Stähle nach kurzzeitigem sensibilisierendem Glühen. Werkst. u. Korrosion **27**, 560–568 (1976).

[202] Wiester, H. J., Schüller, H. J., Schwaab, P.: Ausscheidungsvorgänge in austenitischen Chrom-Nickel-Stählen und ihr Einfluß auf die Neigung zur interkristallinen Korrosion. Arch. Eisenhüttenwes. **30**, 299–309 (1959).

[203] Rocha, H. J.: Die Sensibilisierung stabilisierter und nicht stabilisierter Stähle durch Chromkarbid. DEW – Technische Berichte **2**, 16–24 (1962); Zeitschr. für Schweißtechn. **49**, 98–106 (1962).

[204] Bäumel, A., Bühler, H. E., Schüller, H. J., Schwaab, P., Schwenk, W., Ternes, H., Zitter, H.: Deutung der Ursachen der interkristallinen Korrosion von nichtrostenden Stählen in Zusammenhang mit der Chromverarmungstheorie. Corrosion Sci. **4**, 89–103 (1964).

[205] Stawström, C., Hillert, M.: An Improved Depleted-Zone Theory of Intergranular Corrosion of 18-8 Stainless Steel. J. Iron Steel Inst. **207**, 77–85 (1969).

[206] Bäumel, A.: Korrosion in der Wärmeeinflußzone geschweißter chemisch beständiger Stähle und Legierungen und ihre Verhütung. Werkst. u. Korrosion **26**, 433–443 (1975).

[207] Bäumel, A.: Das Korrosionsverhalten der Schweißnähte an ferritischen Chromstählen. Fachbuchreihe Schweißtechnik, Band 36, 95–105. Düsseldorf: DVS 1964.

[208] Diebold, A., Weingerl, H.: Eigenschaften des schweißbaren, hoch korrosionsbeständigen Stahles X 2 CrNiMoN 25 22. Werkst. u. Korrosion **25**, 175–179 (1974).

[209] Grützer, G.: Über die interkristalline Korrosion stickstofflegierter 18/10-Chrom-Nickel-Stähle. Dr.-Ing.-Dissertation, Techn. Hochschule Aachen 1971.

[210] Grützner, G.: Kornzerfallsanfälligkeit stickstofflegierter austenitischer Chrom-Nickel-Stähle durch Chromnitridausscheidung. Stahl u. Eisen **93**, 9–18 (1973).

[211] Wedl, W., Kohl, H.: Interkristalline Korrosion und Festigkeitswerte austenitischer Stähle mit erhöhtem Chromgehalt und Stickstoffzusatz. Berg- u. hüttenm. Mh. **124**, 508–514 (1979).

[212] Briant, C. L., Mulford, R. A., Hall, E. L.: Sensitization of Austenitic Stainless Steels, I. Controlled Purity Alloys. Corrosion-NACE **38**, 468–477 (1982).

[213] Blazejak, D., Herbsleb, G., Westerfeld, K. J.: Der Einfluß von Stickstoff auf die korrosionschemischen Eigenschaften lösungsgeglühter und angelassener austenitischer 18/10-Chrom-Nickel-Stähle und 18/10-Chrom-Nickel-Molybdän-Stähle. Part I: Ausscheidungsverhalten der Stähle. Werkst. u. Korrosion **27**, 398–403 (1976).

[214] Herbsleb, G., Westerfeld, K. J.: Der Einfluß von Stickstoff auf die korrosionschemischen Eigenschaften lösungsgeglühter und angelassener austenitischer 18/10-Chrom-Nickel- und Chrom-Nickel-Molybdän-Stähle. Part II: Interkristalline Korrosion in Kupfersulfat-Schwefelsäure-Lösungen und in siedender 65%iger Salpetersäure. Werkst. u. Korrosion **27**, 404–415 (1976).

[215] Wessling, W., Bock, H. E.: Kornzerfallsverhalten von gleichartigen und ungleichartigen Schweißverbindungen aus stickstofflegierten und nichtstickstofflegierten austenitischen Stählen. Stahl u. Eisen **91**, 1442–1445 (1971).

[216] Lorenz, K., Médawar, G.: Über das Korrosionsverhalten austenitischer Chrom-Nickel-(Molybdän-)Stähle mit und ohne Stickstoffzusatz unter besonderer Berücksichtigung ihrer Beanspruchung in chloridhaltigen Lösungen. Thyssenforschung **1**, 97–108 (1969).

[217] Edström, H. J., Ljundberg, L.: Rost- und säurebeständige Chrom-Nickel-Stähle mit max. 0,030% Kohlenstoff als Konstruktionsmaterial für die chemische Industrie. Werkst. u. Korrosion **15**, 743–753 (1964).

[218] Kügler, A.: Einfluß der Gefügeausbildung auf das Korrosionsverhalten von hochmolybdänhaltigen austenitischen Chrom-Nickel-Stählen. Dr.-Ing.-Dissertation, Techn. Hochschule Aachen 1976.

[219] Lennartz, G., Oppenheim, R.: Beständigkeit austenitischer Chrom-Nickel-(Molybdän-)-Stähle gegen interkristalline Korrosion nach Dauerglühung bei 500 °C. DEW – Techn. Ber. **13**, 70–74 (1973).

[220] Pohl, M.: Elektronenmikroskopische Untersuchungen zum Ausscheidungsverhalten unstabilisierter vollaustenitischer Chrom-Nickel-Stähle im Temperaturbereich von 500–800 °C. Dr.-Ing.-Dissertation, Techn. Hochschule Aachen 1977.

[221] Bungardt, K., Lennartz, G.: Ausscheidungsvorgänge in einem mit Titan stabilisierten austenitischen Chrom-Nickel-Stahl und ihre Beziehungen zur interkristallinen Korrosion. Arch. Eisenhüttenwes. **27**, 127–133 (1956).

[222] Bungardt, K., Lennartz, G.: Ausscheidungsvorgänge und interkristalline Korrosion in einem nichtrostenden Chrom-Nickel-Stahl mit Niob. Arch. Eisenhüttenwes. **29**, 359–364 (1958).

[223] Bungardt, K., Lennartz, G.: Einfluß des Stabilisierungsgrades auf Ausscheidung und interkristalline Korrosion bei einem hochwarmfesten niobstabilisierten Stahl mit 16% Cr, 13% Ni, Mo und V. Arch. Eisenhüttenwes. **30**, 23–29 (1959).

[224] Bungardt, K., Lennartz, G., Wetzlar, K.: Einfluß des Stabilisierungsgrades auf die Ausscheidungsvorgänge niobhaltiger austenitischer Chrom-Nickel- und Chrom-Nickel-Molybdän-Stähle. Arch. Eisenhüttenwes. **30**, 429–434 (1959).

[225] Reisenhofer, K., Weingerl, H.: Über das Ausscheidungs- und Korrosionsverhalten niobstabilisierter, korrosionsbeständiger Reaktorstähle. Mikrochimica Acta (Wien), Suppl. **7**, 523–530 (1977).

[226] Lennartz, G.: Interkristalline Korrosion und interkristalline Rißbildung an nichtrostenden Stählen und korrosionsbeständigen Nickellegierungen. VDI-Ber. Nr. 235, 169–181 (1975).

[227] Bäumel, A.: Korrosionsverhalten nichtrostender Vergütungsstähle mit rund 13% Chrom. Werkst. u. Korrosion **18**, 289–302 (1967).

[228] Bäumel, A.: Zusammenhang zwischen der Wärmebehandlung und dem Korrosionsverhalten nichtrostender Stähle mit rund 17% Chrom in siedender konzentrierter Salpetersäure. Arch. Eisenhüttenwes. **34**, 135–146 (1963).

[229] Herbsleb, G., Schwenk, W.: Untersuchungen über die Kornzerfallsanfälligkeit eines unstabilisierten 17%igen Chromstahles und ihre Beseitigung durch Stabilglühen. Werkst. u. Korrosion **19**, 103–113 (1968).

[230] Bond, A. P., Lizlovs, E. A.: Neue Erkenntnisse auf dem Gebiet der ferritischen rostfreien Stähle. Werkst. u. Korrosion **21**, 336–342 (1970).

[231] Bäumel, A.: Zusammenhang zwischen dem Ausscheidungs- und Korrosionsverhalten von niobstabilisierten Stählen mit 17% Chrom. Arch. Eisenhüttenwes. **85**, 1462–1476 (1965).

[232] Bäumel, A.: Vergleichende Untersuchung nichtrostender Chrom- und Chrom-Nickel-Stähle auf interkristalline Korrosion in siedender Salpetersäure und Kupfersulfat-Schwefelsäure-Lösung. Stahl u. Eisen **84**, 798–807 (1964).

[233] Fontana, M. G., Green, N. D.: Corrosion Engineering. International student edition, McGraw-Hill Book Comp. 1982.

[234] Lennartz, G., Kiesheyer, H.: Korrosionsverhalten von hochchromhaltigen, ferritischen Stählen. DEW – Techn. Ber. **11**, 230–235 (1974).

[235] Stainless Steel 77. Conference proceedings, London, Sept. 1977. Climax Molybdenum Comp. London: Amax Nickel 1977.

[236] Brezina, P., Erdoes, E., Geiger, T., Habel, L., Lorenz, M., Wintsch, W.: Beitrag zur Methodik der Bestimmung von Phasen in technischen Legierungen. Prakt. Metallogr. Part I: **10**, 343–357 (1973); Part II: **10**, 377–397 (1973).

[237] Folkhard, E.: Neue Werkstoffe auf dem Gebiet des Auftragschweißens. Schweißen u. Schneiden **22**, 245–248 (1970).

[238] Heimann, W., Hook, M.: Ursache der Versprödung weichmartensitischer nichtrostender Stähle. Thyssen Edelst. Techn. Ber. **8**, 126–134 (1982).

[239] Süry, P., Brezina, P.: Kurzzeitprüfmethoden zur Untersuchung des Einflusses von Wärmebehandlung und chemischer Zusammensetzung auf die Korrosionsbeständigkeit martensitischer Chrom-Nickel-(Molybdän-)Stähle mit tiefem Kohlenstoffgehalt. Werkst. u. Korrosion **30**, 341–357 (1979).

[240] Brezina, P., Sonderegger, B.: Wärmebehandlung, Gefüge und Eigenschaften des korrosionsträgen martensitaushärtbaren Stahles X 5 CrNiMoCuNb 14 5 (14-5 PH). Escher Wyss Mitteilungen **1978**, 38–48.

[241] Mende, A. B., Oppenheim, R.: Ausscheidungshärtung eines niedriggekohlten martensitischen 16/5-Chrom-Nickel-Molybdän-Stahles. DEW – Techn. Ber. **13**, 119–122 (1973).

[242] Lindscheid, H., Mayer, H.: G-X 5 CrNiMo 16 5 – eine nichtrostende martensitische Stahlgußsorte mit verbesserten Eigenschaften. Techn. Rundschau Sulzer **4**, 162–166 (1980).

[243] Guiraldenq, P., Vieillard-Baron, B., Hochmann, J.: Influence de la composition et de traitements thermiques sur la structure et les caractéristiques mécaniques des aciers inoxydables martensitiques dérivés du type 16% de chrome et 4% de nickel. Mém. Sci. Rev. Métallurg. **69**, 855–867 (1972).

[244] Van Nassau, L.: The Welding of Austenitic-Ferritic Mo-Alloyed Cr-Ni-Steel. Weld. World **20**, 23–30 (1982).

[245] Herbsleb, G., Schwaab, P.: Precipitation of Intermetallic Compounds, Nitrides and Carbides in AF 22 Duplex Steel and Their Influence on Corrosion Behavior in Acids. Conference at ASM, Int. Symp. on Duplex Austenitic-Ferritic Stainless Steels, St. Louis, U.S.A. 1982, published by American Soc. for Metals, Metals Park, Ohio 44073, catalog No. 83-071074.

[246] Wehner, H., Speckhardt, H.: Zum Ausscheidungs- und Korrosionsverhalten eines ferritisch-austenitischen Chrom-Nickel-Molybdän-Stahles nach kurzzeitiger Glühbehandlung unter besonderer Berücksichtigung des Schweißens. Z. Werkst.-Techn. **10**, 317–332 (1979).

[247] Wessling, W., Bock, E. H., Fuchs, W.: Neuere Entwicklungen von korrosionsbeständigen Sonderstählen für das Chemiewesen. Z. Werkst.-Techn. **4**, 186–195 (1973).

[248] Kohl, H., Hochörtler, G., Kriszt, K., Koren, M.: Sonderstähle für den Apparatebau mit sehr guter Korrosionsbeständigkeit und erhöhter Festigkeit. Werkst. u. Korrosion **34**, 1–6 (1983).

[249] Bäumel, A.: Selektive Korrosion an Verbindungsschweißungen von Stählen. Werkst. u. Korrosion **17**, 299–307 (1966).

[250] Bäumel, A.: Korrosionsprognose für den Bereich der Schweißnaht. Werkst. u. Korrosion **23**, 546–554 (1972).

[251] Stalder, F.: Das Korrosionsverhalten von Schweißnähten bei austenitischen, nichtrostenden Stählen. Z. Schweißtechn. (Basel) **69**, 73–79 (1982).

[252] Schabereiter, H., Folkhard, E., Ablasser, F., Ornig, H., Neff, F.: Schweißen austenitischer Werkstoffe beim Bau großer Kernreaktoren. Berg- u. hüttenm. Mh. **117**, 167–181 (1972).

[253] Schabereiter, H.: Verhalten von austenitischem, Niob-stabilisiertem Schweißgut nach Wärmebehandlung unter besonderer Berücksichtigung der Anwendung für Inneneinbauten von Kernreaktoren. Z. Schweißtechn. (Basel) **62**, 63–79 (1975).

[254] Horn, E. M., Kügler, A.: Entwicklung, Eigenschaften, Verarbeitung und Einsatz des hochsiliziumhaltigen austenitischen Stahls X 2 CrNiSi 18 15. Part II. Z. Werks.-Techn. **8**, 410–417 (1977).

[255] Bäumel, A.: Einfluß des Deltaferrits auf das Korrosionsverhalten von Schweißungen aus austenitischen Zusatz-Werkstoffen. Schweißen u. Schneiden **19**, 264–269 (1967).

[256] Hall, E. O., Algie, S. H.: The Sigma Phase. Metallurg. Reviews (Australia) **11**, 61–88 (1966).

[257] Baerlecken, E., Fabritius, H.: Umwandlungskinetik der Sigmaphase in einer Eisen-Chrom-Legierung mit 48% Chrom. Arch. Eisenhüttenwes. **26**, 679–686 (1955).

[258] Brandis, H., Heimann, H., Schmidtmann, E.: Einfluß des Stickstoffs auf das Ausscheidungsverhalten des Stahles X 3 CrNiMoNbN 23 17. TEW – Techn. Ber. **2**, 150–166 (1976).

[259] Freißmuth, A.: Das Ausscheidungsverhalten stickstofflegierter austenitischer Chrom-Nickel-Stähle. Berg- u. hüttenm. Mh. **112**, 197–203 (1967).

[260] Bungardt, K., Borchers, H., Kölsch, D.: Untersuchung der Bildung der Sigmaphase in ferritischen Chromstählen mit unterschiedlichen Gehalten an Silizium und Aluminium. Arch. Eisenhüttenwes. **34**, 465–476 (1963).

[261] Norström, L. A., Petterson, S., Nordin, S.: Sigmaphase Embrittlement in Some Ferritic-Austenitic Stainless Steels. Z. Werkstofftechn. **12**, 229–234 (1981).

[262] Schwaab, P.: Das Ausscheidungsverhalten des ferritisch-austenitischen Stahles X 2 CrNiMoN 22 5. Praktische Metallographie. Fortschr. Metallogr. **14**, 435–444 (1983).

[263] Schabereiter, H., Rabensteiner, G.: Die Problematik der Spannungsarmglühung von austenitischem Schweißgut. Z. Schweißtechn. (Basel) **63**, 214–230 (1976).

[264] Rabensteiner, G., Schabereiter, H.: Einfluß von Spannungsarmglühungen auf das Korrosionsverhalten und die Zähigkeitseigenschaften von austenitischem Schweißgut. Berg- u. hüttenm. Mh. **120**, 522–528 (1975).

[265] Schabereiter, H.: Beitrag zur Schweißung dickwandiger, vollaustenitischer korrosionsbeständiger Stähle für die Verwendung bei tiefsten Temperaturen und für den chemischen Apparatebau. Z. Schweißtechn. (Basel) **58**, 229–243 (1971).

[266] Rabensteiner, G., Schabereiter, H.: Neuentwickelte hochkorrosionsbeständige austenitische Schweißzusatzwerkstoffe. Z. Schweißtechn. (Basel) **66**, 26–35 (1979).

[267] Perteneder, E., Tösch, J., Schabereiter, H., Rabensteiner, G.: Neuentwickelte Schweißzusatzwerkstoffe zum Schweißen korrosionsbeständiger CrNiMoN-legierter Duplex-Stähle. Schweißtechnik (Wien) **37**, 83–86, 102–104 (1983).

[268] Beckitt, F. R.: The Formation of Sigma-Phase from Delta-Ferrite in a Stainless Steel. J. Iron Steel Inst. **207**, 632–638 (1969).

[269] Baerlecken, E., Fabritius, H.: Einfluß der Auslagerung im Temperaturbereich der 475 °-Versprödung auf einige magnetische Eigenschaften von Eisen-Chrom-Legierungen. Stahl u. Eisen **75**, 1774–1784 (1955).

[270] De Nys, T., Gielen, P. M.: Spinodal Decomposition in the Fe-Cr System. Metallurg. Trans. **2**, 1423–1428 (1971).

[271] Ettwig, H. H., Pepperhoff, W.: Untersuchungen über das metastabile System Eisen-Chrom mit Hilfe des Mössbauer-Effektes. Arch. Eisenhüttenwes. **41**, 471–474 (1970).

[272] Albritton, O. W., Lewis, J. M.: Computer Differentiation of Mössbauer Spectra. Welding J. **50**, Weld. Res. Suppl., 327s–330s (1971).

[273] Solomon, H. D., Levinson, L. M.: Mössbauer Effect Study of 475 °C Embrittlement of Duplex and Ferritic Stainless Steels. Acta Metallurg. **26**, 429–442 (1978).

[274] Heger, J. J.: 885 °F Embrittlement of Ferritic Chromium-Iron Alloys. Metal Progr. **60**, 55–61 (1951).

[275] Bandel, G., Tofaute, W.: Die Versprödung von hochlegierten Chromstählen im Temperaturgebiet um 500 °C. Arch. Eisenhüttenwes. **15**, 307–320 (1942).

[276] Newell, H. D.: Properties and Characteristics of 27% Chromium-Iron. Metal Progr. **50**, 997–1006, 1016–1028 (1946).

[277] Riedrich, G., Loib, F.: Versprödung chromreicher Stähle im Temperaturgebiet von 300–600 °C. Arch. Eisenhüttenwes. **15**, 175–182 (1941).

[278] Grobner, P. J.: The 885 °F (475 °C) Embrittlement of Ferritic Stainless Steels. Metallurg. Trans. **4**, 251–260 (1973).

[279] Trautwein, A., Gysel, W.: Influence of Long Time Aging of CF 8 and CF 8 M Cast Steel at Temperatures Between 300 and 500 °C on the Impact Toughness and the Structure Properties. Techn. Mitt. Georg Fischer Konzern, G. F. Spektrum **5**, 38–48, 49–53 (1981).

[280] Püber, J., Auer, R.: Chromstahlguß GX 5 CrNiMo 13 4 und GX 5 CrNiMo 16 6 – Hinweise über das Verhalten beim Auslagern im Temperaturbereich von 250–400 °C. Berg- u. hüttenm. Mh. **126**, 108–113 (1981).

[281] Malone, M. O.: Sigma and 885 °F (475 °C) Embrittlement of Chromium-Nickel Stainless Steel Weld Metal. Welding J. **46**, Res. Suppl., 241s–253s (1967).

[282] Baker, R. G.: Weld Cracking–A Modern Insight. Brit. Welding J. **15**, 283–295 (1968).

[283] Borland, J. C., Younger, R. N.: Some Aspects of Cracking in Welded Cr-Ni-Austenitic Steels. Brit. Welding J. **7**, 22–59 (1960).

[284] Dokumentation Schweißtechnik. Bibliography of "Hot Cracking in Welds", edited by Wilken, K., Blumendorf, P. Berlin: Fachgruppe Fügetechn. der BAM 1978.

[285] Hemsworth, B., Boniszewski, T., Eaton, N. F.: Classification and Definition of High Temperature Welding Cracks in Alloys. Metal Constr. and Brit. Welding J. **1**, 5–16 (1969).

[286] Apblett, W. R., Pellini, W. S.: Factors, Which Influence Weld Hot Cracking. Welding J. **33**, Res. Suppl., 83s–90s (1954).

[287] Prochorow, N. N., Jakuschin, B. F., Prochorow, N. Nikol.: Theorie und Verfahren zum Bestimmen der technologischen Festigkeit von Metallen während des Kristallisationsprozesses beim Schweißen. Schweißtechn. (Berlin) **18**, 8–11 (1968).

[288] Probst, R., Schirmer, F.: Technologische Festigkeit von Schweißverbindungen an stabilaustenitischen CrNiMo-Stählen. Schweißtechn. (Berlin). Part I: **26**, 322–326 (1976); Part II: **26**, 460–464 (1976).

[289] Klug, P.: Beitrag zur Prüfung von hochlegierten Schweißzusatzwerkstoffen mit dem PVR-Test der VEW-Kapfenberg. Dr.-techn.-Dissertation, Techn. Universität Graz 1980.

[290] Arata, Y., Matsuda, F., Nakagawa, H., Katayama, S., Ogata, S.: Solidification Crack Susceptibility in Weld Metals of Fully Austenitic Stainless Steels (Report III). Trans. Japan Welding Res. Inst. **6**, 197–206 (1977).

[291] Matsuda, F., Nakagawa, H., Ogata, S., Katayama, S.: Fractographic Investigation on Solidification Crack in the Varestraint Test of Fully Austenitic Stainless Steel. Trans. Japan Welding Res. Inst. **7**, 59—70 (1978).

[292] Thier, H.: Delta-Ferrit und Heißrisse beim Schweißen chemisch beständiger austenitischer Stähle. DVS-Berichte **41**, 100—104 (1976).

[293] Homberg, G., Wellnitz, G.: Über die Heißrissigkeit beim Schweißen von Stählen. Schweißen u. Schneiden **27**, 90—93 (1975).

[294] Perteneder, E., Rabensteiner, G., Schabereiter, H.: Einfluß der Primärkristallisation auf das Heißrißverhalten austenitischen Cr-Ni-Schweißgutes. Schweißtechn. (Wien) **33**, 33—39 (1979).

[295] Müsch, H.: Heißrisse und ihre Entstehung. Schweißtechn. Kolloquium „Heißrisse" der SLV-Duisburg 1977.

[296] Van Bemst, A.: Warmrisse und Erstarrungsvorgänge im austenitischen Schweißgut. Z. Schweißtechn. (Basel) **69**, 119—124 (1982).

[297] Baker, R. G., Newman, R. P.: Cracking in Welds. Metal Constr. and Brit. Weld. J. **1**, Nr. 2, 1—4 (1969).

[298] Kulmburg, A.: Private information, Kapfenberg 1975.

[299] Stauffer, W. A., Keller, A.: Rißfreies Schweißen von austenitischem Stahlguß. Z. Schweißtechn. (Basel) **46**, 2—6 (1959).

[300] Honeycombe, J., Gooch, T. G.: Microcracking in Fully Austenitic Stainless Steel Weld Metal. Metal Constr. and Brit. Weld. J. **2**, 375—380 (1970).

[301] Schabereiter, H., Rabensteiner, G., Folkhard, E.: Neue vollaustenitische Schweißzusatzwerkstoffe für den Bau von Harnstoffanlagen. Schweißtechn. (Wien) **30**, 149—153 (1976).

[302] Lundin, C. D., Spond, D. F.: The Nature and Morphology of Fissures in Austenitic Stainless Steel Weld Metals. Welding J. **55**, Res. Suppl., 356s—367s (1976).

[303] Gooch, T. G., Honeycombe, J.: Welding Variables and Microfissuring in Austenitic Stainless Steel Weld Metal. Welding J. **59**, Res. Suppl., 233s—241s (1980).

[304] Lundin, C. D., Chou, C.-P. D., Sullivan, C. J.: Hot Cracking Resistance of Austenitic Stainless Steel Weld Metal. Welding J. **59**, Res. Suppl., 226s—232s (1980).

[305] Klingauf, S.: Theorie zum Entstehen von Heißrissen in hocherwärmten Werkstoffbereichen. Schweißen u. Schneiden **32**, 258—263 (1980).

[306] Klug, P.: Rißgefahr in geschweißten Bauteilen; Beitrag zur quantitativen Beurteilung aus den Ergebnissen der Heißrißprüfung mit dem PVR-Test. Schweißtechn. (Wien) **35**, 33—39 (1981).

[307] Granjon, H.: Informations sur les essais de fissuration. Soudage Techn. Conn. **53**, 290—310 (1963).

[308] Wilken, K.: Aussagefähigkeit von Warmrißprüfverfahren für die Warmrißsicherheit von geschweißten Bauteilen. Chemie Ing. Techn. **11**, 777—783 (1972).

[309] Wilken, K., Schönherr, W.: Description of Four Hot Cracking Tests Without Self-Restraint. Weld. World **13**, 238—249 (1975).

[310] Homberg, G.: Untersuchungen zur Heißrißneigung vollaustenitischer unstabilisierter Stähle des Typs X 3 CrNi 16 16 beim Unterpulverschweißen. Dr.-Ing.-Dissertation, Techn. Hochschule Aachen 1976.

[311] Chène, J. J.: Bestimmung der Empfindlichkeit von austenitischen Stählen gegen Wiedererwärmungsrisse. Z. Schweißtechn. (Basel) **68**, 385—390 (1981).

[312] Halkes, C. L.: Review of Literature Concerning Hot Cracking Tests (Concept Text). Dokument IX-H-107-82 der Unterkommission IX-H des Internationalen Schweißinstituts.

[313] Lundin, C. D., DeLong, W. T., Spond, D. F.: The Fissure Bend Test. Welding J. **55**, Res. Suppl., 145s—151s (1976).

[314] Folkhard, E., Rabensteiner, G., Schabereiter, H., Fuchs, K., Tösch, J.: Der PVR-Test, ein neues Verfahren zur Ermittlung der Heißrißsicherheit von Schweißwerkstoffen mit hoher quantitativer Aussagekraft. Jubiläumsschrift „50 Jahre Böhler Schweißtechnik". Kapfenberg: VEW 1977.

[315] Savage, W. F., Lundin, C. D.: The Varestraint Test. Welding J. **54**, Res. Suppl., 433s–442s (1954).

[316] Senda, T., Matsuda, F., Takano, G., Watanabe, K., Kobayashi, T., Matsuzaka, T.: Fundamental Investigations on Solidification Crack Susceptibility for Weld Metals with Trans-Varestraint Test. Trans. Japan Welding Soc. **2**, 141–162 (1971).

[317] Wilken, K., Kleistner, H.: Der MVT-Test – ein neues universelles Verfahren zur Prüfung der Heißrißanfälligkeit beim Schweißen. Material u. Technik **10**, 3–10 (1982).

[318] Wilken, K.: Heißrißprüfung mit dem MVT- und Heißzugversuch sowie Übertragbarkeit auf Bauteilverhältnisse. Schweißen u. Schneiden **32**, 71–74 (1980).

[319] Rabensteiner, G., Tösch, J., Schabereiter, H.: Heißrißuntersuchungen an chemisch beständigen Schweißgütern mit dem neuentwickelten PVR-Test. Schweißtechn. (Wien) **34**, 213–217 (1980).

[320] Arata, Y., Matsuda, F., Katayama, S.: Solidification Crack Susceptibility in Weld Metals of Fully Austenitic Stainless Steels (Report II). Trans. Japan Welding Res. Inst. **6**, 105–116 (1977).

[321] Folkhard, E., Fladischer, J., Rabensteiner, G., Schabereiter, H., Tösch, J.: Einfluß der Niobstabilisierung, des Umhüllungstyps und des Ferritgehaltes auf die Heißrissigkeit von niedriggekohlten austenitischen 20 Cr/10 Ni-Schweißgutlegierungen. Jubiläumsschrift „50 Jahre Böhler Schweißtechnik". Kapfenberg: VEW 1977.

[322] Lundin, C. D., DeLong, W. T., Spond, D. F.: Ferrite-Fissuring Relationship in Austenitic Stainless Steel Weld Metal. Welding J. **54**, Res. Suppl., 241s–246s (1975).

[323] Tösch, J., Schabereiter, H.: Einfluß der Schweißbedingungen auf den Ferritgehalt und die Heißrissigkeit von austenitischem Cr-Ni-Schweißgut. Schweißtechn. (Wien) **33**, 141–146 (1979).

[324] Asakura, S., Wachi, H., Watanabe, K.: The Effect of Welding Conditions on the Crack Sensitivity of Austenitic Stainless Steel Weld Metal. Trans. Japan Welding Soc. **3**, 264–274 (1972).

[325] Lippold, J. C., Savage, W. F.: Solidification of Austenitic Stainless Steel Weldments. Part III: The Effect of Solidification Behavior on Hot Cracking Susceptibility. Welding J. **61**, Res. Suppl., 388s–396s (1982).

[326] Masumoto, I., Tamaki, K., Kutsuna, M., Ozaki, H.: Einfluß der Primärkristallisation bei peritektischer Reaktion auf die Heißrißneigung von Stahlschweißgut. Schweißen u. Schneiden **27**, 450–454 (1975).

[327] Matsuda, F., Nakagawa, H., Uehara, T., Katayama, S., Arata, Y.: A New Explanation for Role of Delta-Ferrite Improving Weld Solidification Crack Susceptibility in Austenitic Stainless Steel. Trans. Japan Welding Res. Inst. **8**, 105–113 (1979).

[328] Elmer, J. W., Olson, D. L., Matlock, D. K.: The Thermal Expansion Characteristics of Stainless Steel Weld Metal. Welding J. **61**, Res. Suppl., 293s–301s (1982).

[329] Polgary, S.: Crack Sensitivity of Weld Metal from Nb-Stabilized Stainless Electrodes. Svetsaren in Engl. **2**, 10–14 (1965).

[330] Polgary, S.: Einflußgrößen des Ferritgehaltes im Hinblick auf seine Festlegbarkeit im austenitischen rostsicheren Schweißgut umhüllter Elektroden. Z. Schweißtechn. (Basel) **53**, 364–375 (1966).

[331] Ebert, K.-A.: Beeinflussung der Schweißsicherheit bei austenitischen Chrom-Nickel-Stählen durch den Ferritgehalt. Schweißen u. Schneiden **20**, 64–71 (1968).

[332] Hull, F. C.: Effect of Delta Ferrite on the Hot Cracking of Stainless Steel. Welding J. **46**, Res. Suppl., 399s–409s (1967).

[333] Polgary, S.: The Influence of Silicon Content on Cracking in Austenitic Stainless Steel Weld Metal with Particular Reference to 18 Cr-8 Ni Steel. Metal Constr. and Brit. Welding J. **1**, 93–97 (1969).

[334] Berns, H., Killing, U., Thier, H.: Zusammenhang zwischen chemischer Zusammensetzung und Heißrißneigung in austenitischen Bandplattierungen. Schweißen u. Schneiden **33**, 650–655 (1981).

[335] Gueussier, A., Castro, R.: Étude expérimentale des criques de solidification dans les aciers – influence des impuretés. Rev. Métallurg. **57**, 117–134 (1960).

[336] Nakagawa, H., Matsuda, F., Senda, T., Matsuzaka, T., Watanabe, K., Nishino, F., Yaji, M., Hamanaka, T.: Effect of Sulphur on Solidification Cracking in Weld Metal of Steel. Trans. Japan Welding Soc. Report 1: **5**, Nr. 1, 39–46 (1974); Report 2: **5**, Nr. 2, 84–89 (1974); Report 3: **5**, Nr. 2, 26–33 (1974); Report 4: **5**, Nr. 2, 18–25 (1974); Report 5: **6**, Nr. 1, 3–9 (1975); Report 6: **6**, Nr. 1, 10–16 (1975).

[337] Masumoto, I., Tamaki, K., Kutsuna, M.: Über den Einfluß von Schwefel auf die Warmrissigkeit im Schweißgut. ZIS-Mitt. **12**, 624–640 (1970).

[338] Morgenfield, J. I., Solari, M., Ovejero-Garcia, J.: Technical Note: Metallographic Sulphur Determination in Austenitic Stainless Steel Weld Metal. Welding J. **62**, Res. Suppl., 143s–145s (1983).

[339] Cieslak, J., Savage, W. F.: Weldability and Solidification Phenomena of Cast Stainless Steel. Welding J. **59**, 136s–146s (1980).

[340] Jolley, G., Geraghaty, J. E.: Solidification Cracking in 18 Cr-13 Ni-1 Nb Stainless Steel Weld Metal: The Role of Magnesium Additions. Solidification and Casting of Metals. Book 192, S. 411–415. London: The Metals Soc. 1977.

[341] Araki, Y., Sano, H., Kominami, M., Oikawa, H.: Study on the Cr-Ni Austenitic Filler Metal Containing Mn. Trans. Japan Welding Soc. **13**, Nr. 2, 32–40 (1982).

[342] Schmidtmann, E., Homberg, G.: Einfluß der chemischen Zusammensetzung und der Schweißparameter auf die Erstarrungsrißneigung von austenitischem Unterpulverschweißgut. Schweißen u. Schneiden **28**, 470–473 (1976).

[343] Sivonen, S., Suutala, N., Moisio, T.: Elektronenstrahlmikroanalytische Untersuchung der Legierungselementverteilung nahe der Schmelzlinie in Schweißgut aus austenitischem Chrom-Nickel-Molybdän-Stahl. Schweißen u. Schneiden **30**, 211–214 (1977).

[344] Tamura, H., Watanabe, T.: Study on the Behavior of Grain Boundary at Heat-Affected Zone of SUS 347 and SUS 321 Austenitic Stainless Steel. Trans. Japan Welding Soc. **5**, Nr. 1, 47–56 (1974).

[345] Donati, J. R., Lehuede, G., Zacharie, G.: Influence de la tendence ferritique sur la susceptibilité à la fissuration à chaud au cours du soudage d'aciers austénitiques 18-10. Mém. Sci. Rev. Métallurg. **72**, 823–836 (1975).

[346] Honeycombe, J., Gooch, T. G.: Effect of Manganese on Cracking Behaviour of Fully Austenitic Stainless-Steel Weld-Metals. Metal Constr. and Brit. Welding J. **4**, 456–460 (1972).

[347] Arata, Y., Matsuda, F., Katayama, S., Nakagawa, H., Ogata, S.: Solidification Crack Susceptibility in Weld Metals of Fully Austenitic Stainless Steels. Trans. Japan Welding Res. Inst. Report I: **5**, Nr. 2, 35–51 (1976); Report II: **6**, Nr. 1, 105–116 (1977); Report III: **6**, Nr. 2, 37–46 (1977); Report IV: **7**, Nr. 2, 21–24 (1978); Report V: **10**, Nr. 2, 73–84 (1981); Report VI: **11**, Nr. 1, 79–94 (1982).

[348] Rabensteiner, G., Tösch, J., Schabereiter, H.: Hot Cracking Problems in Different Fully Austenitic Weld Metals. Welding J. **62**, Res. Suppl., 21s–27s (1983).

[349] Matsuda, F., Nakagawa, H., Katayama, S., Arata, Y.: Weld Metal Cracking and Improvement of 25% Cr-20% Ni (AISI 310 S) Fully Austenitic Stainless Steel. Trans. Japan Welding Soc. **13**, Nr. 2, 41–58 (1982).

[350] Gooch, T. G., Honeycombe, J.: Microcracking in Fully Austenitic Stainless Steel Weld Metal. Metal Constr. **6**, 146–148 (1975).

[351] Ogawa, T., Tsunetomi, E.: Hot Cracking Susceptibility of Austenitic Stainless Steels. Welding J. **61**, Res. Suppl., 82s—93s (1982).

[352] Husemann, R. U.: Rißbildung in der Wärmeeinflußzone eines geschweißten Formstückes aus dem warmfesten austenitischen Stahl X 8 CrNiNb 16 13. VGB Kraftwerkstechn. **57**, 483—485 (1977).

[353] Lorenz, K., Fabritius, H., Kranz, E.: Vollaustenitische Chrom-Nickel-Molybdän-Stähle mit rund 0,03% Kohlenstoff und 0,1% Stickstoff für den Einsatz im Temperaturbereich von −269 bis 700 °C. Stahl u. Eisen **92**, 393—400 (1972).

[354] Kobayashi, T., Kuwana, T., Kikuchi, Y.: The Nitrogen Contents of Fe-Cr, Fe-Ni and Fe-Cr-Ni Weld Metals. Trans. Japan Welding Soc. **1**, Nr. 4, 36—42 (1970).

[355] Asakura, S., Wachi, H., Watanabe, K.: The Effect of Welding Conditions on the Crack Sensitivity of Austenitic Stainless Steel Weld Metal. Trans. Japan Welding Soc. **3**, Nr. 3, 26—36 (1972).

[356] Bekkers, K., Berkhout, T.: Stickstoffaufnahme und Auswirkungen im Schweißgut von austenitischen Stählen. DVS Bericht **57**, 182—191 (1979).

[357] Killing, R., Thier, H.: Beitrag zur Frage der Stickstoffaufnahme beim Unterpulverschweißen nichtrostender Stähle. DVS Bericht **75**, 10—17 (1982).

[358] Kotecki, D. J.: Welding Parameter Effects on Open-Arc Stainless Steel Weld Metal Ferrite. Welding J. **57**, Res. Suppl., 109s—117s (1978).

[359] Killing, R.: Beeinflussung der Heißrißneigung durch das Schweißverfahren und die Schweißparameter. Schweißtechn. Kolloquium „Heißrisse" der SLV-Duisburg 1977.

[360] Thiessen, W., Poweleit, B.: Einfluß der konstruktiven Gestaltung auf das Entstehen von Heißrissen in Schweißungen. Schweißen u. Schneiden **31**, 148—152 (1979).

[361] Honeycombe, J., Gooch, T. G.: Corrosion and Stress Corrosion of Arc Welds in 18% Chromium-2% Molybdenum-Titanium Stabilized Stainless Steels. Brit. Corrosion J. **18**, 25—34 (1983).

[362] Briggs, J. Z., Parker, T. D.: Die vergütbaren und aushärtbaren 12%-Chromstähle. Bänder, Bleche, Rohre **8**, 695—707 (1967).

[363] Stahl-Eisen-Liste. (Verein Deutscher Eisenhüttenleute, Hrsg.) 7. Aufl. Düsseldorf: Verlag Stahleisen m. b. H. 1981.

[364] Peter, W., Matz, W.: Das Umwandlungsverhalten von Stählen mit 12 bis 14% Chrom. Arch. Eisenhüttenwes. **28** (1957).

[365] Gut, K., Werner, A.: Schweißversuche an 13%igem Chrom-Stahlguß. Z. Schweißtechn. (Basel) **48**, 203—211 (1961).

[366] Gysel, W., Mayer, B.: Ein Beitrag zur Frage der Reparaturschweißung am klassischen 13%-Chromstahlguß im Kraftwerk. Z. Schweißtechn. (Basel) **56**, 163—172 (1969).

[367] Woltron, K.: Beitrag zur Problematik der Schweißsprödigkeit ferritischer Chromstähle. Dissertation Montanuniversität Leoben 1971.

[368] Winkler, F., Tösch, J.: Rostbeständige Schweißzusatzwerkstoffe für Auftragungen an Wasser-, Gas- und Dampfarmaturen. Schweißblitz, Folge 21, 1—5. Kapfenberg: VEW 1980.

[369] Gysel, W., Gerber, E., Trautwein, A.: G-X 5 CrNi 13 4: Neuentwicklungen auf der Basis von 20 Jahren Erfahrung. Konstruieren und Gießen **7**, Nr. 2, 16—33 (1982).

[370] Baggström, G.: New Steels for Power Industry. Water Power **16**, 516—521 (1964).

[371] Cristina, V., Gysel, W.: Hochfester Stahlguß auf der Basis von 13% Chrom mit guter Schweißbarkeit. Z. Schweißtechn. (Basel) **53**, 226—237 (1966).

[372] Souresny, H., Sauer, H.: Chromstahlguß für den Wasserkraftmaschinenbau. Gießerei **58**, 775—782 (1971).

[373] Gysel, W.: Das Zeitstandverhalten von niedriggekohltem martensitischem Stahlguß mit 13 bis 17% Chrom und 1 bis 4% Nickel. International Conference on Properties of Creep-Resisting Steels, Bericht 8.3, Düsseldorf, Mai 1972.

[374] Gysel, W., Gerber, E.: Kaltzäher martensitischer Stahlguß. Konstruieren und Gießen **3**, Nr. 2, 17–25 (1978).

[375] Heimann, W., Strom, F. H.: Eigenschaften und Gefüge einiger weichmartensitischer nichtrostender Chrom-Nickel-(Molybdän-)Stähle. Thyssen Edelstahl Techn. Ber. **8**, 115–125 (1982).

[376] Koren, M., Diebold, A.: Verbesserung der Güteeigenschaften von martensitaushärtbaren chemisch beständigen Stählen. Berg- u. hüttenm. Mh. **121**, 269–274 (1976).

[377] Schmidtmann, E., Kieselbach, M.: Einfluß von Molybdän und Aushärtungsgrad auf die Korrosionseigenschaften chromlegierter martensitaushärtender Stähle. Werkst. u. Korrosion **24**, 185–190 (1973).

[378] Gysel, W.: Die Schweißbarkeit von hochfestem martensitischem Stahlguß mit 13% Cr und 4% Ni. Z. Schweißtechn. (Basel) **55**, 369–393 (1968).

[379] Howard, R. D.: The Weldability of Improved Chromium Steels. Welding Res. Intern. **7**, 482–502 (1977).

[380] Oppenheim, R., Chlibec, G.: Eigenschaften des nichtrostenden Stahles X 2 CrNiMoN 22 5 im Vergleich mit anderen ferritisch-austenitischen Stählen. Thyssen Edelst. Techn. Ber. **8**, 187–193 (1982).

[381] Koren, M., Hochörtler, G.: Eigenschaften des ferritisch-austenitischen Stahles X 3 CrMnNiMoN 25 6 4. Stahl u. Eisen **102**, 509–513 (1982).

[382] Bock, H. E.: Beitrag zum Spannungs- und Schwingungsrißkorrosionsverhalten stickstofflegierter austenitischer und ferritisch-austenitischer Chrom-Nickel-(Molybdän-)-Stähle. Techn. Mitt. Krupp-Forsch.-Ber. **36**, 49–60 (1978).

[383] Herbsleb, G., Poepperling, R. K.: Corrosion Properties of Austenitic Ferritic Duplex Steel AF 22 in Chloride and Sulfide Containing Environments. Corrosion NACE **36**, 611–617 (1980).

[384] Krauß, H., Speckhardt, H.: Beitrag zur dehnungsinduzierten Spannungsrißkorrosion rost- und säurebeständiger Stähle. Z. Werkstofftechn. **10**, 406–416 (1979).

[385] Clark, C. A., Guha, P.: Improvements in Corrosion Resistance, Mechanical Properties, and Weldability of Duplex Austenitic-Ferritic Steels. Werkst. u. Korrosion **34**, 27–31 (1983).

[386] Petersen, W. A., Lang, F. H.: Welding of a High Strength Stainless Steel IN-744. Welding J. **49**, Res. Suppl., 267s–271s (1970).

[387] Blumfield, D., Clark, G. A., Guha, P.: Welding Duplex Austenitic-Ferritic Stainless Steel. Metal Construction **13**, 269–273 (1981).

[388] Nagano, H., Kowaka, M.: Corrosion Resistance of Welded Joints in Some Duplex Alloys. Tetsu to Hagané (J. Iron Steel Inst. of Japan) **66**, 1150–1159 (1980).

[389] Müller-Stock, H. W.: Einfluß der Wärmebehandlung auf die mechanischen Eigenschaften der ferritisch-austenitischen Stähle X 8 CrNiMo 27 5 und X 4 CrNiMoNb 25 7. TEW – Technische Berichte **2**, 47–54 (1976).

[390] Sedriks, J.: Corrosion of Stainless Steels. New York-Chichester-Brisbane-Toronto: John Wiley & Sons 1979.

[391] Nichtrostende Stähle (34 Verfasser), edited by Schierhold, P. Düsseldorf: Verlag Stahleisen m. b. H. 1977.

[392] Werkstoffkunde der gebräuchlichen Stähle, Teil 2 (32 Verfasser). (Verein Deutscher Eisenhüttenleute, Hrsg.) Düsseldorf: Verlag Stahleisen m. b. H. 1977.

[393] Strassburg, F. W.: Schweißen nichtrostender Stähle, 2. Aufl. Düsseldorf: Deutscher Verlag f. Schweißtechnik GmbH 1982.

[394] Castro, R., Cadenet, J. J.: Métallurgie du soudage des aciers inoxydables. Paris: Verlag Dunod 1968.

[395] Lancaster, J. F.: Metallurgy of Welding. London-Boston-Sydney: George Allen & Unwin 1980.

[396] Thier, H.: Aufbau und Eigenschaften austenitischer Stähle. DVS-Berichte **33**, 129–141 (1975).

[397] Thier, H.: Schweißen austenitischer Stähle. DVS-Berichte **76**, 1–8 (1983).

[398] Thier, H.: Schweißen hochlegierter Stähle. DVS-Berichte **33**, 69–75 (1975).

[399] Schabereiter, H.: Verhalten von unstabilisiertem, austenitischem Schweißgut beim Spannungsarmglühen im Temperaturbereich von 550–750 °C. Schweißtechn. (Wien) **29**, 89–95 (1975).

[400] Randak, A., Wessling, W., Bock, H. E., Steinmaurer, H., Faust, L.: Mechanische Eigenschaften und Schweißbarkeit austenitischer stickstofflegierter Stähle. Stahl u. Eisen **91**, 1255–1270 (1971).

[401] Becker, H., Hilbrans, H., Köhlert, G.: Einfluß der Wärmebehandlung auf die Eigenschaften von Schweißverbindungen stickstofflegierter und stickstofffreier austenitischer Chrom-Nickel-Molybdän-Stähle. Stahl u. Eisen **92**, 400–405 (1972).

[402] Herbsleb, G., Westerfeld, K.-J.: Der Einfluß von Stickstoff auf die korrosionschemischen Eigenschaften lösungsgeglühter und angelassener austenitischer 18/10-Chrom-Nickel- und 18/12-Chrom-Nickel-Molybdän-Stähle. III. Loch- und Spannungsrißkorrosionsverhalten in wäßrigen Lösungen von NaCl und $MgCl_2$. Werkst. u. Korrosion **27**, 479–486 (1976).

[403] Gerlach, H., Kautz, H. R., Schäffler, F.: Untersuchungen über die Wirkung von Stickstoff und Niob in austenitischen korrosionsbeständigen Stählen. Bänder, Bleche, Rohre **10**, 418–425 (1969).

[404] Ogawa, T., Aoki, S., Sakamoto, T., Zaizen, T.: The Weldability of Nitrogen-Containing Austenitic Stainless Steel. Part I: Chloride Pitting Corrosion Resistance. Weld. J. **61**, Res. Suppl., 139s–147s (1982). Part II: Porosity, Cracking and Creep Properties. Weld. J. **63**, Res. Suppl., 213s–233s (1984).

[405] Backman, A., Lundquist, B.: Properties of a Fully Austenitic Stainless Steel Weld Metal for Severe Corrosion Environments. Weld. J. **56**, Res. Suppl., 23s–28s (1977).

[406] Gottschalck, H.: Schweißen neuer korrosionsbeständiger Stähle. DVS-Berichte **41**, 91–99 (1976).

[407] Rabensteiner, G., Schabereiter, H.: Die Entwicklung hochkorrosionsbeständiger CrNiMo-Schweißzusatzwerkstoffe mit hoher Festigkeit. Stahl u. Eisen **102**, 577–582 (1982).

[408] Bäumel, A., Horn, E. M., Siebers, G.: Entwicklung, Verarbeitung und Einsatz des stickstofflegierten, hochmolybdänhaltigen Stahles X 3 CrNiMoN 17 13 5. Werkst. u. Korrosion **23**, 973–983 (1972).

[409] Bäumel, A.: Eigenschaften und Einsatzmöglichkeiten des stickstofflegierten Stahles X 5 CrNiMoN 17 13. Berg- u. hüttenm. Mh. **116**, 88–94 (1971).

[410] Advanced Stainless Steels for Seawater Application. Proceedings of the Symposium. The University of Piacenza. Verlag Hugh Morrow III (Climax Molybdenum Comp.) 1980.

[411] Kohl, G., Rabensteiner, G., Hochörtler, G.: Stainless Steels with High Strength and High Corrosion Resistance. Conference proceedings: Alloys for the Eighties, Paper 27, Ann Arbor, Michigan 1980. (R. Q. Barr, ed.) Greenwich, Conn. 1981.

[412] Vicentini, B., Sinigaglia, D., Taccani, G., Rondelli, G., Gherardi, F., Ortali, P. L., Galelli, L.: New Stainless Steels for Sea-Water Application. Part I: Corrosion and Mechanical Properties of Ferritic Stainless Steels. Werkst. u. Korrosion **33**, 132–143 (1982); Part II: Comparison of Corrosion and Mechanical Properties of Commercial and Laboratory ELI Ferritic and Superaustenitic Stainless Steels. Werkst. u. Korrosion **33**, 592–601 (1982).

[413] Garner, A.: Crevice Corrosion of Stainless Steels in Sea Water: Correlation of Field Data with Laboratory Ferric Chloride Tests. Corrosion NACE **37**, 178–184 (1981).

[414] Garner, A.: The Effect of Autogenous Welding on Chloride Pitting Corrosion in Austenitic Stainless Steels. Corrosion NACE **35**, 108–114 (1979).

[415] Garner, A.: Pitting Corrosion of High Alloy Stainless Steel Weldments in Oxydizing Environments. Weld. J. **62**, 27–34 (1983).

[416] Lennartz, G., Kiesheyer, H.: Korrosionsverhalten eines seewasserbeständigen, nichtmagnetisierbaren Chrom-Nickel-Molybdän-Stahles hoher Festigkeit. Arch. Eisenhüttenwes. **43**, 639–642 (1972).

[417] Risch, K.: Nachbehandlung von Halbzeugen und Apparaten aus austenitischen Stählen zur Vermeidung von Spannungsrißkorrosion. Chem.-Ing.-Tech. **51**, 402–406 (1979).

[418] Lundin, C. D.: Dissimilar Metal Welds–Transition Joints Literature Review. Weld. J. **61**, Res. Suppl., 58s–63s (1982).

[419] Faber, G., Gooch, T.: Welded Joints Between Stainless and Low Alloy Steels: Current Position. Weld. World **20**, 87–98 (1982).

[420] Widowitz, H., Folkhard, E.: Schweißzusatzwerkstoffe für Austenit-Ferrit-Verbindungen. Schweißtechnik (Wien) **31**, 101–107 (1977).

[421] Flemming, D.: Schweißverbindungen zwischen ferritischen und austenitischen Stählen. Ind. Anz. **88**, 637–640 (1966).

[422] Hennemann, K., Schütte, H.: Schmelzschweißverbindungen zwischen unlegierten und hochlegierten Stählen. Der Praktiker, Teil 1: **26**, 156–159 (1974); Teil 2: **26**, 177–180 (1974).

[423] Thorneycroft, D. R.: Schweißverbindungen zwischen verschiedenartigen Werkstoffen. Z. Schweißtechn. (Basel) **54**, 267–287 (1967).

[424] Wirtz, H.: Möglichkeiten und Werkstoffprobleme beim Verbinden unterschiedlicher Metalle durch Schweißen. Maschinenmarkt **72**, SP 195–202, Nr. 72, 23–30 (1966).

[425] Gooch, T. G.: Repair Welding with Austenitic Stainless Steel MMA Electrodes. Metal Construction **12**, 622–631 (1980).

[426] Klueh, R. L., King, J. L.: Austenitic Stainless Steel-Ferritic Steel Weld Joint Failures. Weld. J. **61**, Res. Suppl., 302s–311s (1982).

[427] Shifrin, E.: A Review of American Practices in the Welding of Dissimilar Metal Joints in Piping and Tubing for Power Plants. Zváranie **6**, 73–82 (1964).

[428] Rabensteiner, G., Schabereiter, H., Schablauer, C., Brinsky, W., Kadletz, K., Klug, P., Tomann, W., Mösenbacher, H., Mittermeier, J.: Schweißverbindungen zwischen hochlegierten CrNi-Stählen und niedriglegierten Stählen. Berg- u. hüttenm. Mh. **129**, 1–7 (1984).

[429] Matsuda, F., Nakagawa, H., Katayama, S., Arata, Y.: Solidification Crack Susceptibility in Weld Metals of Fully Austenitic Stainless Steels (Report VIII). Trans. Japan Weld. Res. Inst. **12**, 89–95 (1983).

[430] Matsuda, F., Nakagawa, H., Nakata, K., Kohmoto, H., Honda, Y.: Quantitative Evaluation of Solidification Brittleness of Weld Metal During Solidification by Means of In-Situ Observation and Measurement. Trans. Japan Weld. Res. Inst. Report I: **12**, 65–72 (1983); Report II: **12**, 73–80 (1983).

[431] Brooks, J. A., Williams, J. C., Thomson, A. W.: Microstructural Origin of the Skeletal Ferrite Morphology of Austenitic Stainless Steel Welds. Metallurg. Trans. **14 A**, 1271–1281 (1983).

[432] Okagawa, R. D., Dixon, R. D., Olson, D. L.: The Influence of Nitrogen from Welding on Stainless Steel Weld Metal Microstructures. Weld. J. **62**, Res. Suppl., 204s–209s (1983).

[433] Perteneder, E., Tösch, J., Rabensteiner, G.: New Welding Filler Metals for the Welding of Girth Welds on Pipelines of Corrosion-Resisting CrNiMoN-Duplex Steels. International Conference on "Welding in Energy Related Projects". Toronto, Canada 1983.

[434] Slattery, G. F., Keown, S. R., Lambert, M. E.: Influence of Delta-Ferrite Content on Transformation to Intermetallic Phases in Heat-Treated, Type 316 Austenitic Stainless Steel Weld Metal. Metals Technol. **10**, 373–385 (1983).

[435] Perteneder, E., Blöch, R., Wedl, F.: Die Gefügeausbildung des Deltaferrits im Schweißgut hochlegierter Stähle. Prakt. Metallogr. **20**, 588–595 (1983).

[436] Charbonnier, J. C., Jossic, T.: An Electrochemical Study of Intergranular Corrosion Related to the Chromium Depletion Mechanism in the Case of 18-10 Austenitic Stainless Steels. Corros. Sci. **23**, 1191–1206 (1983).

[437] Brezina, P.: Martensitische Cr-Ni-Stähle mit niedrigem C-Gehalt. Härterei-Techn. Mitt. Part I: **38**, 197–214 (1983); Part II: **38**, 251–262 (1983).

[438] Rabensteiner, G.: The Welding of Austenitic CrNi-Stainless Steels with High Mo-Content for Seawater Application. IIW-Document IX-1319/84 (IXH-133/84).

[439] Cieslak, M. J., Ritter, A. M., Savage, W. F.: Chi-Phase Formation During Solidification and Cooling of CF-M 8 Weld Metal. Weld. J. **63**, Res. Suppl., 133s–140s (1984).

[440] Szumachowski, E. R., Kotecki, D. J.: Effect of Manganese on Stainless Steel Weld Metal Ferrite. Weld. J. **63**, Res. Suppl., 156s–161s (1984).

[441] Gooch, T. G.: Stress Corrosion Cracking of Welded Austenitic Stainless Steel. Weld. World **22**, 64–76 (1984).

[442] Van Bemst, A.: Postweld Heat Treatment of Stainless Steel and Nickel Alloys in Thick Sections. Weld. World **22**, 88–106 (1984).

[443] Ritter, A. M., Henry, M. F., Savage, W. F.: High Temperature Phase Chemistries and Solidification Mode Prediction in Nitrogen-Strengthened Austenitic Stainless Steels. Metallurg. Trans. **15 A**, 1339–1351 (1984).

[444] Hammar, O., Svensson, U.: Solidification and Casting of Metals, The Metals Society, London, 401–410 (1979).

[445] Hebble, T. L., Canonico, D. A., Edmonds, D. P., Goodwin, G. M., Nanstad, R. K.: Analysis of Delta-Ferrite Data from Production Welds on Stainless Steel Pipe. Weld. J. **64**, Res. Suppl., 260s–265s (1985).

[446] Olson, D. L.: Prediction of Austenitic Weld Metal Microstructure and Properties. Weld. J. **64**, Res. Suppl., 281s–295s (1984).

[447] Hall, E. L., Briant, C. L.: Chromium Depletion in the Vicinity of Carbides in Sensitized Stainless Steels. Metallurg. Trans. **15 A**, 793–811 (1984).

[448] Vitek, J. M., David, S. A.: The Sigma Phase Transformation in Austenitic Stainless Steel. Weld. J. **65**, Res. Suppl., 106s–111s (1986).

[449] Thomas, R. D.: HAZ Cracking in Thick Sections of Austenitic Steels—Part I: Weld. J. **63**, No. 12, 24–32 (1984); Part II: Weld. J. **63**, Res. Suppl., 355s–368s (1984).

[450] Morishige, N., Okabayashi, H.: Microcracking in Weld Metals of Nickel Alloys and Austenitic Steels. J. Japan Weld. Soc. **51**, 1006–1014 (1982).

[451] Wilken, K.: Gebräuchliche Heißrißprüfverfahren im Vergleich. Schweißen u. Schneiden **37**, 170–174 (1985).

[452] Pohle, C. P.: Möglichkeiten für die Beurteilung der Heißrißanfälligkeit von austenitischem Schweißgut. Schweißen u. Schneiden **37**, 55–59 (1985).

[453] Ritter, A. M., Savage, W. F.: Solidification and Solidification Cracking in Nitrogen-Strengthened Austenitic Stainless Steels. Metallurg. Trans. **17 A**, 727–737 (1986).

[454] Kujanpää, V. P.: Weld Discontinuities in Austenitic Stainless Steel Sheets—Effect of Impurities and Solidification Mode. Weld. J. **63**, Res. Suppl., 369s–375s (1984).

[455] Kujanpää, V. P.: Effects of Steel Type and Impurities in Solidification Cracking of Austenitic Stainless Steel Welds. Metal Constr. **17**, 40 R–46 R (1985).

[456] Lundin, C. D., Chou, C. P. D.: Fissuring in the "Hazard HAZ" Region of Austenitic Stainless Steel Welds. Weld. J. **64**, Res. Suppl., 113s–118s (1985).

[457] Ogawa, T. O., Zaizen, T., Tsunetomi, E.: Hot Cracking Susceptibility of Niobium-Containing Stainless Steel Welds. J. Japan Weld. Soc. **50**, 680–688 (1981).

[458] Nakao, Y., Oshige, H., Koga, S., Nishihara, H., Sugitani, J.: Effects of Nb/C on Sensitivity of Liquation Cracking in 24 Cr-24 Ni-1,5 Nb Fe-Base Heat Resisting Alloy—Study on Weld Cracking in 24 Cr-24 Ni-1,5 Nb Fe-Base Heat Resisting Alloy (Part II). J. Japan Weld. Soc. **51**, 989—995 (1982).

[459] Cieslak, M. J., Savage, W. F.: Hot-Cracking Studies of Alloy CN-7 M. Weld. J. **64**, Res. Suppl., 119s—126s (1985).

[460] Krysiak, K. F.: Welding Behavior of Ferritic Stainless Steels—An Overview. Weld. J. **65**, No. 4, 37—41 (1986).

[461] Gümpel, P., Hook, M., Strom, F. H.: Beitrag zum Versprödungsverhalten des weichmartensitischen Stahles X 5 CrNi 13 4 (Remanit 4313). Thyssen Edelst. Techn. Ber. **11**, 49—56 (1985).

[462] Asakura, S., Nihei, M.: Weldability of High Strength Martensitic Steels. J. Japan Weld. Soc. **50**, 458—464 (1981).

[463] Rabensteiner, G., Tösch, J., Schabereiter, H.: Das Schweißen korrosionsbeständiger CrNiMoN-legierter Stähle. Technica **8**, 37—44 (1985).

[464] Bazzoni, B., Cigada, A., Lazzari, L., Pedeferri, P.: Field and Laboratory Experiences of Sulfide Stress Corrosion of a Duplex Stainless Steel. Werkst. u. Korrosion **30**, 151—156 (1985).

[465] Hoffmeister, H., Mundt, R., Berner, K. L.: Effect of Weld Metal Composition and Welding Conditions on Deltaferrite and CVN Toughness of Austenitic-Ferritic Weld Metal. Steel Research **56**, 163—166 (1985).

[466] Fiehn, H., Spähn, R., Werner, H., Zürn, H.: Duplex-Stähle — ihre Verarbeitung und geeignete Schweißzusätze. DVS-Berichte, Bd. 105, 29—35 (1986).

[467] Bonnefois, B., Catelin, D., Soulignac, P.: Soudage des aciers inoxydables destinés aux milieux corrosifs sévères. DVS-Berichte, Bd. 105, 106—112 (1986).

[468] Perteneder, E., Rabensteiner, G., Tösch, J.: Neue Erkenntnisse zum Schweißen von korrosionsbeständigen CrNiMoN-Stählen im Rohrleitungs- und Apparatebau. DVS-Berichte, Bd. 100, 29—32 (1985).

[469] Duplex Stainless Steels. Conference Proceedings, St. Louis 1982, edited by R. A. Lula, published 1983 by Am. Soc. for Metals, Ohio 44073, Metals Park.

[470] Advanced Corrosion Resistant Steels and Metal Alloys Offshore. Proceedings of the Oslo Conference 1986, edited by The Norwegian Soc. of Engineers and Norwegian Welding Soc., Oslo 1, P.O.Box 9100.

[471] International Conference on Duplex Stainless Steel, The Hague, 1986, publ. by Nederlands Inst. voor Laasttechniek, 2517 AJ Den Haag.

[472] Third International Conference on Welding and Performance of Pipelines. London 1986. Published by The Welding Inst., Abington Hall, Cambridge.

[473] Schläpfer, H.-W., Weber, J.: Austenitic-Ferritic Duplex Steels. Material und Technik 1986, No. 2, 60—69.

[474] Kotecki, D. J.: Extension of the WRC Ferrite Number System. Weld. J. **61**, Res. Suppl., 352s—361s (1982).

[475] Bonnefois, B., Blondeau, R., Catelin, D.: Control of the Ferrite Content in Duplex Stainless Steel Welds. 25. Journées des aciers spéciaux, St. Etienne 1986.

[476] Tösch, J., Rabensteiner, G., Reiterer, P.: Metallurgie der Schweißung korrosionsbeständiger, ferritisch-austenitischer CrNiMoN-legierter Duplex-Stähle. Berg- u. hüttenm. Mh. **132**, 147—154 (1987).

[477] Street, J.: Welding Duplex Stainless Steel for Arduous Service. Metal Constr. **18**, 565—569 (1986).

[478] Gümpel, P., Chlibec, G.: Untersuchungen über das Werkstoffverhalten des ferritisch-austenitischen Stahles X 2 CrNiMoN 22 5. Thyssen Edelst. Techn. Ber. **11**, 3—8 (1986).

[479] Bäumel, A.: Korrosion: Definition, Gliederung, grundsätzliche Einflüsse. DVS-Berichte, Bd. 105, 206—216 (1986).

[480] Rockel, M. B., Renner, M.: Pitting, Crevice and Stress Corrosion Resistance of High Chromium and Molybdenum Alloy Stainless Steels. Werkst. u. Korrosion **35**, 537–542 (1982).

[481] Risch, K.: Maßnahmen gegen die chloridbedingte Spannungsrißkorrosion austenitischer Stähle. Werkst. u. Korrosion **36**, 55–63 (1985).

[482] Herbsleb, G., Pfeiffer, B.: Kritische Werte von Einflußgrößen für die Spannungsrißkorrosion von austenitischen 18 Cr-10 Ni-Stahl in Chloridlösungen. Werkst. u. Korrosion **35**, 245–266 (1984).

[483] Mukai, Y., Murata, M.: Mechanism of Stress Corrosion Cracking in SUS 304 Duplex Weld Metal. Trans. Japan Weld. Soc. **15**, 57–63 (1984).

[484] Kuwana, T., Kokawa, H., Tsujii, H.: The Effect of the Nitrogen Absorption on the TEM Structure of SUS 304 L Austenitic Stainless Steel Weld Metal. Trans. Japan Weld. Soc. **17**, 10–19 (1986).

[485] Gill, T. P. S., Vijayalakshmi, J. B., Gnanamoorthy, J. B., Padmanabhan, K. A.: Transformation of Deltaferrite During the Postweld Heat Treatment of Type 316 L Stainless Steel Weld Metal. Weld. J. **65**, Res. Suppl., 122s–127s (1986).

[486] Rabensteiner, G., Tösch, J.: The Metallurgy of Welding Fully Austenitic CrNiMo Stainless Steels—An Update. Weld. J. **64**, No. 4, 33–38 (1985).

[487] Kurz, W., Fisher, D. J.: Fundamentals of Solidification. Trans. Tech. Publications LTD 1986, Aedermannsdorf, Switzerland and Rockport, MA 01966, U.S.A.

Author Index

The numbers printed in italics refer to the entry in the list of references at the end of this book (page 243), the straight ones to the respective page number in the text.

Subject Index

Fotosatz und Druck:
F. Seitenberg Ges. m. b. H., 1050 Wien